CRIME AND TERRORISM RISK

Crime and Terrorism Risk is a collection of original essays and articles that presents a broad overview of the issues related to the assessment and management of risk in the new security age. These original articles show how researchers, experts, and the public are beginning to think about crime and terrorism issues in terms of a new risk paradigm that emphasizes establishing a balance between threat and resources in developing prevention and response strategies.

Leslie W. Kennedy is University Professor of Criminal Justice at Rutgers University. He is also the Director of the Rutgers Center on Public Security. Dr. Kennedy's current research in public security builds upon his previous research in event analysis and understanding the social contexts in which hazards to society are identified and deterred.

Edmund F. McGarrell is Professor and Director of the School of Criminal Justice at Michigan State University. Dr. McGarrell co-chairs the Michigan State University Risk Research Initiative: Risk, Values, and Decisions. In this role, he is collaborating with faculty from a variety of fields in research on environmental and natural resource crime, supply chain security, and law enforcement intelligence.

CRIME AND TERRORISM RISK

STUDIES IN CRIMINOLOGY AND CRIMINAL JUSTICE

Edited by
LESLIE W. KENNEDY
Rutgers University

EDMUND F. McGARRELL
Michigan State University

Routledge
Taylor & Francis Group
NEW YORK AND LONDON

First published 2011
by Routledge
605 Third Avenue, New York, NY 10017
4 Park Square, Milton Park, Abingdon, Oxon OX14 4RN

Routledge is an imprint of the Taylor & Francis Group, an informa business

Library of Congress Cataloging-in-Publication Data
Crime and terrorism risk: studies in criminology and criminal justice/[edited by] Leslie W. Kennedy, Edmund F. McGarrell.
p. cm.
1. Crime prevention. 2. Terrorism—Prevention. 3. Risk assessment. 4. Criminal justice, Administration of. I. Kennedy, Leslie W. II. McGarrell, Edmund F., 1956–
HV7431.C694 2011
364.4—dc22
2010034563

ISBN 13: 978–0–415–99181–0 (hbk)
ISBN 13: 978–0–415–99182–7 (pbk)
ISBN 13: 978–0–203–89447–7 (ebk)

Typeset in Minion by Swales & Willis Ltd, Exeter, Devon

CONTENTS

About the Contributors x
Preface xvii
Introduction xix

PART I
OVERVIEW OF RISK ASSESSMENT 1

1. **Assessing the Risk in Crime** 6
The Risk of Crime: A Framework
ERIN GIBBS VAN BRUNSCHOT AND LESLIE W. KENNEDY

This chapter examines the different stages of risk: the run-up to an event, the incident itself, and its aftermath. In their piece, the authors suggest that risk operates at all levels of criminological analysis and changes its character at different stages of the crime incident.

2. **Risk and Offender Decision-making** 17
Reducing Homicide Risk
NICHOLAS CORSARO AND EDMUND F. McGARRELL

This chapter explores how risk analysis can be used to identify trends in homicide, the differential risk for homicide, and for assessing risk reduction strategies. The authors describe a "pulling levers" approach that emphasizes a directed intervention plan that identifies and responds to the risk of gun offending and prescribes agency responses directed at potential offenders.

3. **Understanding Scalability of Risk** 29
Global Risk Assessment: The Search for a Common Methodology
LESLIE W. KENNEDY, NEREA MARTEACHE, AND YASEMIN GAZIARIFOGLU

This chapter explores the ways in which major agencies have developed risk assessment procedures in anticipating the negative effects of health crises, disasters, and terrorism. The authors ask whether or not there is a common methodology evident in these approaches that can be applied across all risk domains. In their review, they suggest that there are important similarities in approaches that can be adopted to bring risk assessment into a more standardized and systematic approach.

4. Threats, Vulnerability, and Exposure to Terrorism 54
The Practitioner's View of the Terrorist Threat
MARK ANARUMO

In looking at the extent to which we can know about the real threat of terrorism, this chapter focuses on three factors: local actors, local targets, and local agency. These comprise a trilogy that is a cornerstone of opportunity-based theories of crime. There is a logical partnering of these three elements and the elements tested via a survey to discover "why" chiefs of police feared certain groups over others.

5. Terror Groups and Radicalization 80
Radicalization and Risk Assessment
MICHAEL SUTTMOELLER, STEVEN CHERMAK, JOSHUA D. FREILICH, AND SARAH FITZGERALD

One critical task for risk assessment is to reliably characterize extremist groups by their propensity for violence. The authors focus on right-wing terrorist groups and compare them by their involvement in violence. This, in turn, provides a window to the processes in which these groups radicalize their members. Continuing to build this type of database, on white supremacist groups and other extremist groups, is critical to effective risk assessment and ultimately to risk reduction.

6. Prospective Rather than Reactive Risk Assessment 97
Forecasting Global Maritime Piracy Utilizing the Risk Terrain Modeling Approach to Spatial Risk Assessment
JOEL M. CAPLAN, WILLIAM D. MORETO, AND LESLIE W. KENNEDY

In looking at the results from a study of international piracy, the authors report an increase in the predictive capability of risk-based mapping over the more conventional forecasts offered by retrospective hot spot mapping, reinforcing the idea that these approaches provide a greater focus for hazard detection and response.

7. Risk Assessment Models in the Security Sector 116
Product Counterfeiting: Developing a Risk Assessment Model
JOHN SPINK

Although the financial risks to businesses are apparent, this chapter documents that counterfeiting also poses significant risk to human health and that the range of products subject to counterfeiting covers virtually all sectors of the economy. Given the global nature of twenty-first-century trade, the development of risk assessment models becomes essential for enterprise risk management as well as for countering this emerging form of transnational crime.

PART II
PRINCIPLES OF RISK MANAGEMENT 131

8. **Setting Law Enforcement Priorities** 133
Intelligence-led Policing as a Framework for Responding to Terrorism
EDMUND F. MCGARRELL, JOSHUA D. FREILICH, AND STEVEN CHERMAK

This chapter examines the challenges facing law enforcement in managing the risk of terrorism. The authors look at this problem through the lens of intelligence-led policing, and draw on the lessons learned in programs developed to respond to crime problems, that emphasizes the need to coordinate information within the law enforcement agency in a way that connects threats to response.

9. **Police Organizations and the Adoption of Risk Management Strategies** 151
A Model of Risk Assessment, Management, and Decision Making Strategies Used by the Turkish National Police
SULEYMAN HANCERLI, LESLIE W. KENNEDY, AND A. RAHMI KIRKPINAR

This chapter examines how risk is used in the organizational decision-making of an integrated police agency, the Turkish National Police (TNP). In investigating the ways in which different offices within the TNP conceptualize and act on risk, the authors speculate that a more rigorous approach to the planning of risk management strategies would improve current practice while still conforming to what is currently being done procedurally.

10. **Sharing Information to Manage and Reduce Risk** 166
Reducing Risk through Suspicious Activity Reporting
DAVID L. CARTER

Effective risk management and governance requires sharing of information across agencies, between the public and private sectors, and from citizens to law enforcement officials. In this chapter, the author explores how this process can be facilitated through the use of interagency reporting techniques, in particular, suspicious activity reports (SARs) in monitoring illegal activities.

11. **Coordinating Counter-terrorism Risk Assessment and Management** 183
Law Enforcement Intelligence: Suspicious Activity Reporting and Preventing Self-radicalized Terrorism
JEREMY CARTER

This chapter discusses the role that intelligence plays in defining the risks at all jurisdictional levels and outlines the most effective ways of responding to these risks. This effort includes a strategy that would educate officers at all levels to both understand what the intelligence means and how the terrorist risk can be manifested. The chapter draws upon several contemporary cases of terrorist threats in the United States.

12. **Emergency Response: Public–Private Sector Cooperation for Preparedness** 196
Public–Private Partnerships for Crisis Management: A Community Solution
BRIT R. WEBER AND SUSAN J. GADE

Although we often think of the public sector's role in emergency preparedness and response, the fact that the private sector controls most of the critical infrastructure points to the need for public–private cooperation. In this chapter, the authors examine private–public partnership in crisis management, showing how this collaboration can work effectively and reduce the long-term negative consequences of emergencies.

PART III
CASE STUDIES OF RISK ASSESSMENT AND RISK MANAGEMENT IN AN ALL HAZARDS ENVIRONMENT **211**

13. **Global Trade and Crime** 213
JIE XU

This chapter reviews the role that risk plays in influencing the security of the global supply chain.

14. **Counterfeit Pharmaceuticals: Examining the Multi-dimensional Risk to African Nations** 227
ROY S. FENOFF AND JEREMY M. WILSON

The authors review the threat to African nations of the introduction of counterfeit pharmaceuticals into the local marketplace.

15. **Case Study: Counterfeit Product Risk Model** 242
JOHN SPINK

This chapter presents a risk assessment model that was developed in collaboration with a corporation seeking to manage counterfeit risk.

16. **Food Frauds: Counterfeit and Adulteration in Food Supply Chain Networks** 246
ROBYN R. MACE

The author explores the aspects of food security and the ways in which protections against fraud and adulteration can be breached.

17. **Privatized Cybersecurity and the Challenges of Securing the Digital Environment** 255
PANAYOTIS A. YANNAKOGEORGOS

This chapter examines different aspects of the challenges faced in maintaining cybersecurity.

18. The Risk Posed by Zebra Mussels 268
DAVID L. CARTER

The author looks at the threat posed by invasive species used as an instrument by a terrorist threat group.

PART IV
SUMMARY AND CONCLUSIONS **273**

Index 276

ABOUT THE CONTRIBUTORS

Mark Anarumo earned his PhD in Criminal Justice in 2005 from Rutgers Graduate School, Newark. His dissertation topic was terrorist threat forecasting, a topic sponsored by the United States Air Force (USAF) Fellowship Program. Mark is currently serving as a Lieutenant Colonel in the USAF Security Forces and is stationed at Kunsan Air Base, Gunsan City, Republic of Korea. During his 20-year career he has worked in the United States and foreign nations including Iraq, Afghanistan, Kyrgyzstan, Kuwait, and Saudi Arabia. In these positions Mark has served as Chief of Anti-Terrorism, Chief of Police, and Chief of Information Protection, protecting United States interests and personnel from terrorism, crime, and cyber attacks. He is a graduate of the FBI National Academy and three-time graduate of Rutgers University.

Joel M. Caplan is an Assistant Professor at the Rutgers University School of Criminal Justice and Associate Director of the Rutgers Center on Public Security. He specializes in spatial analysis and advises many public safety agencies on how to use geographical information systems (GISs) for strategic decision-making and evidence-based practices. Joel's work has been disseminated in books, peer-reviewed journal articles, and at national and international conferences. His recent book, *GIS for Public Safety* (2010, Rutgers Center for Public Security), is an annotated guide to ArcGIS that presents many of the tools and techniques commonly used by public safety researchers, analysts, and practitioners. In addition to his academic accomplishments, Joel has professional experience as a police officer, 911 dispatcher, and an emergency medical technician.

David L. Carter is a Professor in the School of Criminal Justice and Director of the Intelligence Program at Michigan State University. A former Kansas City, Missouri police officer, Dr. Carter has served as a trainer, consultant, and advisor to many law enforcement agencies throughout the US, Europe, and Asia on matters associated with officer behavior, community policing, law enforcement intelligence, and computer crime. Dr. Carter is the author of the Community Oriented Policing Services-funded publication, "Law Enforcement Intelligence: A Guide for State, Local and Tribal Law Enforcement," Project Director for multi-million dollar national intelligence training programs funded by the Department of Homeland Security, and co-Project Director of a National Institute of Justice grant to do a nationwide study on best practices and efficacy of law enforcement intelligence initiatives. He is an Academic Fellow of the Foundation for the Defense of Democracies wherein he studied terrorism in Israel. He is the author or co-author of five books and numerous articles and monographs on policing issues and is a member of the editorial boards of various professional publications.

Jeremy Carter is a doctoral student (ABD) in the School of Criminal Justice at Michigan State University. He is also a Research Specialist and Instructor for the Intelligence Program—a national law enforcement training program for state and local agencies funded by the Department of Homeland Security. His work has appeared in journals such as *Criminal Justice and Behavior*, *Criminal Justice Policy Review*, and *Police Practice & Research*. He is also the Editor-in-Chief for the *International Association of Law Enforcement Intelligence Analysts Journal*. His research interests include: policing, law enforcement intelligence, homeland security, risk prevention, and organizational behavior.

Steven Chermak (PhD, State University of New York at Albany) is a Professor in the School of Criminal Justice at Michigan State University (MSU). Dr. Chermak is also a lead investigator affiliated with the National Consortium for the Study of Terrorism and Responses to Terrorism (START). Dr. Chermak's terrorism research has focused on four general areas. First, he has received funding from START to study the criminal and terrorist activities of domestic far-right extremists. Second, he received funding (with Drs. David Carter and Edmund McGarrell—MSU) from the National Institute of Justice to study the intelligence practices of state, local, and tribal law enforcement agencies. This project includes a national survey of fusion centers and local police agencies to identify the major obstacles of effective intelligence gathering and information sharing. Third, he is studying the sources of funding used by terrorist organizations, with a particular emphasis on examining their use of counterfeited products. Finally, he has studied the media's role in relation to crime and terrorism issues.

Nicholas Corsaro is an Assistant Professor in the Department of Criminology and Criminal Justice at Southern Illinois University Carbondale (SIUC). Prior to joining SIUC, he received his PhD in 2007 from the School of Criminal Justice at Michigan State University. Corsaro's research interests include strategic approaches to reducing crime, environmental criminology, program evaluation, and quantitative analytical modeling. His recent work appears in the *Journal of Urban Health*, *Journal of Quantitative Criminology*, *Crime & Delinquency*, and the *Journal of Experimental Criminology*.

Roy S. Fenoff is a forensic document examiner and an expert in forgery detection who provides expert scientific advice and forensic examinations for individuals, businesses, law enforcement, and law firms throughout the United States and overseas. Roy received a Bachelor of Science in 2002 and a Bachelor of Art in 2004, both from the University of Georgia, and a Master of Science from the University of Wyoming in 2007. Presently, Roy is pursuing a Doctorate degree in Criminal Justice at Michigan State University (MSU) where he is a research associate with MSU's Anti-Counterfeiting and Product Protection Program (A-CAPPP). Roy's primary research interests are in anti-counterfeiting, forgery, document fraud, and white-collar crime.

Sarah Fitzgerald is currently a Master's student in the criminal justice program at Michigan State University, with plans to obtain her doctorate in the coming years. She has her Bachelor's degree in Criminal Justice with a certificate in Law Enforcement Intelligence Analysis from Michigan State University. Her research interests include terrorism, religion, and cybercrime.

Joshua D. Freilich is the Acting Executive Officer (Director) of the Criminal Justice PhD program and a member of the Criminal Justice Department at John Jay College, the City University of New York. He is a lead investigator for the National Consortium for the Study of Terrorism and Responses to Terrorism (START), a Center for Excellence of the US Department of Homeland Security (DHS) and a member of the Terrorism Research and Analysis Project (TRAP), sponsored by the Federal Bureau of Investigation's (FBI) Behavioral Science Unit (BSU). Freilich's research focuses on domestic terrorism as well as criminological theory. His research has been funded by DHS directly as well as through START. Freilich is currently the principal investigator (with Steven Chermak) on the United States Extremist Crime Database (ECDB) study, a large-scale data-collection effort that is building the first-of-its-kind relational database of crimes committed by far-right, Al-Qaeda-inspired, and animal rights and environmental rights extremists in the United States reported in an open source.

Susan J. Gade is a crime analyst for the Milwaukee Police Department. She received her Bachelor's and Master's degrees from Michigan State University where she also worked on the Critical Incident Protocol Public–Private Partnerships for Emergency Preparedness Program.

Yasemin Gaziarifoglu is a PhD student in the Criminal Justice program at Rutgers University, where she received an MA in Criminal Justice. Prior to undertaking her graduate study at Rutgers, Ms. Gaziarifoglu received her MA degree in Forensic Science at Istanbul University Institute of Forensic Sciences with her thesis titled "Fear of Crime and Perceived Risk of Crime." She holds her BA degree in Sociology (with a minor in Psychology) from Middle East Technical University, Turkey. Her research interests include systemic and local analysis of corruption, activity spaces of offenders, victimization, and risk assessment methodologies.

Suleyman Hancerli is a Chief Superintendent in the Turkish National Police (TNP). He has been in the agency for the last 17 years. He worked as an instructor in the TNP Academy between 1994 and 2000, and as the deputy director at the Turkish International Academy Against Drugs and Organized Crime (TADOC) between 2000 and 2009. He has currently been working as the Chief of Police of the Yenisehir Police Department, which is located in Diyarbakir/Turkey. He has a Bachelor's degree from the TNP Academy in Turkey, a Master's degree in the Criminal Justice Department and Doctoral degree from the Department of Information Science at the University of North Texas in the US. Chief Superintendent Dr. Hancerli shared his information and knowledge about his policing experience with more than 30 nations' law enforcement professionals during his teaching experience at the TNP

Academy, TADOC, and other police agencies abroad. He is specifically interested in researching terrorism and hostage/crisis sieges.

Leslie W. Kennedy is University Professor of Criminal Justice at Rutgers University. He is also a core faculty member of the Rutgers Division of Global Affairs. Dr. Kennedy's current research in public security builds upon his previous research in event analysis and understanding the social contexts in which hazards to society are identified and deterred. He is the co-author, with Erin Van Brunschot, of *Risk Balance and Security* (2008, Sage), which explores risk from an all hazards approach and *Risk in Crime* (2009, Rowman and Littlefield), which looks at how risk is conceptualized in criminology. In addition, he has published (with Cynthia Lum and Alison Sherley) an extensive review of counter-terrorism studies in the *Journal of Experimental Criminology.* Dr. Kennedy is currently the Director of the Rutgers Center on Public Security, whose primary focus is to conduct original analytical, objective research on how risk assessment can be applied to the study of security and to inform public policy discussions in this area.

A. Rahmi Kirkpinar is a Police Chief at the Turkish National Police (TNP). He has 16 years of professional experience as a police administrator in the cities of Ankara and Bursa in Turkey, and the city of Prague in the Czech Republic. He has served in the TNP at different ranks and capacities including Executive Officer and Advisor of the General Director of the TNP. He worked as the Police Attaché at the Turkish Embassy in Prague, the Czech Republic, between 2001 and 2004. He was responsible for advising the Ambassador of Turkey to the Czech Republic on security-related issues and served as a liaison officer between the Turkish Police and the Czech Police. A. Rahmi Kirkpinar earned his Master's in Criminal Justice at the John Jay College of Criminal Justice, and his undergraduate Bachelor of Criminal Justice from TNP Academy. He is currently a PhD candidate at the Rutgers School of Criminal Justice. His research interests include spatial analysis, risk assessment and prediction, environmental criminology, and police administration.

Robyn R. Mace is an academic specialist at the School of Criminal Justice at Michigan State University. Her research focuses on risk, security, and crime issues including food protection and defense, counterfeiting, policing, and environmental crime and enforcement. She teaches courses on terrorism, security management, and risk. She received her PhD from the School of Criminal Justice at Rutgers University.

Nerea Marteache has been a PhD student at Rutgers School of Criminal Justice since 2008. She has worked for the Department of Justice of Catalonia, Spain, first as a legal advisor in the Data Management and Strategy Development Unit (2005–2007), and then as a researcher at the Criminological Research Unit of the Center of Juridical Studies (2007–2008). She has also participated in several comparative European research projects in the area of corrections and transnational crime. Her research interests include the areas of situational crime prevention, criminal policy making, corrections, and global risks. Marteache's participation in the research project reported in her chapter was supported with funding from the Foundation Caja Madrid.

Edmund F. McGarrell is Professor and Director of the School of Criminal Justice at Michigan State University. Dr. McGarrell co-chairs the Michigan State University Risk Research Initiative: Risk, Values, and Decisions. In this role, he is collaborating with faculty from a variety of fields in research on environmental and natural resource crime, supply chain security, and law enforcement intelligence. Dr. McGarrell is the Principal Investigator (PI) of a project involving a national research team studying Project Safe Neighborhoods (PSN). PSN is a major Department of Justice program intended to reduce the risk of firearms violence in the United States. He is also the Co-PI for projects involving law enforcement intelligence and food supply chain security, both sponsored by the Department of Homeland Security.

William D. Moreto is a doctoral student at the Rutgers School of Criminal Justice. He holds a Bachelor of Arts in Criminology from Simon Fraser University. Versed in environmental criminology, the routine-activity approach and situational crime prevention, his research focuses on examining the interaction between individuals and their surrounding environment while also investigating the geo-spatial and temporal characteristics of various criminal events. He is also interested in theory-based risk assessment. As a research assistant for the Rutgers Center on Public Security (RCPS), he combines theory with practicality by utilizing geographic information systems in the analysis of specific crimes. Currently, he is working with colleagues at RCPS on further expanding the application of Risk-Terrain Modeling. His current research interests include wildlife crime and maritime piracy.

John Spink is an Assistant Professor and Associate Director for the Anti-Counterfeiting and Product Protection Program (A-CAPPP) at Michigan State University. Spink has 15 years experience in industry, 12 of which were with Chevron Corporation. Spink directs executive education and research programs within A-CAPPP. Before helping merge a number of research and outreach activities into the A-CAPPP concept with the School of Criminal Justice, Spink was an Instructor in the Michigan State University Online Master of Science in Food Safety program and developed and teaches on-line graduate classes including "Anti-Counterfeit and Product Protection," the packaging module of "Food Protection and Defense," and "Packaging for Food Safety." He serves on the State of Michigan's Agriculture and Food Protection Strategy Team and as the Chair of the US Technical Advisory Group for the ISO/TC247 Fraud Control and Countermeasures (which includes anti-counterfeiting and product fraud).

Michael Suttmoeller is a doctoral student in the School of Criminal Justice at Michigan State University. His research interests include extremist groups, domestic terrorism, organizational behavior, and policing.

Erin Gibbs Van Brunschot is Associate Professor in the Department of Sociology at the University of Calgary. Her primary research interests are in the realms of crime, security, and risk, with specific interests in how individual, organization, and state orientations to related issues both diverge and converge. In 2008, she authored *Risk Balance and Security* (Sage) with Les Kennedy. She recently coauthored a second book with Les Kennedy entitled, *Risk in Crime* (2009, Rowman and Littlefield).

Brit R. Weber is the director of Michigan State University's School of Criminal Justice Critical Incident Protocol–Community Facilitation program, which is a no-cost, grant-funded opportunity for communities across the nation that builds public and private partnerships for joint critical incident preparedness. Previously, he served in the United Nations Peacekeeping Mission in Kosovo, Yugoslavia as a project manager and instructor with the United Nations' Police Force. He developed and directed various training programs for the newly created Kosovo Police Service and the United Nations' Police. Prior to that he was the Manager of Safety and Human Resources responsible for training, inspection, and regulatory compliance, along with planning, project management, and policy development at the R. W. Mercer Company. Previously, he served with the Michigan State Police holding operational, specialty, and management positions in the Uniform Division. He worked on numerous community-related initiatives, in addition to local, state, and federal multi-agency cooperatives. Brit received his Bachelor of Science degree in Public Safety from Eastern Michigan University and is working on his Master's degree in Security Management at Michigan State University.

Jeremy M. Wilson is the Associate Director for Research and an Associate Professor in the School of Criminal Justice at Michigan State University (MSU). He recently founded and directs the MSU Anti-Counterfeiting and Product Protection Program (A-CAPPP). Prior to joining MSU, Jeremy was a Behavioral Scientist at the RAND Corporation, where, in addition to directing many local, national, and international law enforcement projects, he led the development of several successful initiatives, including the Center on Quality Policing and the Police Recruitment and Retention Clearinghouse. He is a visiting scholar in the Australian Resource Council's Centre of Excellence in Policing and Security at Griffith University, and he recently held the Willett Chair in Public Safety in the Center for Public Safety at Northwestern University and was an adjunct professor of public policy at Carnegie Mellon University. Jeremy has collaborated with police agencies, communities, task forces, and governments throughout the US and the world on many of the most salient public safety problems. He has written broadly in the areas of police administration, violence prevention, and internal security. His research and commentary have been featured in numerous books, professional journals, and in various forms of national and international media. Jeremy received his PhD in Public Administration from the Ohio State University.

Jie Xu is a PhD candidate at the Rutgers School of Criminal Justice. She earned an MS in Statistics from Rutgers University-New Brunswick. She received her Bachelor of Laws from Central University of Finance and Economics, Beijing, China. Her current research employs advanced statistics and spatial analysis techniques to the study of crime risks.

Panayotis A. Yannakogeorgos is an expert on issues pertaining to international relations, counterterrorism, global cybersecurity cooperation, cyberwarfare, and other areas of cyberspace policy. He is Founder and President of Cyrtone Consulting, a firm offering strategic cyber services. At Rutgers Division of Global Affairs (DGA), Pano taught graduate-level courses while managing DGA's research arm, the Center for Emergent Threats in the 21st Century (ET21). He has worked in cybersecurity at the United Nations, which included an appointment to the International Telecommunications Union (ITU) as a delegate-expert by the Greek Ministry of Foreign Affairs. He also served as an Advisor in for the United Nations Security Council desk on issues related to non-proliferation, the Middle East, Al-Qaeda, and Internet misuse. Pano received his PhD (2009) and MS (2007) in Global Affairs from Rutgers University, and his ALB in Philosophy from Harvard University, *cum laude*, in 2005.

PREFACE

When we began this project, there was a great deal of interest in the role that police agencies were playing in the general response to the terrorism threat in the United States and worldwide. In examining this, researchers began to realize that there was a dearth of good information about how the interface between law enforcement and national security agencies worked, what major challenges were being faced, and how successful were some of the initiatives that were being developed to combat crime and terrorism. The new initiatives that had been developed in the last decade, including new laws, inter-agency cooperation, new data collecting procedures, innovations in analytical techniques, and so on were rolled out but not carefully scrutinized. Also, as the key players in the security arena began discussing requirements for response and prevention they discovered that they did not share a common language of threat and hazard.

So, it is into this arena that we have entered. This reader is intended to provide a framework for analyzing the state of the art in crime and terrorism studies approached from a risk perspective. We believe that introducing risk analysis and management strategies into the discussion provides the language and the assessment tools needed to address the new security realities. We have reached out to leading scholars in the areas of crime and terrorism to provide new perspectives on the ways in which researchers and policy makers are tackling this new reality. We have, in addition, reached out to practitioners outside of the conventional criminal justice community who are interested in the applicability of these risk approaches to hazards in areas that impinge on security but not in the strict sense of how this has been studied in the past by criminologists and criminal justice experts.

We think that this reader will be of particular interest to students involved in global studies of crime and terrorism and also to those who are looking at the issues of global threats from a security studies perspective. Indeed, the chapters include applications to traditional street crime, to emerging transnational crimes, and to terrorism. Practitioners will also find the material of use in their review of their current practices and applications. Included are concrete examples and descriptions of the development of risk assessment models and risk management processes that will be of value to law enforcement and security officials.

In putting together this reader, we would like to thank the people at Routledge, in particular Steve Rutter and Leah Babb-Rosenfeld, who shepherded us to a final product. Our students and colleagues at Rutgers and Michigan State universities have provided ongoing support to our efforts and helped us with ideas that have been incorporated into the book. We would also like to especially thank our contributors who came through with great work in a short period of time. They have made this a high-quality book that we believe will receive a positive response in the community.

Leslie W. Kennedy
Edmund F. McGarrell

INTRODUCTION

This book represents a broad overview of the issues related to the assessment and management of risk in the new security age. We include original work to show how researchers and policy experts are beginning to think about crime and terrorism issues in terms of a new risk paradigm that emphasizes establishing a balance between threat and resources in developing prevention and response strategies.

There is a fast-growing interest in risk-based issues as the models from public health and the private sector are increasingly applied in the public sector response to crime and terrorism. The college audience for this book is broad, including students enrolled in criminal justice courses focused on homeland security, risk assessment and crime, law enforcement, terrorism, and the administration of justice. Beyond criminal justice, there is a potential audience in related fields of public administration, business courses that focus on corporate security and supply chain management.

This reader begins with an overview of the literature relating risk perspectives to an "all hazards" approach, which has become a central focus in modern policing. This approach suggests that while crime dominates most of the work of police, they are now involved directly in counter-terrorism activities. In addition, most catastrophes have a component to them that includes the need to maintain order, a role that police must assume. But, this is not just about policing. Criminal justice approaches are expanding to consider how risk assessment can apply to broader issues that threaten social order including terrorism, natural disasters, and so on. We demonstrate how there has been a movement into the mainstream literature of risk-based concepts that have joined up with the work that has already been done in applied disciplines on risk management. This literature is particularly interesting, we feel, for criminal justice specialists who have been confronted with the new demands to research and explain the origins of terrorism and transnational crime. There is increased attention paid to the connections between crime and terrorism. Agencies at the local, national, and international levels are engaged in ongoing discussions about how they can address these different types of problems using common methods (with mixed success).

Bringing risk into the equation in the study of "all hazards" is not uncontroversial and we explore some of the concerns that have been raised about using these types of metrics in prevention and response strategies. Looking at the problem in this way, we can set more light on how the pressures that are being put on law enforcement, national security, and corporate security agencies must be tempered with a clear understanding of how risk is calculated and its uses and misuses in assessment and management.

We intend, in this collection, not only to bring the criminal justice perspectives to the risk audience but, in addition, we hope that this will raise awareness in the criminal justice community of the applicability of risk approaches to the study of all manner of crimes, terrorism, and justice issues. As these cross-disciplinary efforts continue, we anticipate a rich debate around the merits of these approaches and the efficacy of the prescriptions that they offer. In this regard, the emphasis here on both assessment and management underlines the importance that we put on the idea that research must be practical and that solutions that are offered to practitioners are supported through solid research.

We are strong advocates, not only of solid empirical studies (which can be qualitative and quantitative), but also we subscribe to the view that agency programs should be routinely subjected to evaluation. In this way, we can have greater certainty that the interventions that are planned and implemented actually work. Given our emphasis on empirically based findings underlying programs we have found that the risk-based approach provides a strong framework in which this type of research can be performed, with an emphasis on standardization, reliability, and scalability of findings. In addition to generating new knowledge about risk assessment and management within security organizations, a broader goal of the book is to inform scholarship surrounding decision-making, risk, and uncertainty by examining the organizational dimensions of collective decisions.

Examination of these issues within the security domain may be particularly fruitful for scholarship around risk due to the sense of urgency within the organizations and the inherent sociopolitical context surrounding these organizations. We explore all of these ideas (and more) in this reader and finish with a section that provides case studies that are illustrative of risk applications. In these, we have intentionally included examples from different countries to underline the global nature of this type of approach and the many different types of problems that are faced by agencies around the world.

PART I

OVERVIEW OF RISK ASSESSMENT

In the years since the terrorist attacks of September 11, 2001, the continuing threat has raised significant issues related to risk source assessment, analysis, management, communication, decision making, and values in the context of uncertainty. Yet, to a significant extent, law enforcement agencies confronting these hazards do not have the level of experience with risk assessment and management as have organizations in domains such as environment, food, and health. More accurately, although security agencies deal with risk continually, the tendency has been to do so from a closed-system perspective. In contrast, multidisciplinary research has consistently demonstrated that risk assessment is a highly complex process, dependent on the effective integration of scientific and human dimensions and involving sociopolitical factors that constrain or facilitate risk-based practices and policies (National Research Council, 1983; Stern and Fineberg, 1996).

As we have indicated, the interest in risk in the new research on public security can be divided into two analytical parts: risk assessment and risk management. In risk assessment (covered in Part I) attention is paid to the threats that are perceived in the environment, the opportunities that provide targets, and the calculations that are made by individuals and agencies to balance risk against the demands of routine daily activities. This assessment is influenced by the stage at which we judge risks relative to an actual incident where prevention, preparedness, response, and recovery all affect our risk balance calculation.

Risk management, which we cover in Part II, addresses the ways in which agencies plan for and react to threats and incidents, in particular, in the judgments that they make about resource allocations. This process has important effects, as well, depending on the level of impact of incidents where different levels of jurisdiction may influence the ways in which resources are managed, the ways in which information about threat is used, and the authority that is applied in responding to and coordinating response.

Risk analysis is probabilistic but also must account for *uncertainty*, where we have more difficulty in make firm assessments of what might occur. The extent to which we can remove uncertainty will influence how well we are able to anticipate hazards and threats and their consequences. Probabilistic risk analysis, based on Bayesian inference, is a well-developed method that has been shown to provide risk management models despite limited data. Similarly, stochastic risk modeling has been used to develop warning systems that can generate testable hypotheses. The National Infrastructure Advisory Council also recommends consideration of the application of the COSO financial risk management model to security risks (see www.coso.org). The COSO model includes risk assessment, risk response, and governance and the development of metrics to assess risk and the integration of the model throughout the enterprise (Marsh and Noonan, 2005).

The RAND Corporation advocates event-based models that include detailed analysis of vulnerability and consequence of specific terrorist attack scenarios (Willis et al., 2005). In this context of uncertainty, the goal of event-based models is to reduce "egregious errors" in risk estimates as opposed to identifying optimal estimates. The point is that risk modeling is a well-developed approach in the private sector and comparative research assessing the use of these models in security organizations offers significant value to policymakers and leaders of organizations such as the Department of Homeland Security.

There is an extensive literature on risk and risk management that has dealt with a range of topics from health security (Glass and Schoch-Spana, 2002), to property protection (O'Malley, 1992), to crime control (Simon, 1988), to crime prevention (Bradley and Morss, 2002), and to environmental threats (Douglas and Wildavsky, 1982; Dimitrov, 2002). Increasingly, researchers are beginning to apply the risk models to the study of terrorism (Cummins and Lewis, 2003; Viscusi and Zeckhauser, 2003). These approaches can be typified as concentrating on applications of a perspective that has become a more prevalent part of criminal justice thinking, beginning with the idea that we are living in a "risk society", where hazards and threats become a focus of agencies to control through selective application of resources (see Beck, 1992). These controls are calculated through an application of probabilistic reasoning that incorporates the value of information, surveillance, and insurance (or some other form of mitigation) as key components in risk management of crime and terrorism (Simon, 1988). What this means is that agencies become more aware of the need to determine the dangers in an environment and marshal resources to repair damage that come from these dangers. Agencies must also enlist the support of the public in managing their own risks and in allowing the institutional interventions that attack the threats, sometimes at the cost of privacy and freedom of movement.

The calculus for risk management has not had an explicit focus of discussions around counter-terrorism until recently. This has emerged in the United States primarily through the ongoing political debate surrounding the distribution of homeland security funds across the country. While the discussion began as one of fairness, arguing that the areas most vulnerable should receive most funding, this debate has become more focused around how vulnerability can actually be calculated. In these deliberations, US government agencies have offered the formulation that risk is a product of threat, vulnerability, and consequence. In examining these factors, risk is profiled as "relative," with the direct acknowledgment that resources must be applied unevenly and that not all threats will be addressed. In fact, some have argued that too much protection against loss creates loss. Baker and Simon (2002) state that society should embrace some risks. This view flies in the face of thinking that "we need to be right all the time, the terrorist needs only be right once," a form of thinking that says that we cannot afford to make a mistake and must take action to impede all possible attacks. This strategy is used in airport screening but not applied successfully in any other instance of broad-based terrorism control (see Lum, Kennedy, and Sherley, 2006; Mueller, 2006).

In articulating risk management procedures, there has been much talk about prevention, response, and recovery (these phrases appear in many different forms in documents prepared by a range of agencies involved in crime prevention, emergency preparedness, terrorism response, and so on). In the case of terrorism, these procedures are based on a view that if attacks can be anticipated, we can quickly respond and recover from these events. But many problems emerge in the prediction of terrorism acts, the most difficult being the fact that these are low incident occurrences that are difficult to plan for. Despite this, the cases where events have occurred have provided important models for response and recovery and are giving us some important insights into how we might better prevent these events (as we have seen in the aftermath of the London transit bombings and the break up of the threatened attacks on airlines in Britain).

In the discussions of risk assessment, there have been conversations about the importance of values and institutional expertise in influencing the success of programs addressing threats and balancing risks (Van Brunschot and Kennedy, 2008). An interesting example of the contrast across societies concerning the tension that develops between experts and the public concerning risks (characterized as the "precautionary principle") is the aversion of Europeans to genetically modified food, despite continuous claims about its safety, a reaction not reflected in North America. In similar ways, discussions about risk management in responding to terrorism, on the face of it, varies dramatically from country to country. Part of this is determined by the local experiences with threats but it is in part a function of the degree of trust that the public puts in law enforcement and its ability to counter dangers in an effective way. The judgment about the success of risk management in counter-terrorism, then, is dependent on not just reducing incidents but also by delivering programs in which the public has confidence. This success also depends on the degree to which the public becomes a partner in the risk management process, working with agencies to deter terrorism. This can take on many different forms but two examples that illustrate this are the actions taken within ethnic communities that are vulnerable to terrorism recruitment to address these threats and the voluntary participation of industry in securing their supply chains in collaboration with customs and border patrol (Closs and McGarrell, 2004).

An important element in this cultural approach relates to the social construction of risk and the management of risk communications (Wildavsky, 1995; Glassner, 1999). Social construction relates to the ways in which risk and security are discussed in the media, by politicians, in local communities, and in small circles of family and friends. How risk is constructed can have important implications for the steps that we take to mitigate it. In addition, the ways in which we manage risk communication can influence our views of threat. Agencies play an important role in this communication and this needs to conform to their actions in preparing for and responding to crime and terrorism.

In sum, in developing risk assessments strategies, we have to address a number of issues. We can summarize the major ones by posing questions that will be the subject of the articles that we have included in this book:

1. What are the common features of risk assessments of crime and terrorism, natural disasters, and health events and how do they differ?
2. How do we assess the risks of "all hazards" in the ever-changing global environments in which we now live?
3. How do we go about finding out what are the threats to security?
4. How do we keep it from getting too complicated to understand?
5. How do we make sure that we leave no stone unturned in identifying threats to security?
6. How do we determine that what we are told about security and threats is correct and valid (trust)?
7. How do we make sure that everyone understands our results?
8. How do we address different levels of security by circumstance (I'm safe, are you?)?
9. How do we establish trust in prescribing solutions or strategies based on our results?
10. How do we account for changes in security over time?
11. How do we contain the consequences of threats so that we minimize the damage that comes from hazardous events such as terrorist attacks?
12. How do we know that the things that we do actually change the final outcome of security (efficacy)?
13. How do we avoid attributing blame to those who are not involved in compromising our security?
14. How do we get people to return to "normal", i.e., resume regular routines, without getting too comfortable in their security?

In setting the context for examining risks in modern society, we must also look at the changes that have occurred in global security where the shifting character of national borders and the forces of globalization have brought on new challenges for agencies responsible for addressing threats and hazards.

References

Baker, Tom and Jonathan Simon, eds. 2002. *Embracing Risk: The Changing Culture of Insurance and Responsibility.* Chicago: University of Chicago Press.

Beck, U. 1992. *Risk Society: Toward a New Modernity.* London: Sage.

Bradley, Ben S. and John R Morss. 2002. "Social Construction in a World at Risk: Toward a Psychology of Experience." *Theory & Psychology* 12:509–532.

Closs, D. J. and E. F. McGarrell. 2004. *Enhancing Security Throughout the Supply Chain.* Washington, DC: IBM Center for the Business of Government.

Cummins, J. David and Christopher M. Lewis. 2003. "Catastrophic Events, Parameter Uncertainty and the Breakdown of Implicit Long-Term Contracting: The Case of Terrorism Insurance." The *Journal of Risk and Uncertainty* 26:153–178.

Dimitrov, Radoslav S. 2002. "Water, Conflict, and Security: A Conceptual Minefield." *Society & Natural Resources* 15:677–691.

Douglas, Mary and Wildavsky, Aaron. 1982. *Risk and Culture.* Berkely, CA: University of California Press.

Glass, Thomas A. and Monica Schoch-Spana. 2002. "Bioterrorism and the People: How to Vaccinate a City against Panic." *Clinical Infectious Diseases* 34:217–223.

Glassner, Barry. 1999. *The Culture of Fear: Why Americans Are Afraid of the Wrong Things.* New York: Basic Books.

International Risk Governance Council. 2005. *Risk Governance: Towards an Integrative Approach.* Geneva, Switzerland: International Risk Governance Council.

Lum, C., L. W. Kennedy, and A. Sherley. 2006. "Are Counter-terrorism Strategies Effective? The Results of the Campbell Systematic Review on Counter-terrorism Evaluation Research." *Journal of Experimental Criminology* 2(4):489–516.

Marsh, M. H. and T. E. Noonan. 2005. *Risk Management Approaches to Protection: Final Report and Recommendations by the Council.* Washington, DC: National Infrastructure Advisory Council.

Mueller, J. 2006. *Overblown: How Politicians and the Terrorism Industry Inflate National Security Threats, and Why We Believe Them.* New York: Free Press.

O'Malley, Pat. 1992. "Risk, Power and Crime Prevention." *Economy and Society* 21:252–275.

Simon, Jonathan. 1988. "The Ideological Effects of Actuarial Practices." *Law and Society Review* 22: 772–800.

Van Brunschot, E. and L. W. Kennedy. 2008. *Risk Balance and Security.* Thousand Oaks, CA: Sage.

Viscusi, W. Kip and Richard J. Zeckhauser. 2003. "Sacrificing Civil Liberties to Reduce Terrorism Risks." *Journal of Risk and Uncertainty* 26:99–120.

Wildavsky, Aaron B. 1995. *But Is it True?: A Citizen's Guide to Environmental Health and Safety Issues.* Cambridge, MA: Harvard University Press.

Willis, H. H., A. R. Morral, T. K. Kelly, and J. J. Medby. 2005. *Estimating Terrorism Risk.* Santa Monica, CA: RAND.

1.
ASSESSING THE RISK IN CRIME

Putting these risk approaches in context, the Van Brunschot and Kennedy chapter examines the different stages of risk: the run-up to an event, the incident itself, and its aftermath. In their piece, the authors suggest that risk operates at all levels of criminological analysis and changes its character at different stages of the crime incident. In addition, depending on the level of analysis (individual, institution, or state) risk analysis has been applied in various ways to identify, control, respond to, or prevent crime. The authors remind us that we are often confronted with uncertainty about how and why crime may occur but risk models provide tools for monitoring through surveillance and direct identification of hazards that can lead to bad outcomes.

There is, in addition, a realization that stems from a risk perspective, that is, risk can never be zero. There is always a chance that bad things will happen and, despite all good efforts to take crime down to new lows, we are confronted with the ongoing reality that we will always be facing crime of some sort. The authors emphasize, however, that the confusion that derives from a myriad of explanations and prescriptions about crime can be reduced somewhat when we draw on the common elements of risk. Understanding that we use risk ideas in different ways, in different contexts (we take risks, we lead risky lifestyles, we measure risks), provides a common theme that can go through our analysis when approaching crime problems in this way.

Risk provides a metric (albeit influenced by both subjective and objective factors) that can help tie different parts of the crime problem together. It offers a probabilistic interpretation to crime analysis that suggests that certain things are likely to happen and others can be prevented based on our risk assessments. It provides a framework, a common structure to our approach that is missing in previous work on crime. It also makes sense to all of the participants. Offenders know that they take risks (and talk about them in explaining their behavior); victims understand that they need to reduce risks and insure themselves against them; police consider risks in doing their jobs but also in understanding how certain areas and certain behaviors can be considered riskier than others.

THE RISK OF CRIME

A Framework

Erin Gibbs Van Brunschot and Leslie W. Kennedy

In January 2009, the BBC[1] reported on the Davos World Economic Forum where a panel of experts discussed the threat of cybercrime and how it is rising sharply. The experts determined that there were a number of threats associated with the internet—threats that not only come from different components of the world wide web, but also threats that differentially target particular users and user groups. The first threat identified is the threat of crime. It was noted that internet crime threats are wide-ranging, from individual hackers who might attempt to steal identities or crash systems, to the more recent threat emerging from organized crime groups who have been able to control and redirect many millions of dollars:

> in the United States, a "virtual" group had managed to hijack and redirect the details of 25 million credit card transactions to Ukraine. The group used the data to buy a large number of goods, which were then sold on eBay.[2]

The second threat is a design flaw emanating from the internet system itself. The experts noted that the internet system is disproportionately based upon a system of trust. This means that responsibility for catching and punishing those who misuse the system for fraudulent or criminal purposes is difficult to establish. For example, when an internet service provider (ISP) mistakenly made access to a particular YouTube video impossible for all viewers around the globe, it was unclear who was to take responsibility for such an occurrence. Ultimately, a group of volunteers remedied the situation.

The final threat identified is that of "cyber warfare." A recent political dispute between two countries, Russia and Estonia, resulted in internet services being cut off, effectively disabling Estonia's banking and electricity system. The experts note that the threat of cyber warfare becomes even more problematic as there has been a "multiplication of web-enabled devices,"[3] from phones to environmental sensors. While individual loss could potentially be dealt with, the damage done at the societal level through cyber warfare is grave: "systemic risk, where fraud and attacks undermine either trust in or the functionality of the system" could result in systems becoming unusable.

This brief discussion of internet threats highlights particular aspects of the discussion that will follow in this chapter—a discussion that focuses on the multidimensional concept of risk as it applies to the study of crime. First, threats associated with a system may be directed toward a number of targets. We see that not only are individuals potentially threatened by forms of internet-related crime, but so too are institutions and even entire political systems at risk from crime threats. Second, beyond being directed toward a variety of targets, threats also emanate from various sources. Individuals, in other words, may threaten other components of the system, and at the same time institutions or states may also pose threats. Third, the nature of the threat varies. Some threats may be more readily identifiable than others. It may be difficult, for example, for individuals to identify fraudulent internet activity—often our identification of this activity requires technical knowledge that we simply may not possess. Police forces may be similarly hampered in their ability to pursue particular types of crimes. These threats therefore remain at a level of abstraction unlike more familiar and concrete threats, such as traditional types of fraud. Fourth, the issue of responsibility is central to the determination of threat. It may be that threats are defined as less serious, or ignored, if we cannot afford or do not have the means to address them. The determination of threat may be heavily impacted by available resources. Finally, threats associated with the internet illustrate the dynamic and evolving nature of many types of crime threat. As noted above, cybercrime has changed and has evolved beyond individual perpetration to the inclusion of more organized crime groups. Similarly, threats may evolve and become non-threats due to the advent of technology, or simply due to prevention methods that occur in the early stages of threat identification.

Individual, Institutional, and Societal Risk

In everyday life, individuals make numerous risk assessments. They consider their environments, interpret them based on what they know and what they have experienced, and proceed accordingly. Particular individual characteristics may be associated with increased sensitivity to an assortment of cues in the environment, which may, in turn, be associated with both perceived and real vulnerabilities. Of course, individuals may assess mistakenly—they may register cues that have less impact than they believe, or fail to register cues that may actually cause them harm. Because individuals and institutions are differentially attuned to cues in the environment, vulnerabilities will also vary, presumably influencing the likelihood of victimization and harm. The elderly, for example, may view themselves as more likely to be criminally victimized than those who are younger, and may interpret environmental cues in light of this perception. Perceived vulnerability is also impacted by how individuals assess their likelihood of recovery and how quickly it is believed that recovery from victimization may take. Individual characteristics, along with associated environmental and contextual factors, result in individuals inhabiting different risk positions: individuals are more or less vulnerable and/or impervious to harm according to a host of factors.

Risk assessments are not only associated with individual circumstances. Risk can also play an important role in influencing the actions, as well as the policies and procedures that institutions, such as law enforcement agencies, may develop. For example, at its most basic, policing is about response. Police agencies are structured to be attentive to problems, to react quickly to disorder and crime, and to provide assistance to people who need it (not necessarily in terms of crime). Despite innovations that have encouraged policing to be more proactive, the methods by which police do their jobs have stayed predominantly reactive. At the same time, police leaders have begun to move their organizations toward more strategic planning and greater accountability of administration and officers for the crime that occurs. The emphasis on accountability has made police increasingly likely to explore ways in which they can do their jobs more effectively.

At the end of the twentieth century, two major changes took place in policing that brought about key modifications to the ways in which policing is done. These changes can be framed in terms of risk assessment—a consideration of probabilities of particular outcomes, both positive and negative. The first shift was the introduction in the late 1970s of the 911 phone system that provided the public with immediate access to the police in emergencies (although it took until the late 1990s for this to be designated as a nationwide emergency system in the United States). The impact was that 911 calls broadened the focus of police from patrol and random surveillance to one that was based on the demands by the public for assistance. Initially, all calls for assistance were treated equally despite substantial evidence that policing, to be effective, needed to be targeted depending on the seriousness of the incident being reported. For example, calls for burglaries long after the intruder had left the premises were considered as important as calls for assaults where the assailant might still be in proximity. After finding that requests for more resources to respond to this ever-growing demand were very difficult to meet, police soon began to develop methods of prioritizing calls. They began a practice of using dispatchers to estimate call importance, employing a calculation of the caller's ability to deal with the problem on her or his own; the amount of danger the situation posed to potential victims; the consequences of not arriving in a timely way. This estimation of risk was central to the strategies that police began to employ and decisions made with respect to the relative seriousness of incident.

A second major innovation enabled police to make sense of where to place resources in responding to particular types of offenses and also to monitor where police are called to for more serious crimes. This involved the building of information gathering systems, often referred to as CompStat programs, which monitored the distribution, in time and space, of criminal action and minor delinquency (Willis, Mastrofski, and Weisburd, 2003). CompStat programs became important diagnostic tools allowing police to determine if and how their actions influenced crime; allowing them to use resources gained by reducing the constraints of 911 systems; and enabling redeployment to areas in need of the most help. Most importantly, it taught police that, when concentrating their attention on problem areas, they had to follow up on issues that may have initially resisted intervention. Though still based on reactions to crime occurrences, this approach gave policing agencies the opportunity to adopt new elements of risk management helping to define, albeit on a short time horizon, how police resources were to be managed.

Risk also has a more societal and global aspect, as the cybercrime example at the beginning of the chapter suggests. Efforts have been directed at creating legislation to reduce the likelihood of crime. The European Institute for Crime Prevention and Control, for example, undertook a study to "reduce the probability that an economic/

financial/organized crime [would occur] due to opportunities arising from legislation."[4] Further, the probability of crime is reduced by enforcing product codes that require manufacturers to sell electronic goods that are hard to steal. This "crime proofing" assumes that, by changing the characteristics of consumer goods with crime risk in mind, less theft will occur.

At the state level, we also have examples of the ways in which private sector security operations address the threats that exist while conducting business on a global scale. In their ongoing efforts to update their employees and clients about the risks to the corporation on an international level, CISCO Systems, the world's largest provider of internet technology, produces a widely distributed document outlining the problems that they face. Recently, the authors of the report identified seven key threats to security (primarily related to cybercrime but also to issues related to global business). These threats include: vulnerability of the internet; threats to physical assets; lack of legal protections; lack of trust; identity protection; human error; and geopolitical conflict. The report illustrates the new thinking in corporations about crime and other forms of risk that are posed in doing business, domestically and internationally. The challenges in meeting these types of threats set up new requirements within these companies and heavily influences both how they do business and the expectations they place on governments and law enforcement agencies to provide them with protection and support in their efforts to mitigate these risks.

Nature of Threat/Hazard

A discussion of the varying nature of threats is facilitated by reference to the example of surveillance cameras or closed-circuit television (CCTV). CCTV has been promoted as a method of situational crime prevention. If would-be offenders know there are cameras recording their movements, this may serve as a deterrent since the likelihood of getting caught for a crime is increased. CCTV is also promoted as a crime-solving method—those who commit crimes are more likely to be identified after the fact than they might be otherwise. Yet despite the deterrence and crime-solving aspects associated with CCTV, there are also limitations. In terms of deterrence, cameras are focused on a specific and limited geographic area. Crime may still occur in the immediate vicinity of a camera simply due to the fact that the camera may not be pointed in the direction in which a crime could take place. The use of CCTV as a means of facilitating identification after the fact is simply that—after the fact. A crime may take place and CCTV becomes a responsive technique, rather than a preventative measure.

At the same time, CCTV also highlights how there are abstract threats that impact on the perceived likelihood of crime. While CCTV is geared toward specific areas—cameras are aimed at and located in particular directions—this method can only address abstractions such as particular categories of people who might offend. For example, setting up a camera at a bus station can only serve as a general deterrent—specific offenders are not necessarily identified, but rather, the idea is that the cameras simply address the abstract threat that someone (we don't know who) will commit a crime and be deterred. As with random checks (if they are, in fact, random), these are not used to identify specific offenders but rather address the location—a subway—as being the type of site in which these abstract threats ("those kinds of people", whoever they may be) are realized.

While cameras may address abstract threat, they may also help to relieve or address abstract anxiety. CCTV may provide the public, in some instances, with feelings of reassurance that whatever threatens is in some way reduced by the mere presence of the cameras. Whether cameras and security measures inspire confidence and reassurance of diminished threats, or simply serve to remind the public that threats prevail, is a matter of some debate (Welsh and Farrington, 2003).

Threats may therefore be abstract—with few specific identifiers attached, but characterized by an associated possibility of harm, nonetheless. One of the more significant abstract threats in the twenty-first century is the "terrorist threat." The identities of terrorists are not necessarily known and also expand and contract. When threats are abstract, the means by which deterrence may be achieved is wide-ranging and possibly ineffective. Airports, for example, apply their general deterrence policies and procedures to all travelers in the hope of identifying an individual who is a specific threat. There are huge costs associated with this type of approach. At the same time, it is possible to identify specific threats—threats that have been recognized through various methods as constituting either a source of harm or a potential source of harm. Many communities receive police alerts when particular offenders are released into the community, for example. Such offenders have been known to be a source of harm in the past, and the expectation is that they may continue to threaten (if not actualize) harm in the future. Dealing with risk means that the nature of the threat—abstract versus specific—must be considered.

The Dynamic Nature of Threat

The introduction regarding the internet and crime suggests that the potential for harm does not remain static. Instead, threats may recede or come to the forefront; they may be removed, or they may become more threatening. By receding, there may be other threats that become priority concerns. Threats that perhaps initially appear more benign may be altered over time and become more potentially harmful. New information may become accessible that either eliminates various threats or suggests that abstract threats are now specific (abstract terrorist threat may be revealed as a specific plot by specific individuals to blow up a particular facility or structure, for example).

Along with the changing nature of the threat itself, threats also have a temporal dimension associated with them. In the first instance, in the precursor stage, threats are a prediction of a future possibility—they anticipate a particular type of outcome. In the second instance, threats may be realized—a crime is committed, a bomb goes off, or a fraud occurs. In the next stage, there is a response to the actualization of a particular type of incident. What are the consequences of an incident occurring? Do the consequences spur change in order to prevent a similar occurrence? For example, in the aftermath of the bombing of the New York World Trade Center towers in the fall of 2001, many measures were put into place at airports to prevent a similar event from occurring again. Importantly, the aftermath of these bombings is really the anticipation of a future incident: every aftermath is a precursor to another event. Individuals, organizations and states are continually dealing with a variety of threats in a variety of stages.

Responsibility for Threat

Our consideration of "threat" has made clear that there is a chance that harm or negativity might occur. (Likewise, opportunities suggest a chance that something positive may occur.) Although threats are directed to the possibility of a future event, at the same time, threats may be identified because of what has occurred in the past. We know, for example, that cybercrime and cyber warfare have occurred, are occurring, and threaten (are likely) to occur in the future. The identification of threats draws our attention to how it is that threats may be minimized or eradicated before any harm or damage may be done. All of this begs the question as to how threats are identified in the first place. Clearly there are a number of factors, ranging from information and data, to resources, to political pressures—some of which make some more "evident" than others.

Associated with the identification of threat is responsibility. Once a threat has been identified, often the implication of such identification is that something will be done to address the threat. For example, if we note that the parking lot we use has become a hangout for gangs, having identified this as threatening, we may change our behavior and park elsewhere. If, however, we have little means or opportunity to change our parking place we may "re-negotiate" how it is that we frame this "threat." Rather than being seen as threatening, we may identify these gangs as "groups of kids" and therefore not imbue the group with a moniker that is suggestive of threat (if we suppose that "gang" is indicative of a threat). Another course of action may be that we weigh the risk of harm from that group with the costs associated with parking much farther away. The convenience of the parking spot may outweigh the threat posed by the gang, reducing the threat to something that might be considered acceptable. Once a threat is identified, what is the cost of addressing it?

The challenge of dealing with crime risk is essentially the need to balance the cost of response against the likelihood of victimization. In other words, there are times when positions regarding "acceptable" risk must be taken, realizing that a certain amount of loss may occur in the normal course of business. We recognize, deal with, and define acceptable risks during the course of our everyday lives. But acceptable risks must be defined at all levels—not just by individuals. Institutions, governments, and states define levels of acceptable risk and weigh whether applying resources to reduce risks is cost effective, including both social and economic costs. Certainly, when there are scarce resources and limited funds, there may be an implicit decision that a certain amount of acceptable risk will have to be tolerated in certain areas.

A Framework of the Three Stages of Risk

In examining crime from a risk perspective, we are cognizant of the many different factors we need to examine to see how probabilities, propensities, and likelihoods connect the important factors used to combat and control crime (Kennedy and Van Brunschot, 2009). We have identified three major facets of risk assessment where the risk dynamic relates to these different dimensions of risk (see Table 1.1). First, we identify the nature of the hazard under consideration and whether it constitutes a more general and abstract threat (suggesting impending or potential harm) or a specific danger/hazard. More abstract threats may result in free-floating anxiety that can cause harm. The uncertainty creates challenges as the efforts to prevent abstract harm can be costly and is often hard to justify if the threat is undefined or indeterminate. On the other hand, when facing specific hazards, our framework accounts for the ways in which we prepare for, respond to, and recover from these incidents. Second, we consider the level of analysis (individual, institution, or state) associated with the threat or hazard under consideration. We recognize that people and institutions face threats and hazards that are sometimes abstract and indirect in addition to those that are very specific and readily identifiable. Depending upon the level of analysis, threats will not vary—states are not necessarily threatened by the same things as individuals, for example; nor are individuals, institutions and states comparably threatening.

On a third dimension we offer a spectrum that runs from the precursors of crime, to realization of threat (the incident), to the aftermath (Sacco and Kennedy, 2002). As

Table 1.1 Risk in Crime Matrix

Stage	*Nature of threat/hazard*	*Level of analysis*	*Example*
1. Precursor	General/abstract threat [uncertainty]	State	Surveillance
		Institution	Quality of life policing
		Individual	Fear
	Specific hazard [certainty]	State	Public security
		Institution	Situational crime prevention
		Individual	Criminal careers
2. Transaction	General/abstract threat [uncertainty]	State	Crime waves
		Institution	Managing threat information
		Individual	Multiple victimization
	Specific hazard [certainty]	State	Crime rates
		Institution	Hot spots
		Individual	Scripts
3. Aftermath	General/abstract threat [uncertainty]	State	Crime trends and statistics
		Institution	CompStat/punishment
		Individual	Hate crime
	Specific hazard [certainty]	State	Border control
		Institution	Sex offender registries
		Individual	Domestic violence

Vince Sacco and Leslie Kennedy (2002) make clear, crime explanations can accommodate different factors, and the importance of each explanation depends on what stage of the criminal event is under consideration. The event perspective allows us to tie different components together to explain individual, group, and institutional influences and impacts on crime. The criminal event perspective reminds us to anticipate the effects of motivation, vulnerability, and prevention on crime. From this vantage point, what is left implicit is the likelihood of an event beginning, continuing, and ending in a certain way.

This three-dimensional matrix serves as an organizational tool for locating a multitude of crime concepts that draw on the notion of risk and suggest how particular ideas may be used to explain the different types of issues that are relevant along particular dimensions. The examples on the right-hand side of Table 1.1 illustrate how risk comes into play as suggested by the three dimensions of risk under consideration (nature of the threat, level of analysis, and time frame or stage).

Part of the dilemma in our understanding of crime comes from the imprecision in our understanding of complex systems that we encounter and that actions based on individual decision-making, institutional practice, and state-level policies have important impacts on and for the other levels. In fact, the processing and understanding of risk at these different levels may actually come into conflict with one another. This reality means going beyond simple or single factor solutions. Rather, it requires that we think probabilistically as we face uncertainty and that we learn from the effects of our decisions as we set out to cope with what may be threatening and disorganized environments. As Paul Slovic (2000: 135) observes: "Uncertainty about facts and uncertainty about values both imply that determining the acceptability of a hazard must be an iterative process, partly because, as time goes on, we learn more about how a hazard behaves and how much we like or dislike consequences." The bottom line is that we learn about managing risk through trial and error.

From this matrix, we can see how the concept of risk is central to our thinking about crime in a variety of ways, including: surveillance, quality of life policing, fear, scripts, sex offender registries, border control, and so on. The wide array of risk-related concepts shows how deeply imbedded these ideas are in and to our thinking about crime.

Conclusion

In presenting our model, we suggest that the concept of risk operates at all levels of crime analysis and changes its character depending on whether or not we look at it before, during, or after the crime incident. In addition, depending on the level of analysis (individual, institution, or state), we are able to see that risk analysis has been used in different ways to identify, control, respond to, or prevent crime. Even though we are often confronted with uncertainty about how and why crime occurs, risk models provide tools for the identification of hazards that can lead to negative outcomes. There is, in addition, a realization that stems from a risk perspective: the probability of a negative outcome can never be zero. There is always a chance that bad things will happen—but we can reduce the probability of negative events by anticipating their occurrence and by taking measures to reduce their likelihood. Despite such efforts, the reality is that we will likely always face some crime probability.

Our discussion emphasizes that the confusion that derives from a myriad of explanations and prescriptions about crime can be reduced somewhat when we draw in the common elements of risk. Understanding that we use risk ideas in different ways, in different contexts (we take risks, we lead risky lifestyles, and we measure risks), there is still a common theme that runs through our analysis when approaching crime problems in this way. Risk, as we have said, provides a metric (albeit influenced by both subjective and objective factors) that can help us tie different parts of the crime problem together. It offers crime analysis probabilistic interpretations that allow us to suggest that certain things are likely to happen while others can be prevented based on our risk assessments. It provides a framework, a common structure that has been missing in previous work on crime. It also makes sense to all of the participants. Offenders know they take risks (and talk about them in explaining their behavior); victims understand they need to reduce risks and insure themselves against them; and police consider risks in doing their jobs but also in understanding how certain areas and certain behaviors can be considered riskier than others. There is a common script (if not a complete language) found in risk that can govern our study of crime. The intent of our framework is to provide a means to understand how it is that seemingly disparate approaches and philosophies may be, in effect, broken down and replaced by a more integrated, contextual approach to crime through an elaboration based on risk.

Notes

1. http://news.bbc.co.uk/2/hi/business/davos/7862549.stm, accessed January 2010.
2. http://news.bbc.co.uk/2/hi/business/davos/7862549.stm, accessed January 2010.
3. http://news.bbc.co.uk/2/hi/business/davos/7862549.stm, accessed January 2010.
4. http://marc.unicatt.it/MARC_Annex_I_March_2006.pdf, accessed October 2010.

References

Kennedy, L. W. and E. G. Van Brunschot. 2009. *The Risk in Crime.* New York: Rowman and Littlefield.

National Research Council. 1983. *Risk Assessment in the Federal Government: Managing the Process.* Washington, DC: National Academy Press.

Sacco, V. and L. W. Kennedy. 2002. *The Criminal Event: Explorations in Time and Space.* Belmont, CA: Wadsworth.

Slovic, P. 2000. *The Perception of Risk.* London: Earthscan.

Stern, P. and H. Fineberg. 1996. *Understanding Risk: Informing Decisions in a Democratic Society.* Washington, DC: National Academy Press.

Welsh, B. and D. P. Farrington. 2003. "Effects of Closed-Circuit Television on Crime." *The ANNALS of the American Academy of Political and Social Science* 587(1):110–135.

Willis, J., D. Mastrofski, and D. Weisburd. 2003. *Compstat in Practice: An In-depth Analysis of Three Cities.* Washington, DC: The Police Foundation.

Questions

1. How do we determine what are the major threats to global security in the coming decade?
2. What are the similarities and differences in threats to global security posed by crime versus terrorism?
3. In considering uncertainty when comparing crime to terrorism, how do we deal with the major differences in the number of times that incidents occur? How does this affect the ways in which we identify and respond to these problems?
4. The degree of uncertainty in events can be affected by the presence of willing collaborators. How do risk assessments look at this co-offending problem differently for terrorism than for crime?

2.
RISK AND OFFENDER DECISION-MAKING

Moving past the assessment of risk of particular incidents and making plans for response and recovery, analysts are also deeply involved in identifying systemic aspects of modern hazards. There are many different factors that contribute to risk of harm from these dangers and form into definable patterns of harm. Criminologists, for example, talk about risky lifestyles, hot spots, drug markets, crime rates, and crime waves. These patterns become important considerations in the ways in which we approach explanations of the origins of crime and our analysis of how crime responds to prevention and law enforcement. Within these regularities, we can view the potential of crime or the concentration of criminality. In addition, we can use patterns to define the relationship that these behavioral outcomes have to characteristics of the offenders and victims that are involved in day-to-day interaction. Changes in society that come from shifting demographics, economic booms and busts, and movement of people through migration all affect the patterns of risk.

The Corsaro and McGarrell chapter explores how risk analysis can be used to identify trends in homicide and for assessing risk reduction strategies. They describe a "pulling levers" approach that emphasizes a directed intervention plan that identifies and responds to the risk of gun offending and prescribes agency responses directed at potential offenders. As they describe it, from a theoretical perspective, pulling levers interventions can be considered focused deterrence strategies based on the characteristics associated with offending and the criminal justice system response to offending. So, from the beginning the strategy is specifically focused on the problem of gun crime as a way of addressing violent outcomes. Second, this approaches works from the assumption that a small number of offenders accounts for a disproportionate number of crimes.

Third, the strategy that emerges takes advantage of the fact that much of this chronic offending occurs within a group setting. As a result, pulling levers relies on gangs and networks of crews or offenders to warn others of the risks that await them from gun offending and victimization. Thus, the involvement of many agency members focused on a specific problem should increase the severity and certainty of penalties, leading to alterations in perceptions about sanctions and risk. The importance of clearly defining the risks and connecting these to punishment is clearly communicated in this approach. The meetings that are held with offenders provide the types of new information that may cause offenders to reassess the risks of committing gun crime.

REDUCING HOMICIDE RISK

Nicholas Corsaro and Edmund F. McGarrell

Introduction

Risk is a term frequently used in criminological research. A cursory search of the National Criminal Justice Reference Service or Criminal Justice Abstracts using the term "risk" yields thousands of "hits." Yet few of the associated articles, books, and reports conceptualize risk from the risk assessment, management, and governance framework presented in this volume and observed in fields such as environmental science, food security, business, and public health. In criminology, risk is often focused on individual-level factors associated with the risk of being victimized, the risk of committing criminal or delinquent acts, and the risk of re-offending. Another body of research examines the perceived risk of being victimized and the perceived likelihood of the detection and punishment of offending. Some research comes closer to the risk assessment and governance framework when examining security risks and emerging crimes such as terrorism, transnational crime such as money laundering and internet-based fraud, and identity theft. Yet, rare are the applications of the risk framework to traditional street crime.

This chapter reports on a planned intervention that was designed to reduce levels of homicide in Indianapolis, Indiana. Although conceptualized as a problem-solving intervention, review of the initiative suggests it followed a risk assessment, management, and governance approach. Specifically, the initiative began with a careful analysis of the variation in homicide, or the risk terrain for homicide. The intervention, or risk management strategy, focused on the factors associated with elevated risk for homicide and a risk communication strategy directed at high-risk groups. The risk governance component included ongoing assessment of the impact of the intervention and the creation of partnerships, or a risk management network, responsible for implementing, refining, and sustaining the risk management program.

Boston's Operation Ceasefire

In the mid-1990s, the Boston Police Department designed an intervention strategy to address the city's growing gun violence and homicide problem (Kennedy, Piehl, and Braga, 1996). The program known as Operation Ceasefire was initiated by a multi-agency working group involving the US Attorney's Office, local prosecutors, the Boston Police Department, probation, youth service workers, and a team of researchers from Harvard's Kennedy School of Government. Data analysis indicated that Boston began experiencing a significant increase in the number of youth homicides, and as a result a set of strategies was employed that was intended to reduce this heightened level of youth victimization.[1]

The program was designed to reduce violent crime by using a deterrent-based message targeted specifically at high-risk offenders while also relying on the involvement of pro-social groups to provide desisting offenders access to legitimate opportunities (e.g., employment, housing, mentorship, community service, and drug treatment programs). Impact analysis indicated a significant reduction of 63 percent in the monthly number of youth homicides in the city following the intervention (Braga et al., 2001).[2] Officials in Indianapolis became aware of the Boston strategy, and building upon the momentum from a recently implemented "directed patrol" deterrence-based policing intervention (see McGarrell et al., 2001), key personnel decided to replicate the Ceasefire program.

Setting and Background

Indianapolis is the county seat for Marion County and the capital of the Midwestern state of Indiana. The city had an estimated population just over 780,000 people according to the 2000 US Census, making it one of the top 20 most populated cities in the nation and the third largest city in the Midwest, behind Chicago, IL and Detroit, MI. At the turn of the century, Indianapolis ranked just inside the top ten in violent crime rates (ninth), murder and aggravated assault rates (eighth), and robbery rates (seventh) among cities with a population between 500,000 to 999,999 (see Pastore and Maguire, 2000).

Related to the context of the city's violent crime rate, the Office of Juvenile Justice and Delinquency Prevention conducted the National Youth Gang Survey in 1988. In this particular study, the questionnaire contained a variable that allowed researchers to make distinctions in terms of defining each city's localized gang problem (Spergel and Curry, 1993). "Chronic gang cities" were defined as cities that had developed gang problems prior to 1980, while emerging gang cities were noted to have developed gang problems after 1980. Importantly, the citywide homicide rate in Indianapolis rose from 10.2 (per 100,000) in 1990 to 20.1 in 1998 (Chermak and McGarrell, 2004). Like many US cities during this period, Indianapolis began experiencing rises in violence and gun crime that were associated with the evolving crack cocaine drug markets and subsequent gangs involved with the illegal distribution of narcotics (Blumstein and Rosenfeld, 1998).

In response to this record increase in lethal violence, officials decided to go beyond simply trying to boost homicide clearance rates as a way to address the problem. In December 1997, the Indianapolis Violence Reduction Partnership (IVRP) was founded as a coalition of criminal justice agencies (i.e., police, prosecution, probation, and parole), social service providers, and community groups.

Law Enforcement Response

One risk for police, criminal justice officials, and political leaders is that pressure to respond swiftly at a time when concerns about crime control are at an extremely high level may result in unintended and negative consequences that include (but are not limited to) perceived corruption, procedural injustice, and enforcement inconsistency that may generate community resistance and even rebellion (Morris and Hawkins, 1970). Research shows that many police crackdown initiatives (i.e., saturation-based enforcement approaches) are either unsuccessful or when they impact crime they most likely will have short-term success (Sherman, 1990). The literature suggests a thoughtful and deliberate set of approaches focused on specific high-risk contexts offers the greatest promise for sustained crime reduction and program impact (National Research Council, 2004).

One of the initial key decisions made by the IVRP task force was to follow a framework that centered on data analysis, target identification, and crafting a number of strategic approaches to address the problem (see Goldstein, 1990). Retrospective bi-monthly homicide incident reviews examining 1997 incidents (onward) began in January 1998, and continued well beyond the time frame of the evaluation presented herein. Detailed incident and actor information were discussed among coalition members that centered on who was involved in each incident, the networks of association that existed among offenders (as well as victims), and the context of the neighborhoods where events occurred.

By mid-year 1998, it became exceedingly clear that a vast majority of homicide victims and offenders were routinely involved in loosely structured gang networks (i.e., either were specific gang members or were part of a known group of gang associates). In addition, the reviews indicated that most homicide participants were in their late teen to early adult years and were almost exclusively male. Another consistent theme that emerged was a clear nexus between drug use, drug distribution, and homicide incidents (e.g., robbery of a known crack-house, retaliation for a drug deal gone wrong, etc.). Firearms were involved in the vast majority of homicides and most offenders and victims had prior felony convictions that legally prohibited them from possessing firearms. Finally, geographic data showed that a vast majority of the drug/gang/gun related incidents occurred in a relatively small number of areas. In sum, the lengthy risk analysis identified the salient and consistent themes of homicide incidents in Indianapolis. In 1999, IVRP officials began to implement a series of specific strategies to address the growing lethal violence problem.

The Indianapolis Intervention

The Indianapolis intervention (outlined in Table 2.1) followed many of the tenets of Ceasefire, which was a strategy designed to reduce youth and gun violence in Boston that was comprised of the following key elements: (1) law enforcement crackdowns on illegal gun traffickers; (2) notification to gang members that continued involvement in violent crime would no longer be tolerated; and (3) in the event notified offenders continued to engage in violent crime, the multi-agency response team would impose all possible sanctions on chronic offenders residing within or found near the area where the incident occurred, or with individuals involved in violence (Braga et al., 2001; McGarrell and Chermak, 2003a, 2003b).

The first component of the strategy involved a range of standard law enforcement practices, which included canvassing homicide incident locations, conducting detailed investigations, and attempting to build and strengthen community networks. In essence, saturation patrols, police crackdowns, and community-oriented policing approaches were all a part of the strategy in high-risk neighborhood contexts (Weisburd and Braga, 2006).

The second component of the IVRP initiative began in 1999 when the multi-agency task force conducted a series of notification hearings called "pulling levers" meetings with high-risk individuals in high-risk contexts (Kennedy, 1997). Between 1999 and 2000, a total of 30 such meetings occurred (including three in early 1999 in violent crime hot spots). In essence, the risk communication message that emerged was a targeted deterrence approach combined with linkages to pro-social and legitimate opportunities. Persons identified through the rigorous investigation phase were summoned to the group-based meetings. Once there, state and federal law enforcement officials, state and federal prosecutors, and probation and parole agents warned high-risk individuals (i.e., currently on probation or parole, extensive criminal histories, prior gun offenses, known gang associates in high-crime neighborhoods, etc.) that continued offending would lead to the heaviest possible sanctions, or that "every possible legal-lever (sanction) would be pulled" that was available to the working group (for a lengthier review, see Braga et al., 2001). These individuals were also informed that they were at high-risk of being a victim of lethal violence. In addition to the deterrence message, notified individuals were provided enhanced access to social service, counseling, and community-service provisions in an effort to both facilitate a transition out of offending the offending lifestyle, and to reduce potential program "defiance" (Sherman, 1993).

The third component of the strategy was implemented on a case-by-case basis with specific groups and individuals who continued to offend despite the notification. This was also the case with the use of social service provisions as a way to assist offenders out

Table 2.1 Major Elements of the IVRP Intervention

Component	*Selected risk reduction strategies*
Law enforcement	• Community outreach with members of neighborhood watch programs. • Contacting local "weed and seed" community partnerships. • Visiting (and presenting at) area middle and high schools. • Partnership with a group of faith-based ministers from Indianapolis. • Multi-agency response to each homicide incident. • Increased directed patrols in homicide hot-spot areas.
Notification sessions	• A total of 30 notification meetings held between 1999 and 2000 including those with high-risk probationers and parolees associated with gang networks. • Meetings held in three violent crime areas (February and March, 1999). • Notification to male juveniles in detention facility (September 1999).
Imposition of sanctions	• When violence after a meeting did occur, the working group followed through with the maximum possible sanctions against those in violation. • Brightwood investigation and crackdown (April 1999). • Task force members in subsequent meetings used sanctions levied against the Brightwood gang as part of the deterrent message.
Social service provisions	• Individual offenders were provided a range of specifically crafted social service opportunities in order to assist with their transition, which included: mentoring from ex-offenders, employment, housing, substance abuse, education, and vocation training.

of the offending lifestyle. Chermak and McGarrell illustrated, "when violence occurred after a meeting, the members of the working group would pull ever available lever in response to violent acts and then specifically describe why such levers were being pulled" (2004: 166). It is also worth noting that in April 1999 officials in Indianapolis conducted the Brightwood investigation, which was a drug and gang market crackdown. As a result, officials arrested 16 individuals (known as the "Brightwood gang"), seized 78 firearms, 500 grams of crack, 12 kg of powered cocaine, and over US$150,000 in cash. The arrests and subsequent prosecution of the gang members served as an illustration of the seriousness of the deterrent message (for more detail, see McGarrell and Chermak, 2003a, 2003b; McGarrell et al., 2006).[3]

Assessing Impact

A number of rigorous evaluations have been conducted to assess the changes in violent crime patterns associated with the implementation of the IVRP intervention. In this section, we provide a detailed review of the scholarly evaluations that have been conducted measuring changes in both process and outcome measures. Next, we follow the summary of the research findings with a more comprehensive analysis that focuses on the change in homicide risk patterns among the most vulnerable population in the city between the pre- and post-intervention period. Ultimately, we conclude with an outline of directions for future homicide research from a risk perspective.

Awareness

Kennedy (1997) contends that offender awareness of the increased risks associated with continued illegal behavior is central to the pulling levers initiative. Consequently, Indianapolis researchers added a series of "IVRP awareness" questions as an addendum to the local Arrestee Drug Abuse Monitoring survey, which can best be described as a methodological instrument (employed by approximately 35 US cities during the period examined here) to assess recent and lifetime drug usage behavior patterns among recent detainees (Chermak and McGarrell, 2004; Decker, Pennell, and Caldwell, 1997). These panel data served as indicators of perceived effectiveness (from an evolving population of offenders) of the local criminal justice community to respond to violent crime.

Across eight waves of data collection that included the added measures of IVRP awareness,[4] results indicated a steady, statistically significant, and progressive increase among new arrestees perceptions' regarding the likelihood of enhanced sanctions for future violent crime activity (Chermak and McGarrell, 2004: 176). More specifically, when asked about the likelihood of local criminal justice responses for homicide as well as robbery, an average increase of nearly 16 percent (i.e., increasing from roughly 72 percent of respondents to nearly 88 percent of respondents) believed future involvement in these offenses would yield arrest, charges, conviction, and receiving enhanced prison sentences. The greatest increase regarding changes in perceived sanctions between any two time periods of data collection occurred between wave 1 and wave 2 in 1999, which corresponded with the aforementioned April 1999 Brightwood investigation.

Comparatively, when asked about the likelihood of sanctions against "drug sales" over this same period, an average increase of roughly 3.2 percent (i.e., increasing from roughly 58 percent to 61.2 percent of all respondents) believed future involvement in illegal drug distribution would yield higher levels of punishment (Chermak and McGarrell, 2004). Given that the IVRP intervention focused specifically on violent crime, these data indicate that samples from the offending population between 1999 and 2001 became increasingly more likely to believe that enhanced criminal justice sanctions were more likely to follow violent crime activity.

Effectiveness

Ultimately, IVRP was intended to reduce firearm related offenses, lethal violence, and gang crime. McGarrell and Chermak (2003a, 2003b) also showed that homicides experienced a linear decline between 1997 and 2001, and that the percentage of incidents involving a firearm declined from 76 percent to 69 percent between pre- and post-intervention periods.[5] Thus, preliminary evidence indicated that IVRP was successful at altering the process of homicide investigations and subsequent criminal justice responses.

In addition, a study relying on a quasi-experimental time series research design showed that Indianapolis experienced a statistically significant decline of 34.3 percent in homicides after the April 1999 Brightwood gang implementation date (McGarrell et al., 2006: 222). Homicide data for six similar Midwestern cities (i.e., Cincinnati (OH), Cleveland (OH), Columbus (OH), Kansas City (MO), Louisville (KY), and Pittsburgh (PA)) were compared over the same period (i.e., 1997 to 2001) and only Indianapolis experienced a statistically significant decline in homicides. Importantly, study findings also indicated that the homicide trends in Indianapolis followed an abrupt, permanent decline rather than a gradual short-term impact, which has been referred to as a "light switch" program effect (see Kennedy, 2006).

Another study that relied on a "within-city" time series research design indicated that gang homicides (i.e., incidents that involved either a confirmed gang member or a confirmed network associate of a gang) were significantly more likely to involve younger, non-white, male actors than were non-gang homicides (Corsaro and McGarrell, 2009). In addition, gang homicides experienced a statistically significant decline of 38.1 percent after the April 1999 intervention, compared with an 8.6 percent (and non-statistically significant) decline in non-gang homicides (seen in Figure 2.1). Thus, the vehicle behind the decline in Indianapolis homicides was the specific decline in gang homicides, which was the major focus of the IVRP initiative.

Perhaps most consistent with a risk-reduction framework, Corsaro and McGarrell (2010) conducted a study that assessed the changes in homicide patterns for the highest risk populations as well as changes in the highest risk communities (i.e., target neighborhoods where many of the community-based programs were implemented). Growth curve models indicated that citywide homicides involving 15–24 year old black and white males experienced the greatest reduction (by 59 percent and 62 percent respectively) between the pre-intervention period (1997–1998) and post-

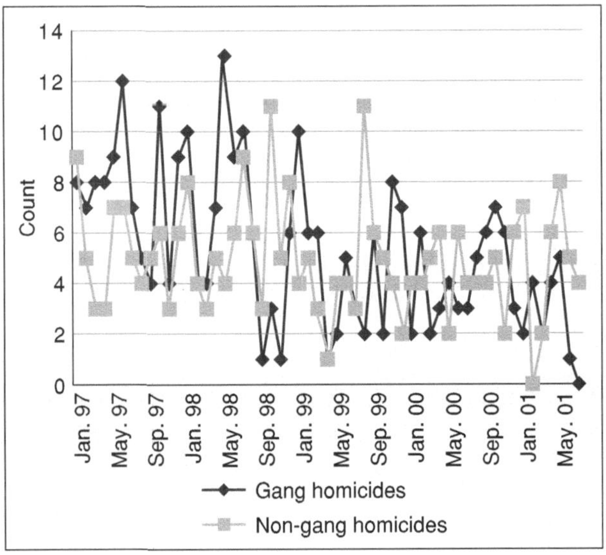

Figure 2.1 Trend in Indianapolis homicides

intervention period (1999–2000).[6] In addition, 15–24 year old black male homicides in the target neighborhoods declined by roughly 44 percent, indicating the statistically significant decline in Indianapolis homicides were risk and place specific. Figure 2.2 displays the bivariate change in risk-specific homicide patterns in Indianapolis during this period.

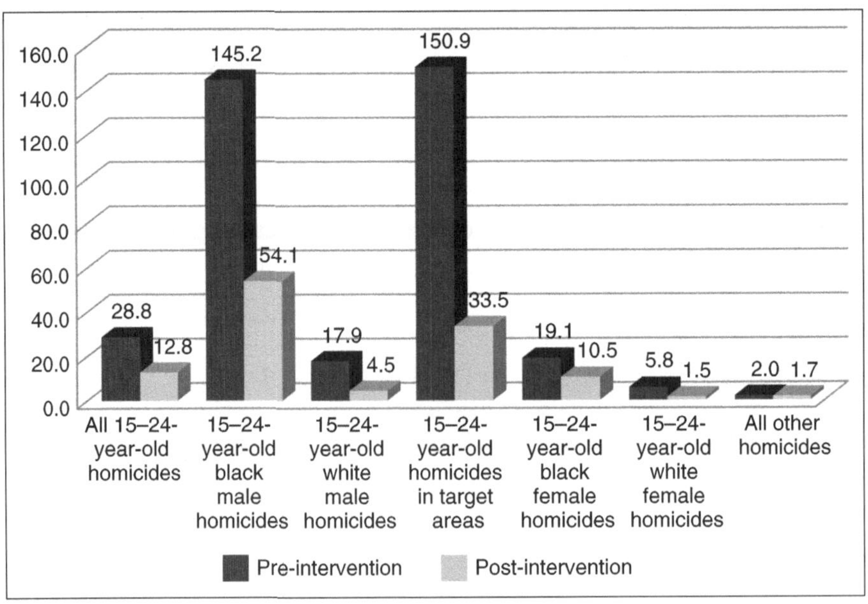

Figure 2.2 Homicide risk changes (per 10,000 at-risk population)

Discussion

The systematic review of the Indianapolis evaluation literature indicates that homicides across the city declined at a time and in a manner that corresponded with the IVRP intervention. The reduction in lethal violence was abrupt, immediate, and very consistent with the key elements of the risk-based pulling levers framework. Both geographic areas and targeted populations at greatest risk for homicide experienced the most substantive reductions following the multi-agency strategy.

In a larger context, criminal justice agencies have a long history of working independently from one-another and often without coordinating their efforts (Kennedy, 2006), and in some cases in an atmosphere of both dislike and distrust (Kennedy, 2002). The formulation of "networks of capacity" (i.e., the collaboration between criminal justice and social service agencies) appears to be extremely vital to effective communication and partnership building (Braga and Winship, 2006; Moore, 2002). The IVRP and Boston's Operation Ceasefire have been regarded as illustrations where these types of partnerships were forged and sustained. Additional sites that have replicated the strategy in an effort to reduce violent crime and have had similar levels of program impact, which include (but are not limited to) Chicago, IL (Papachristos, Meares, and Fagan, 2007), Eastern Los Angeles, CA (Tita et al., 2003), Lowell, MA (Braga et al., 2008), and Stockton, CA (Braga, 2008). These networks of capacity are also illustrative of risk governance networks responsible for risk assessment, risk management, and risk communication.

Along these lines, Braga and Winship (2006: 184) also note that replication sites such as Baltimore, MD, Minneapolis, MN (for a lengthier review, see Kennedy and Braga, 1998), and San Francisco, CA did not have sufficient networks of capacity in place prior to implementation of the Ceasefire program. In essence, they argue that this type of intervention is "relationship intensive" and must be based on a foundation of understanding and trust among inter-agency participants. Indeed, the Project Safe Neighborhoods (PSN) nationwide gun and violent crime reduction initiative was based on the foundation of the aforementioned studies. In a national evaluation of PSN, results indicated that sites with higher levels of "PSN dosage" (i.e., collaborative implementation, research integration, and enhanced federal prosecution) experienced a significantly greater reduction in violent crime than cities with less intense implementation (McGarrell et al., 2010).

In summary, the IVRP, and the Boston Ceasefire intervention upon which it is based, reflect a risk assessment, management, and governance approach. For decades, homicide was largely handled through a reactive police strategy that attempted to increase response times, clearance rates, and prosecution of offenders. The problem-solving approach (Goldstein, 1990), and we argue, the risk framework, fundamentally shifts the approach to a problem such as homicide. Systematic risk analysis was used to develop a risk terrain illustrating the concentration of homicide risk by geography, age, gender, race, network affiliation, and behavior (e.g., illegally carrying firearms, involvement in illicit markets). Risk management strategies, including risk communication, were then directed at these risk terrains. Risk governance included the ongoing assessment of the impact of these strategies as well as the formation of a risk governance network ultimately responsible for developing, managing, and sustaining the overall program. The fact that evaluations show both an overall decline in homicides but even more dramatic declines when focused on the risk terrain (gang, firearms, race, age, gender, and geography), suggest the potential power of the risk framework when applied to major social problems such as urban homicide.

Notes

1. A review of the literature indicated that "youth homicides" in Boston were originally defined as incidents involving those 21 years and younger (Kennedy et al., 1996), but this definition was later altered to include all incidents involving 24 year olds and younger as the strategy progressed (Braga et al., 2001).
2. It should be noted that Rosenfeld, Fornango, and Baumer (2005) performed a national trend analysis using growth curve models and controlled for factors shown to affect city-level violent crime rates. In their findings, they report less robust impact of the Boston gun project, though Berk (2005) raised a number of methodological limitations with the Rosenfeld et al. critique.
3. In a review of the deterrence literature and more specifically of the pulling levers strategy, Kennedy (2008) does not specifically indicate that a public illustration of enhanced sanctions against a group in violation is a necessary component of an intervention. However, it is equally important to note that such public illustrations have taken place in Boston (i.e., "Intervale Posse" gang) and Minneapolis (i.e., "Bogus Boyz" gang) as well as Indianapolis. Thus, future research focusing on problem-oriented policing/pulling levers should attempt to disentangle the degrees to which crackdowns on gangs that continue to engage in violence actually impacts the strategy.
4. The eight waves of Indianapolis surveys occurred in a quarterly fashion (i.e., roughly every three months) between 1999 and 2001.
5. The pre-intervention period was operationalized as the monthly count of homicide incidents between January 1997 and April 1999, while the post-intervention period was defined as April 1999 through July 2001 (i.e., just over a two-year pre-/post-intervention period).
6. The vast majority of the IVRP intervention studies have used April 1999 as the "intervention date" as the public demonstration of increased risks of criminal justice sanctions given the use of monthly homicide count data. Importantly, the Corsaro and McGarrell (2010) study used annual homicide counts as outcome variables in order to integrate neighborhood-level measures into corresponding hierarchical statistical models. The authors re-examined the published statistical models relying on annual count measures that centered on April 1999 (i.e., year 1 = April 1997 to March 1998; year 2 = April 1998 to March 1999, etc.), and the results were virtually identical.

References

Berk, Richard. 2005. "Knowing When to Fold 'Em: An Essay on Evaluating the Impact of Ceasefire, Compstat, and Exile." *Criminology and Public Policy* 4:451–466.

Blumstein, Alfred and Richard Rosenfeld. 1998. "Explaining Recent Trends in US Homicide Rates." *Journal of Criminal Law and Criminology* 88:1175–1216.

Braga, Anthony A. 2008. "Pulling Levers Focused Deterrence Strategies and the Prevention of Gun Homicide." *Journal of Criminal Justice* 36:332–343.

Braga, Anthony A., David M. Kennedy, Elin J. Waring, and Anne M. Piehl. 2001. "Problem-oriented Policing, Deterrence, and Youth Violence: An Evaluation of Boston's Operation Ceasefire." *Journal of Research in Crime and Delinquency* 38:195–226.

Braga, Anthony A., Glenn L. Pierce, Jack McDevitt, Brenda J. Bond, and Shea Cronin. 2008. "The

Strategic Prevention of Gun Violence among Gang-involved Offenders." *Justice Quarterly* 25:132–162.

Braga, Anthony A., and Christopher Winship. 2006. "Partnership, Accountability, and Innovation: Clarifying Boston's Experience with Pulling Levers." Pp. 171–190 in *Police Innovation: Contrasting Perspectives*, edited by David Weisburd and Anthony Braga. New York: Cambridge University Press.

Chermak, Steven M. and Edmund F. McGarrell. 2004. "Problem-solving Approaches to Homicide: An Evaluation of the Indianapolis Violence Reduction Partnership." *Criminal Justice Policy Review* 15:161–192.

Corsaro, Nicholas and Edmund F. McGarrell. 2009. "Testing a Promising Homicide Reduction Strategy: Re-assessing the Impact of the Indianapolis 'Pulling Levers' Intervention." *Journal of Experimental Criminology* 5:63–82.

Corsaro, Nicholas and Edmund F. McGarrell. 2010. "Reducing Homicide Risk in Indianapolis between 1997 and 2000." *Journal of Urban Health* 87: 851–864.

Decker, Scott H., Susan Pennell, and Ami Caldwell. 1997. *Illegal Firearms: Access and Use by Arrestees.* NIJ Research in Brief, Office of Justice Programs, Washington, DC: US Department of Justice.

Goldstein, Herman. 1990. *Problem-oriented Policing.* New York: McGraw-Hill Publishing.

Kennedy, David. 1997. "Pulling Levers: Chronic Offenders, High-crime Settings, and a Theory of Prevention." *Valparaiso University Law Review* 31:449–484.

——. 2002. "A Tale of One City: Reflections of the Boston Gun Project." Pp. 229–261 in *Securing our Children's Future: New Approaches to Juvenile Justice and Youth Violence*, edited by Gary Katzmann. Washington, DC: Brookings Institution Press.

——. 2006. "Old Wine in New Bottles: Policing and the Lessons of Pulling Levers." Pp. 155–170 in *Police Innovation: Contrasting Perspectives*, edited by David Weisburd and Anthony Braga. New York: Cambridge University Press.

——. 2008. *Deterrence and Crime Prevention: Reconsidering the Prospect of Sanction.* New York: Routledge.

Kennedy, David and Anthony A. Braga. 1998. "Homicide in Minneapolis: Research for Problem Solving." *Homicide Studies* 2:263–290.

Kennedy, David, Anne Piehl, and Anthony Braga. 1996. "Youth Violence in Boston: Gun Markets, Serious Youth Offenders, and a Use-reduction Strategy." *Law and Contemporary Problems* 59:147–196.

McGarrell, Edmund F. and Steven Chermak. 2003a. "Problem Solving to Reduce Gang and Drug-related Violence: Final Report on the Indianapolis Violence Reduction Partnership." Final report submitted to the National Institute of Justice. Retrieved April 23, 2009, from http://www.ncjrs.org/pdffiles1/nij/grants/203976.pdf.

McGarrell, Edmund F. and Steven Chermak. 2003b. "Problem Solving to Reduce Gang and Drug-related Violence in Indianapolis." Pp. 77–101 in *Policing Gangs and Youth Violence*, edited by Scott H. Decker. Belmont, CA: Wadsworth Publishing.

McGarrell, Edmund F., Steven Chermak, Alexander Weiss, and Jeremy M. Wilson. 2001. "Reducing Firearms Violence through Directed Police Patrol." *Criminology and Public Policy* 1:119–148.

McGarrell, Edmund F., Steven Chermak, Jeremy M. Wilson, and Nicholas Corsaro. 2006. "Reducing Homicide through a 'Lever-Pulling' Strategy." *Justice Quarterly* 23:214–231.

McGarrell, Edmund F., Nicholas Corsaro, Natalie K. Hipple, and Timothy S. Bynum. 2010.

"Project Safe Neighborhoods and Violent Crime Trends in US Cities: Assessing Violent Crime Impact." *Journal of Quantitative Criminology* 26:165–190.
Moore, Mark H. 2002. "Creating Networks of Capacity: The Challenge of Managing Society's Response to Youth Violence." Pp. 338–385 in *Securing Our Children's Future: New Approaches to Juvenile Justice and Youth Violence*, edited by G. Katzmann. Washington, DC: Brookings Institution Press.
Morris, Norval and Gordon Hawkins. 1970. *The Honest Politician's Guide to Crime Control.* Chicago: University of Chicago Press.
National Research Council. 2004. *Fairness and Effectiveness in Policing: The Evidence.* Washington, DC: National Academies Press.
Papachristos, Andrew V., Tracey L. Meares, and Jeffery Fagan. 2007. "Attention Felons: Evaluating Project Safe Neighborhoods in Chicago." *Journal of Empirical Legal Studies* 4:223–272.
Pastore, Ann L. and Kathleen Maguire. 2000. *Sourcebook of Criminal Justice Statistics—1999.* US Department of Justice, Bureau of Justice Statistics, Washington, DC: Government Printing Press.
Rosenfeld, Richard, R. Fornango, and Eric Baumer. 2005. "Did Ceasefire, Compstat, and Exile Reduce Homicide?" *Criminology and Public Policy* 4:419–450.
Sherman, Lawrence W. 1990. "Police Crackdowns: Initial and Residual Deterrence." *Crime and Justice: A Review of Research* 12:1–48.
——. 1993. "Defiance, Deterrence and Irrelevance: A Theory of the Criminal Justice Sanction." *Journal of Research in Crime and Delinquency* 30:445–473.
Spergel, Irving A. and G. David Curry. 1993. "The national youth gang survey." Pp. 361–402 in *The Gang Intervention Handbook*, edited by Arnold Goldstein and Ronald Huff. Champaign-Urbana, IL: Research Press.
Tita, George, K. J. Riley, Greg Ridgeway, Clifford Grammich, Allan F. Abrahamse, and Peter W. Greenwood. 2003. *Reducing Gun Violence: Results from an Intervention in East Los Angeles.* Santa Monica, CA: RAND Corporation.
Weisburd, David and Anthony A. Braga. 2006. "Introduction: Understanding Police Innovation." Pp. 1–27 in *Police Innovation: Contrasting Perspectives*, edited by David Weisburd and Anthony A. Braga. New York: Cambridge University Press.

Questions

1. When we consider how offenders make decisions about whether to act, what are the factors that we would consider lead them to decide that the risk of success are high?
2. What are the most effective ways of raising the risk of detection and punishment of offenders?

3.
UNDERSTANDING SCALABILITY OF RISK

Risk analysis must be scalable. We intend that the risk model should have elements that work at the smallest level of social interaction (in explaining the hazards that come from crime, for example) to the highest level (in addressing the effects of globalization on public security).

We have argued that the lessons learned in the areas of public health, disaster response, and environmental impacts can be applied in the study of crime and terrorism. The conception of risk as an integral part of planning and responding to the threats and hazards appear to have common elements across these topics of concern. Equally, we would argue that the risk models provide us with the opportunity to provide frameworks of analysis that are scalable; in other words, the application of central elements of the risk approach works equally well at the local, community, or global level. Now, this scalability allows us to accept certain basic principles in risk analysis to be true at all levels with the operationalization of the model varying depending on what we are studying.

As an example, the elements included in our risk model presented above suggest that the key elements of risk assessment include uncertainty, vulnerability, and exposure in conditions where there are identifiable hazards and threats. If we look at the scalability of uncertainty, we can see that this can pose quite different challenges in specificity and measurability, but uncertainty is relevant at both the local and global levels in assessing risk. The means available for reducing uncertainty certainly differ from level to level and the successes achieved in dealing with its consequences will also vary, but this does not reduce the importance of using this as a concept at both levels. And of course, the scalability of these ideas extends to the reality that individual-level assessments can influence reactions at the global level and vice versa (although this connection may be quite tenuous and hard to specify).

Kennedy, Marteache, and Gaziarifoglu explore the ways in which major agencies have developed risk assessment procedures in anticipating the negative effects of health crises, disasters, and terrorism. The authors ask whether or not there is a common methodology evident in these approaches that can be applied across all risk domains. In their review, they suggest that there are important similarities in approaches that can be adopted to bring risk assessment into a more standardized and systematic approach.

GLOBAL RISK ASSESSMENT

The Search for a Common Methodology

Leslie W. Kennedy, Nerea Marteache, and Yasemin Gaziarifoglu

Introduction

There have been many efforts in the past to confront the uncertainties and vagaries of future threats, dangers, and catastrophes that emerge in public health, in the aftermath of disasters, in controlling crime and terrorism, and in protecting our material well-being. Agencies such as the World Health Organization (WHO), Centers for Disease Control, United Nations (UN), Interpol, and others have been struggling to develop means by which they can plan for these outcomes and reduce their negative consequences. In this chapter, we will examine the literature on risk analysis that provides the framework for the study of threats. We will then review case studies from three areas of interest (health, disasters, and crime) to review how relevant agencies have addressed these problems through a risk perspective. We will then conclude with a discussion of the merits of a common methodology for risk assessment that can be used across these areas, identifying what can be shared and what is unique to the specific areas of threat.

Background

Since the establishment of first civilizations risks and adaptive responses to risks have been the foremost concern of all societies. The earliest and simplest analyses and management of risks exercised by the priests of Mesopotamia in BC 3200 have experienced many structural shifts with the development of numerical systems, probability, game theory, and cost-benefit understanding of risks. Yet the very same problems of hunger, drought, famine, and conquest that dominated the early societies as a byproduct of increased production, population, and settlement (McDaniels and Small, 2004: 1–11) have subsisted until now. Nonetheless the changes in the modes, means, and relations of production and the ongoing process of globalization have manifested and are still manifesting themselves in a constantly transforming risk discourse.

In defining risk as a global concern, the Global Risk Network describes systemic risk as the

> [p]otential loss or damage to an entire system as contrasted with the loss to a single unit of that system. Systemic risks are exacerbated by the interdependencies among the units often because of weak links in the system. These risks can be triggered by sudden events or built up over time with the impact often being large and possibly catastrophic. (World Economic Forum, 2010: 10)

The systemic nature of risk adds up to the homogeneity of risk. With globalization, many risks that are considered as developing countries' priorities become global priorities with the extent and speed of international movements (Wooldridge, 2001; World Economic Forum, 2010: 4). In risk assessment and management it is much more preferred to focus on risks on a one-by-one basis since the expectations of different stakeholders limit the analysis and response to the threats that are most immediate and most evident. Especially when risk assessment targets an agreement among national and international experts on issues where governments hesitate to take action, parties prefer to elucidate a given amount of uncertainty, rather than channeling their efforts towards extreme events that will detract people from concrete policy actions (Patt, 2006). However as stated earlier, the systemic nature of risks makes it necessary to understand the relationship between risks as it eases the risk assessment, decision-making and risk management processes. As the Global Risk Network phrases it: "Identification of individual global risks is only one part of the story as global risks never exist in isolation. Accordingly in today's world risks are neither constrained by geographical nor systemic boundaries" (World Economic Forum, 2006: 6).

Over the last decade various threats have attracted the public's, policy-makers', and other stakeholders' attention to the vulnerabilities of the global system. However, in the course of time, the sole identification of immediate and emerging risks started to be perceived as half-way solutions to risks. This has been a consequence of the tension that has developed between the ideal and real assessments that have emerged.

In the ideal risk assessment process authorities (UN/ISDR, 2004: 63):

- identify the risk factors, vulnerabilities, and capabilities;
- estimate the risk level;
- evaluate the risks;
- perform the socio-economic cost-benefit analyses;
- establish priorities;
- establish acceptable levels of risks;
- elaborate on the scenarios and measures; and
- all agree on how to respond.

But in the real world assessment processes differ from the ideal. These are now discussed from various perspectives.

From an information and communication perspective:

- In many countries there are few local organizations with adequate resources or capacity for continuous analysis and assessment of threats and vulnerabilities to threats.
- Although in many countries there are different organizations with different sets of risk data, such information may not be accessible across different parties for security reasons or power relations.
- Production and communication of risk data may not be the first priority of many organizations.

From an ownership and governance perspective:

- Although we foresee cooperation on the local, regional, national, and international levels, it's often hard to identify the actors (governments, public and private organizations, and individuals) who will take the responsibility of the risks (World Economic Forum, 2008: 39).
- In most circumstances any government, business, or international institution is incapable of addressing these risks independently (World Economic Forum, 2006: 1).
- For the emergence of a global risk management paradigm, there is an increasing need for long-term dedication from the political sphere but this becomes unlikely due to pressures of electoral cycles and executive tenures (World Economic Forum, 2010: 6). In most instances governments are expected to guarantee zero loss, where zero risk is impossible to reach (Wooldridge, 2001).

From a perception perspective:

- Every nation's definition and prioritization of risks is different. Accordingly some governments are relatively more risk-averse.
- In most situations it's not possible to find management and mitigation strategies that will satisfy the needs and requirements of every stakeholder (involved in the risk assessment process). There'll always be parties profiting or suffering more from a specific response (World Economic Forum, 2008: 5).

Agencies have offered the formulation that risk is a product of threat, vulnerability, and consequence. In examining these factors, risk is profiled as "relative," with the direct acknowledgment that resources must be applied unevenly and that not all threats will be addressed. In fact, some have argued that too much protection against loss creates loss. Baker and Simon (2002) state that society should embrace some risks but this does not suggest that steps should not be taken to take account of these risks and to mitigate the damage that they inflict.

In articulating risk management procedures, there has been much talk about prevention, response, and recovery (these phrases appear in many different forms in documents prepared by a range of agencies involved in crime prevention, emergency preparedness, terrorism response, and so on). Despite this, the cases where events have occurred have provided important models for response and recovery and are giving us some important insights into how we might better prevent these events from occurring or having catastrophic effects.

In applying the risk perspective to diverse areas of threat, we find an extensive literature on risk and risk management dealing with a range of topics from health security (Glass and Schoch-Spana, 2002), to property protection (O'Malley, 1999), to crime control (Simon, 1988), to crime prevention (Bradley and Morss, 2002), and to environmental threats (Douglas and Wildavsky, 1982; Dimitrov, 2002). In addition, researchers are beginning to apply the risk models to the study of terrorism (Cummins and

Lewis, 2003; Viscusi and Zeckhauser, 2003). These approaches are set in the context of the idea that we are living in a "risk society," where hazards and threats become a focus of agencies to control through selective application of resources (see Beck, 1992; Van Brunschot and Kennedy, 2008). These controls are calculated through an application of probabilistic reasoning that incorporates the value of information, surveillance, and insurance (or some other form of mitigation) as key components in risk management of the negative outcomes of these threats (Simon, 1988). What this means is that agencies become more aware of the need to determine the dangers in an environment and marshal resources to repair damage that come from these dangers.

Risk analysis is probabilistic but also must account for *uncertainty*, where we have more difficulty in making firm assessments of what might occur. The extent to which we can remove uncertainty will influence how well we are able to anticipate hazards and threats and their consequences. Risk assessment must be considered in terms of *events* that occur over time. The precursors might include the hazards or threats that we face but also the exposure and vulnerability that we experience relative to these hazards. Turning our attention away from single incidents and on to the broad evaluation of factors that precede and follow incidents allows us to incorporate a more contextual view into our assessment of risk. The incident itself had a significant impact on its victims and the area surrounding the attack. But, this incident cannot be understood outside of the more general assessment of the factors that led up to the incident and the monumental changes that it spawned in its aftermath. The retrospective analysis of the precursors to the incident showed up not only a failure to prevent the breach of safe environments but also illustrated the failures by agencies that were tasked to assess the risk of dangers and plan ways of averting these outcomes.

In this chapter we have chosen to analyze the approach to risk assessment and management that is made in different sectors according with the following structure: (1) risk assessment; (2) decision-making; and (3) risk management (see Figure 3.1). Different pieces of literature have different versions of this structure, sometimes including the decision-making stage under risk management. In our case, this distinction has

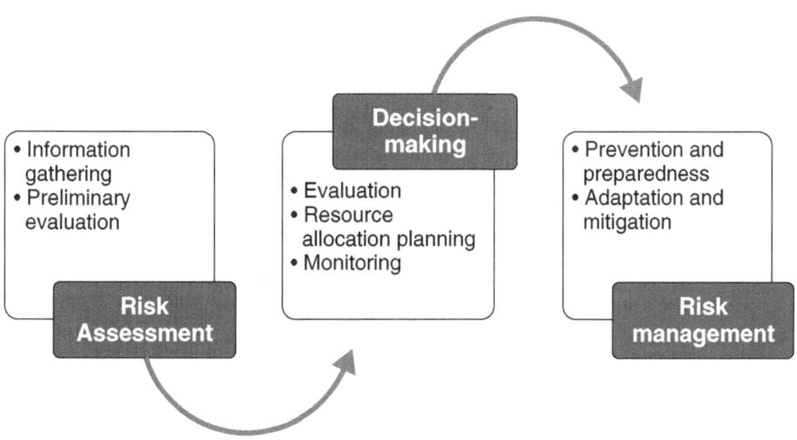

Figure 3.1 Structure used to analyze the approach to risk assessment and management

been made according to the goal pursued and the type of activity that is required at every stage. The risk assessment process is oriented to the description and evaluation of the risk at hand. It is important to gather as much information as possible, in order to move to the second stage: decision-making. We know now the scope and characteristics of the problem: what resources do we have to address it? How do we plan to allocate them? How are these decisions taken and who is responsible for them? Finally, the risk management process involves specific actions in order to bring to life the decisions that have been made, that can be directed either to the prevention of the risk or to the mitigation of its effects, once the threat has become a reality.

In order to further describe what exactly is included under each one of those stages, we list the questions that should be answered in every step:

1. *Risk assessment.* The aim of the risk assessment process is to evaluate the threat and answer the following questions:

   a. What type of threat or hazard are we facing?
   b. What types of data are needed?
   c. What are the sources of information available?
   d. What is the level of certainty that the event will occur? (uncertainty)
   e. How vulnerable are we? Why? (vulnerability)
   f. How exposed are we? Why? (exposure)
   How does the flow of information from the local to a higher level work? Are there specific protocols to follow?

The information is gathered at a local level (for this reason it is useful to have protocols of data gathering) and a preliminary assessment of the situation is made (according to certain models used for risk analysis). If the situation is judged to be a real threat, then the data are sent to the authorities in charge, the leaders, who are responsible for the decision-making process. For this reason we can say that the risk assessment is a bottom-up process.

2. *Decision-making.* The institution or organization responsible for deciding how to face the threat receives the information from the local level. The questions that need to be addressed are the following:

   a. Who is in charge of evaluating the risk at this level and making decisions? How is that authority established? (centralized vs. diffuse leadership)
   b. What is the risk assessment that has been received from the local level? Has the risk been reassessed at a higher level? With what result?
   c. What are the priorities?
   d. What are the available resources? How can they be allocated effectively?
   e. Once the decisions about how to face the threat are made, how are they communicated? To whom? How is that established?
   f. Is monitoring foreseen as a source of data for constant decision-making? Are there protocols of re-examination of decisions according to new information in place?

Decision-making is usually performed at a higher level, by the institutions or organizations with authority to determine what is the adequate course of action.

3. *Risk management.* The risk management process includes all specific actions addressed to the avoidance of the risk (prevention and preparedness), as well as everything that is done to manage the consequences after the event has happened (adaptation and mitigation). Some of the important questions in this area are the following:
   a. Once the decisions are communicated to the institutions or individuals responsible to carry them out, is further communication to lower levels required? How is this done? Are protocols designed with that goal?
   b. Are the specific responsibilities of each level (national, regional, local) established? How?
   c. Are data gathered about the effective implementation of the measures adopted? (monitoring) How? Who is responsible to do that? Is that information passed on to the authorities? How?
   d. How is the effectiveness of the measures established? By whom? Is this communicated to higher authorities? How?

Risk management is a top-down process, since the decisions tend to be made at a higher level. Those decisions are communicated to the regional authorities, and then actual resource allocation takes place, either to prevent or to mitigate the risk.

Methods

Data

Consistent with the view expressed by Beck that the essential elements of public security revolve around health, economic well-being, and rights (Beck, 1992) we selected three general topic areas to compare: health, natural disasters, and crime. The aim is to analyze how risk assessment and management process is conducted in each of these three areas regarding a threat of global dimensions. However, due to the fact that these topics are too generic, we have decided to narrow them down by picking one specific example of each of them for our analysis. In the area of health we have chosen pandemic influenza; for natural disasters, we are focusing on floods; and, finally, we have picked terrorism as the example of crime. Obviously, these are illustrative, not exclusive, examples. We could draw others from other topic areas, such as piracy and its effects on supply chain or the threat from hunger on civil unrest. The topics chosen are supported by a comprehensive response network and have been subject to extensive discussion concerning risks and prevention. For each one of these subtopics we carried out an extensive review of risk assessment and management processes. Multiple sources were consulted, both at the governmental and at an academic level. The summaries of those reviews will serve as the base for the study.

Common Structure

In order to be able to compare how the literature on specific areas addresses global risk assessment and management, the first thing we need is a common structure. By classifying all the information offered by the agencies about their risk assessment procedures according to the same structure we are able to analyze the different approaches and to compare, in a cross-sectional analysis, if some of the areas have more developed risk assessment and management tools than others. In a second step of the analysis we formulate hypothesis of the reasons why such differences are shown.

Findings Regarding the Development of a Common Methodology of Risk Assessment and Management

We will describe our findings from two different angles. First, we will explain how the risk assessment and management process is conducted in each of the three areas mentioned above (pandemic influenza, floods, and terrorism). Then we will do a cross-sectional analysis comparing how the risk assessment, the decision-making, and the risk management processes differ across topics.

Examples

Public Health (Pandemic Influenza)

Overview

From the topics proposed, pandemic influenza is the one where the common structure proposed in the previous section is most developed at an international level. The WHO centralizes the authority to declare the status of the pandemic, and gives the general guidelines about what needs to be done in every phase of the pandemic, but national governments are responsible for planning and organizing the concrete actions. This is the "cleanest" example of gathering information at a local level and communicating it to the international agency in charge of the decision-making at a global level, which in turn issues the guidelines for the implementation of measures at a national and local level. The risk assessment and management process is carried out at a local, national, and international level. Monitoring and decision-making are done continuously throughout the different phases.

Another trait that characterizes the risk assessment and management of the risk of pandemic influenza is the level of organization of the whole process. As we will see, a number of protocols are in place before the threat even exists, and they are a step-by-step guide of the procedures that have to be followed from the moment when the first case is discovered until the pandemic is completely over.

Risk Assessment

Risk assessment is done in a very structured and homogeneous way across all the countries. The WHO has a protocol of communication of the events that may constitute a public health emergency of international concern. If such an event is detected by the national surveillance system, it must be reported to WHO under the International Health Regulations (WHO, 2005), which are an international legal instrument adopted by the World Health Assembly in 2005. They are legally binding upon 194 States[1] around the world and provide a global legal framework to prevent, control, or respond to public health risks that may spread between countries (WHO, 2009). The level of discretion at a local level is very small, since the WHO guidelines describe in detail in which cases the detected events must be notified. However, each country establishes its own "national surveillance system." Once the information has reached the WHO, they assess the level of alert by using a scale from 1 to 6. Each of these phases is accurately described by the WHO (2009) and, as we will see, specific actions are linked to the different levels of alert (see Table 3.1).

Decision-making

The authority in charge of making decisions related to the risk governance of pandemic influenza is the WHO. However, due to the high level of organization and planning in the area of pandemic flu, not many decisions are left once the threat becomes a reality. Decision-making is done primarily through the establishment of protocols that detail what actions must be taken in each level of alert. Similarly, each country is supposed to develop national and sub-national influenza pandemic preparedness and response plans, as well to enact legislation that allow for all proposed interventions.

Table 3.1 Pandemic Phase Descriptions

	Description
PHASE 1	No animal influenza virus circulating among animals has been reported to cause infection in humans.
PHASE 2	An animal influenza virus circulating in domesticated or wild animals is known to have caused infection in humans and is therefore considered a specific potential pandemic threat.
PHASE 3	An animal or human-animal influenza reassortant virus has caused sporadic cases or small clusters of disease in people, but has not resulted in human-to-human transmission sufficient to sustain community-level outbreaks.
PHASE 4	Human-to-human transmission (H2H) of an animal or human-animal influenza reassortant virus able to sustain community-level outbreaks has been verified.
PHASE 5	The same identified virus has caused sustained community-level outbreaks in two or more countries in one WHO region.
PHASE 6	In addition to the criteria defined in Phase 5, the same virus has caused sustained community-level outbreaks in at least one other country in another WHO region.
POST-PEAK PERIOD	Levels of pandemic influenza in most countries with adequate surveillance have dropped below peak levels.
POSSIBLE NEW WAVE	Level of pandemic influenza activity in most countries with adequate surveillance rising again.
POST-PANDEMIC PERIOD	Levels of influenza activity have returned to the levels seen for seasonal influenza in most countries with adequate surveillance.

Source: WHO (2009).

When all those protocols and guidelines are in place, decision-making becomes a very automatic process that saves time in the event of an outbreak. All the decisions about allocations of resources and courses of action are taken before hand, so when the risk becomes real it is just a matter of following them.

Risk Management

Again, the specific actions that have to be taken to prevent and mitigate the risk have been previously described exhaustively in protocols that cover global, regional, national, and sub-national levels. The WHO establishes the duties of the countries, and within each country a response plan must be in place, detailing how those duties will be carried out and who is responsible for doing so. Actions taken during pandemic phases 1–3 are aimed at strengthening pandemic influenza preparedness and response capacities at global, regional, national, and sub-national levels (see Table 3.2).

An example of national risk management protocols is the document created by the US Department of Health and Human Services (HSS) (2005), the HHS Pandemic

Table 3.2 Example of Actions that Must be Taken by the WHO and by the Countries

WHO actions	***National actions***
• Provide technical support to Member States in the preparation of national pandemic preparedness plans.[26] • Provide leadership in coordinating the prioritized activities for epidemic and pandemic preparedness.[27] • Advocate new partnerships with organizations of the United Nations systems, bilateral development agencies, nongovernmental organizations, and the private sector.[28] • Facilitate and encourage the operability of national pandemic plans through preparedness activities, including exercises.	• Establish and activate a cross-governmental, multi-agency national pandemic preparedness committee that meets regularly. • Assess capacities and identify priorities for pandemic preparedness planning and response at national and sub-national levels. • Advise sub-national governments on best practices in pandemic planning; monitor and evaluate the operability and quality of their plans. • Develop, exercise, and periodically revise national and sub-national influenza pandemic preparedness and response plans in close collaboration with human and animal health sectors and other relevant public and private partners with reference to current WHO guidance. • Establish, as needed, full legal authority and legislation for all proposed interventions. • Anticipate and address the resources required to implement proposed interventions at national and sub-national levels including working with humanitarian, community-based, and non-governmental organizations. • Develop an ethical framework to govern pandemic policy development and implementation.[29] • Integrate pandemic preparedness and response plans into existing national emergency preparedness and response programmes. • Provide to public and private sectors the key assumptions, guidance and relevant information to facilitate their pandemic business continuity planning. • Identity and address trans-border issues, including interoperability of plans across borders. • Consider providing resources and technical assistance to resource-poor countries with foci of influenza activity. • Participate, when possible, in regional and international pandemic preparedness planning initiatives and exercises.

Source: WHO (2009).

Table 3.3 Example of Actions that Must be Taken at a National Level in the US for the Different Pandemic Phases Declared by the WHO

	WHO phases 1 and 2: interpandemic phase actions
Planning and coordination	■ Assess preparedness status and identify actions needed to fill gaps. ■ Coordinate with federal, state and local, tribal, and private-sector authorities and organizations, and with media and the public. ■ Develop and conduct tabletop and field exercises to evaluate and improve preparedness plans and response capabilities. ■ Coordinate completion of pandemic preparedness and response plans at federal, state local, tribal, and private-sector levels. ■ Update HHS Pandemic Influenza Plan as needed. ■ Consider indemnification and liability protection issues for affected entities, including pandemic vaccine manufacturers, pandemic vaccine distributors, and healthcare providers who administer pandemic vaccines.
Surveillance, investigation, and protective public health measures	■ Prepare reagents for identification of new influenza strains in animals and other strains with pandemic potential. ■ Assist in domestic and international influenza outbreak investigations. ■ Develop strategies to enhance domestic surveillance and collaborate with international organizations to improve global surveillance to allow earlier detection of novel influenza viruses. ■ Enhance collaborations with international organizations and governments to facilitate surveillance and reporting and the ability to investigate disease outbreaks and implement containment measures that could prevent a pandemic ■ Develop guidance for outbreak control measures in healthcare settings and other institutions such as long-term care facilities. ■ Develop strategies to prevent spread of infections to or within the US form affected areas (e.g., travel advisories or precautions, asessment of travelers returning from affected areas). ■ Assess pathogenicity, antiviral susceptibility, and other characteristics of novel influenza strains.

Source: US Department of Health and Human Services (2005).

Influenza Plan, which serves as a blueprint for all HHS pandemic influenza preparedness planning and response activities. In that document the specific actions that need to be taken are detailed, and the roles and responsibilities of HHS agencies and offices are identified. Again, this high level of organization allows for a rapid response once the pandemic has been declared (see Table 3.3).

Natural Disasters (Floods)

Overview

According to the European Space Agency (ESA) (www.esa.int):

> Flooding is the world's most costly type of natural disaster. Across the developing world floods can strike with deadly regularity, destroying housing, agriculture and communications.

> Developed nations are hardly immune: the floods that struck Europe in 2002 cost dozens of lives and billions of Euros.

The risk assessment and management of natural disasters and, in particular, of floods, is less straight forward and much more complicated that in the case of pandemic influenza. Not only are there many factors that intervene in the generation of the risk, but also the lack of a clear leadership at an international level complicates the flow of information and the decision-making process. Furthermore, when a flood occurs it elicits other risks, affecting the economy (i.e., the supply chain), disrupting the communications, threatening public health (i.e., spread of water-borne and vector-borne diseases) (WHO, n.d.).

Also, very often natural disasters tend to affect poor communities that are repeatedly affected by these phenomena. Disaster prevention might not be a priority until it has happened, since those communities have lots of other things to take care of.

Risk Assessment

There are three main types of floods. Flash floods have an accelerated runoff and are often caused by factors such as dam failure or breakup of ice jams. River floods have a slower build-up and are usually seasonal in river systems. Coastal floods are usually associated with tropical cyclones, tsunami waves, and storm surges. Flood forecasting depends on seasonal patterns, capacity of drainage basins, flood plain mapping and surveys by air and land. Warning is possible well in advance for seasonal floods, but only minutes before in case of flash floods (UN OCHA, n.d.). There are a number of monitoring tools available, such as the Dartmouth flood observatory website (http://floodobservatory.colorado.edu), the NOAA Flash Flood Monitoring and Prediction (http://www.nws.noaa.gov/mdl/ffmp/index.htm), and the ESA Flood Monitoring (http://www.esa.int/esaEO/SEMQFF3VQUD_environment_0.html).

In general we can say that the assessment of the risk of certain types of floods can be done quite accurately, while in other cases (i.e., flash floods) that risk is extremely difficult to foresee and therefore, to assess.

Decision-making

The biggest issue with the risk of floods at a global level is the lack of "coordination leaders," which makes the decision-making process more chaotic. Unlike the case of pandemic flu, where international leaders are clearly identified, and responsibilities and duties are detailed and attributed to each agency, in the case of floods there is not only no international agency in charge, but there is also a lack of coordination of the flow of information. As we have seen in the previous section, to be able to make decisions about the effective use of available resources it is necessary to have an accurate assessment of what is the risk at hand.

In this sense there are some interesting initiatives such as the Global Disaster Alert and Coordination System (GDACS), which is a cooperation framework of the European Union and the UN. According to its website, its aim is

> to consolidate and strengthen the network of providers and users of disaster information worldwide in order to provide reliable and accurate alerts and impact estimations after sudden-onset disasters and to improve the cooperation of international responders in the immediate aftermath of major natural, technological and environmental disasters. (GDACS, 2009)

Another initiative worth mentioning is the Global Pulse, now being developed by the UN. Global Pulse

> is intended to help fill [the information] gap by linking together existing early warning systems and making better use of new innovative ways of collecting real time data. The system is both intended to show impact (i.e. what's happening right now) and raise alarm bells as to potentially dramatically worsening vulnerabilities (i.e. what could happen if we don't act). Its main purpose is to ensure that we have the information and analysis needed to protect our most vulnerable populations against crisis.

Until now we have already identified several sources of problems in the risk assessment and decision-making process: floods are not always possible to predict, they can affect communications, and the scarce local information available is not transmitted to a unique agency in charge of making the adequate decisions.

Risk Management

The lack of leadership also affects the risk management process in case of floods, since there is no unique chain of commands to address the situation. The UN Office for the Coordination of Humanitarian Affairs (OCHA) is in charge of mobilizing and coordinating humanitarian assistance delivered by international and national partners to populations and communities in need. However, as we can see in Figure 3.2 in the next section, there are many agencies and organizations involved in disaster mitigation and humanitarian aid.

It is important to highlight that OCHA is in charge of the "coordination" of all these efforts, but its functions are not of the type of "command and control." For these reasons, even with this level of coordination sometimes there are situations in which a certain level of chaos doesn't allow for the humanitarian aid to be delivered in the most effective way (i.e., Haiti). Regarding the measures to be taken in order to prevent the event from occurring, in the area of floods preventive measures involve macro indicators (poverty, climate change) that are very difficult to change.

Crime (Terrorism)

Overview

From the three topics under study, terrorism is the one that, although having a great impact at a global level, it is addressed by every country independently. It is true that there are some international cooperation agreements, but there is no international agency in charge of gathering all the information from every country and of making decisions about how to address the issue. There are very little data available and, when they are, they tend to be kept at a national level.

Risk Assessment

The most important problem that one faces when trying to assess the risk of terrorism is the difficulty of foreseeing human actions. Unlike pandemic flu or floods that are natural events and have certain patterns, terrorist attacks are the result of planning and organization by human beings. For this reason, assessing the level of threat is much more difficult in this area.

However, there are certain warning signs that can and must be taken into account when assessing the risk. Because estimating the threat is especially difficult, the emphasis in this case is in the evaluation of the vulnerability and exposure of selected targets. For example, the US Department of Homeland Security bases heavily the estimation of risk in the vulnerability and consequence aspects of risk assessment (80 percent), and not that much in the aspect of threat (20 percent) (Masse, O'Neil, and Rollins, 2007) (see Figure 3.2).

Although nothing is said explicitly in the literature, the efforts of improvement in the risk assessment process are not limited to the search of estimators or evidence-based models. The information necessary to feed all those risks models needs to be communicated from the local level to the decision-makers, which in the case of the US happens at a federal level.

Decision-making

The biggest issue that must be addressed when talking about decision-making in the area of terrorism is the fact that, as stated above, decision-making is done almost always

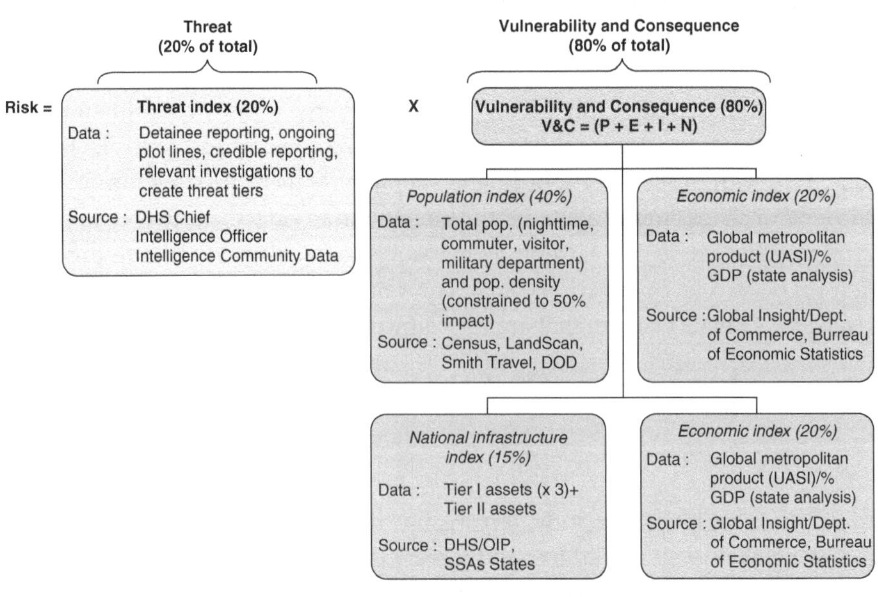

Figure 3.2 Risk formula by the Department of Homeland Security for the fiscal year 2007
Source: DHS, 2007.

at a country level by the agency in charge of national security and defense. In the United States, this agency is the Department of Homeland Security. Each country can have different approaches to risk management, prioritizing different elements such as preventive measures, or mitigation plans. The lack of a supranational organization that coordinates these efforts is translated in the fact that terrorism risk management is purely a national matter.

Recent literature highlights the need of the allocation of resources based on risk. However, as we have seen, estimating the time, place, or scope of the next terrorist event is difficult. Until that intelligence is available, the approach that is being used is, as we will see next, the allocation of resources not only in preventive, but also in mitigation strategies.

Risk Management

The idea behind risk management of terrorist threats is to prevent the attackers' ability to kill, injure, or cause other damage, whether or not their attack is successful (Jackson, 2008). This means that the emphasis is not only on the prevention of the events from occurring, but also on implementing the necessary measures to mitigate the effects of the attack. Although this approach is shared by the other topics analyzed, the difference lays in the increased weight that mitigation has in the area of terrorism, which makes sense if we put this in relation to the increased emphasis that is done in vulnerability and consequence in the risk assessment process.

This "hybrid" approach is not unique to the US Department of Homeland Security as other countries, such as the UK, share the same idea. For example, the Home Office counter-terrorism strategy is divided into four strands (the four Ps): *pursue* terrorists wherever they are and stop terrorist attacks; *prevent* people from becoming terrorists or supporting violent extremism; *protect* the UK by strengthening our defenses against terrorism; and *prepare* to respond to an attack to lessen its impact (Home Office, n.d.).

Counter-terrorism strategies are decided at the higher national level. Just like local information needs to follow the bottom-up chain of communication, those strategies have to be communicated top-down, as the only way to conduct a unified and homogeneous strategy for the prevention and mitigation of terrorist threats.

Comparisons

Each topic has different characteristics that explain why some parts of the risk assessment and management process are more developed in some cases than in others.

Risk Assessment

- Based on previous experiences (Spanish flu, avian flu), a detailed protocol of information gathering and risk assessment is in place. This has been possible because each pandemic might differ in seriousness and geographic scope, but the characteristics of the threat are the same and the analysis of previous experiences has been used to build a very effective risk assessment process.

- In the case of floods, sometimes it is possible to foresee them and therefore adopt preventive measures in order to minimize the impact. But some of the worst disasters are those that cannot be foreseen or those of such magnitude that, if predicted, the main target of the measures adopted will be to save lives, but the infrastructures and economic loss will be extremely high anyway.
- Terrorist threats are very difficult to predict, and this is why the emphasis is on the vulnerability and consequences part of the risk assessment. Historical data, although compiled in databases such as the Global Terrorism Database (n.d.), are very limited, and furthermore terrorist attacks are adaptive, the modus operandi changes in response to the measures implemented to prevent them. For this reason, learning opportunities from previous events are also limited.

Decision-making

- Pandemic flu is likely to spread across countries, therefore the supranational approach makes sense and countries are willing to "obey" those commands. And even in that case, the particular measures taken to meet the goals specified by the WHO are decided at a national level.
- Natural disasters affect one country (or a few) at a time. However, the magnitude of the humanitarian aid needed requires some sort of international approach, and that's what the UN does through GDACS or Global Pulse. But there is no international organization that has the power to make decisions; each country implements the measures that they see fit. This is done mostly on a case-by-case approach.
- Terrorism involves the defense departments of the countries affected. The information is not widely shared with other countries, and no global approach exists. This is a topic that is "national" by nature.

Risk Management

- The centralized decision-making in the area of pandemic influenza facilitates and organizes the risk management process. Each particular measure is discussed and prepared in advance, so when there is a pandemic flu outbreak everybody knows what to do and who is responsible for doing what.
- Natural disasters operate in a case-by-case basis, with some general disaster preparedness at a national level. However, the lack of a supranational approach leaves most of the risk management to the national authorities, and the international community intervenes when the magnitude of the event is an important humanitarian disaster. Even in those cases, the organization that does most to coordinate the efforts of different countries and agencies is the UN, but it doesn't have competences beyond the simple coordination.
- In the case of terrorism, management is completely done at a national level. International agreements such as the one that Spain has with France regarding the terrorist group ETA are in place. The lack of information about the threat

displaces the emphasis of the risk management toward a hybrid approach of prevention and mitigation.

Is a Common Methodology Possible and, if so, Desirable?

The analysis displayed above shows that the risk assessment and management process in the three topics suggested share some commonalities, but also have many differences. If we want to search for a common methodology, one of the best ways to approach this idea would be to take the "best practices" in risk assessment and management, and suggest that each sector applies them in exactly the same way. But the question is: would that be possible? And the next thing we must think about is: would it even make sense?

Table 3.4 summarizes the findings of the cross-sectional analysis. According to this table, the way that the threat of pandemic influenza is handled should serve as an example to other sectors, since every step of the process is highly detailed and organized, leaving little room for chaos when the threat becomes a reality and a rapid response is necessary. However, as we have seen, the different characteristics of each one of the topics make it very difficult to aim for a unified methodology.

If we take the structure discussed at the beginning, we can see that risk assessment is a step previous to, and essential for, the decision-making process. Having accurate up-to-date information is key to be able to make any decisions about how to manage the risk. That level of information is just completely unavailable in the case of terrorism, and might be only partially available in the case of floods, but without a system that organizes that information, the use that can be made of it is very limited. In this case, we could say that "whenever the information is available, a system should be put in place that allows for the information from the local to the higher level to be communicated." But our next question would be: what is that "higher level"?

A system such as the one established for the pandemic flu makes sense only when there is one centralized agency in charge, and that is only likely to happen in those topics where the countries are ready to "obey" a supranational agency. Some disasters involving critical national infrastructures or any type of terrorist attack are certainly not among those. However, the same type of organization and hierarchy is also possible at a national level, as we can see in the case of terrorism. In every country it is clearly delimitated what are the agencies in charge of national security. So, country-wise, it would be possible to establish a detailed plan of action to face threats such as natural disasters or terrorist attacks. But there is an added difficulty: more often than not, those types of threats operate as a trigger for further risks. The first event is a

Table 3.4 Cross-Sectional Analysis of the Three Topics: Pandemic Influenza, Floods, and Terrorism

	Public health: pandemic influenza	*Natural disasters: floods*	*Crime: terrorism*
Risk assessment	✓	✓/×	×
Decision-making	✓	×	✓
Risk management	✓	×	✓

flood or some sort of attack, and the country might be ready for that but what about the risks that follow including infectious diseases, disruption of the supply chain, interruption of communications, and so on. To what extent can we find a methodology to assess and manage those risks, in common with those that have no impact beyond the event per se?

An interesting idea that takes into account multiple risks converging in time and space is the concept of composite measures of risks, which we analyze in the next section.

A Step Further: Developing Composite Measures of Risks

Interconnectivity of Risks

As indicated in the introduction, the relationships between risks increase the odds of systemic risks. The interconnectivity that is seen as the main driving force of global prosperity also acts as a driving force for increased vulnerability to risks (World Economic Forum, 2006: 6). Furthermore this interconnection may damage different global networks in a way the cumulative risk events outsize the individual contribution of each risk event.

So how does this interconnection nature of risks (or how it's likely to) reflect itself in the twenty-first-century risk assessment methodology? As a rule of thumb, whether a specialist or an ordinary citizen, nearly everyone has the tendency to fear the most unanticipated events (Patt, 2006). Accordingly, for an effective management of globalization the uncertainty of global risks must be decreased to the maximum extent. Additionally for sustaining credibility of risk assessment efforts, the scientific risk methodology should also account for the probability of "perfect storms," where "cumulative risk events cause damage in far excess of the sum of each individual risk event."

This is particularly true as current risk assessment and decision-making practices have started to shift towards systemic methodologies where ideal risk assessment and decision-making is seen as a dynamic and continuous process where relationships, specifically the direction of relationships, the strength of relationships, and the nature of relationships between risks, are analyzed.

The Evolution of Risk Assessment and Decision-making

In conventional methodology, risk assessment is defined as a "methodology to determine the nature and extent of risk by analyzing potential hazards and evaluating existing conditions of vulnerability that together could potentially harm exposed people, property, services, livelihoods and the environment on which they depend" (UN/ISDR, 2009: 26).

Using the idea of "risk conflation," we can see a transition towards a more holistic risk assessment and decision-making approach where identification and prioritization of risks is based on the assumption that "risks are rapidly transmitted across geographical and systemic boundaries" (World Economic Forum, 2006: 6) (see Figure 3.3).

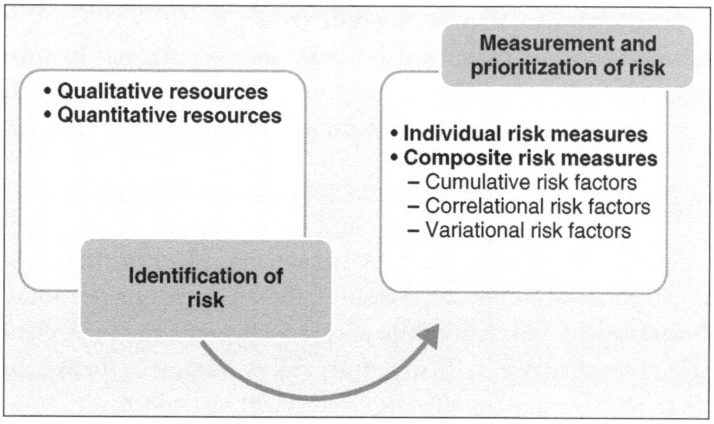

Figure 3.3 Evolution of risk assessment and primary decision-making

Identification of the Risk

In any risk management practice before any counter-measure is implemented, a risk analysis should be implemented. All risk assessment and decision-making processes, whether from an individualistic or holistic approach, starts with the identification of the threat or hazard focusing on the source of the problem or the "problem" itself. This identification is established through data collection from various resources, therefore the nature of risk identification can be qualitative and/or quantitative.

Prioritization of the Risk

The main aim of risk prioritization is to prioritize the risks identified for risk mitigation as resources for risk mitigation are limited. Similar to the risk identification phase, prioritization of risks can be qualitative and/or quantitative in nature. However unlike the risk identification phase, today's risk prioritization procedures have started to show a drastic evolution toward a holistic approach in response to the increasing systemicity between risks and increasing homogeneity of risk across the globe. Accordingly, from this point on we'll discuss real-life applications of risk prioritization to highlight this methodological transition.

Individual Measures

In this very first stage, risk prioritization is established through a process in which the probability of the risk event and the consequence of the relevant risk's occurrence are used to create a risk factor. Accordingly this risk prioritization application may be seen as an elimination process in which relative importance and impact of each risk is calculated to exclude the irrelevant risks from a pool of risks. In this simple stage of risk prioritization the interactions between risks are not taken into consideration.

A classical example for this type may be a simple supply chain risk assessment and evaluation process in which the risk in the extended supply chain is tried to be identified through prioritization among different types of risks such as: supply risks; operational risks; demand risks; security risks; macro risks etc. In this example—assuming a supply chain is vulnerable to many risks—the risks to which the supply chain is most vulnerable are tried to be identified (Manuj and Mentzer, 2008).

Composite Measures

This stage constitutes a big leap in risk assessment and risk prioritization as we start to move along categories of composite measures, the existence of a relationship between risks and the strength of the relationships are gradually more and more elaborated. This characteristic is pivotal in twenty-first-century risk assessment as the systemic nature of risk requires an ongoing and iterative risk assessment procedure.

Cumulative Calculation of Risks

The basic difference of this category from our first individual measures category is its acceptance of the relationships between different risks. Therefore in this category basically every new condition is assigned the same risk factor and by adding up the ratings of different risk indicators, the final risk is calculated. The classification of the risk ranges is often subjective where risk analysts and experts identify the thresholds for certain risk categories (such as low risk, moderate risk, and high risk).

An example for this type is the Failed States Index (Foreign Policy, 2009) in which countries are ranked on their global security levels (such as critical, in danger, borderline, stable, and most stable) according to the total scores of 12 indicators indicated in the index (each indicator is assigned a value between 0–10, accordingly the total score is the sum of the 12 indicators and is on a scale of 0–120). Although this index is highly influential as with its structure it attracts attention to the additive effect of risks (rather than the individual effect of each risk); the index doesn't reflect the direction and strength of relationships between different indicators. Additionally with regards to the scoring system: a score of 60 (or any other score) in this scale neither indicates the weight of indicators nor an existence of a relationship between indicators.

Identification of the Relationship between Risks

As indicated in the previous section, one of the biggest problems in relation to initial composite measures of risks is their deficiency to detect relationships between different risk indicators. Taking this point into consideration some organizations have proposed new risk assessment models in which the relationships of risks are elaborated upon. The World Economic Forum's Risk Correlation Matrix provides an index in which the interconnectedness of different risks is assessed in terms of correlations. This correlation matrix shows the strength of the macro correlations perceived by the experts of the relevant global risk report. Although this example represents a further step in risk assessment with regards to the emphasis on the interconnectedness of risks through a correlation analysis in which existence of a relationship, and the strength of relationships are taken into consideration, still we cannot understand the dynamics of the relationship between risks and causal relationships. As these maps only show the positive relationships between risks, the negative relationships, which might be important for mitigation efforts, are also omitted from the analysis (World Economic Forum, 2008: 25).[2]

The Nature of Relationship between Risks

As indicated in the previous section although the correlation analysis of risks through correlation matrixes revolutionizes the way risk assessment is performed, the methodology still lacks some essential components for identifying the dynamics of relationships between risks. The Global Risk Network's Risk Interconnection Maps (World Economic Forum, 2010: 35–37) and UN's GIVAS Initiative, renamed in 2010, Global Pulse, (GIVAS, 2009) are the most recent examples for a more sophisticated risk assessment approach that point to the complexity and interconnectedness of the risks in the twenty-first century. The main aim of GIVAS as indicated on the UN website was "a representative but manageable set of global indicators that can provide powerful signals of changing vulnerabilities on the ground across regions" (GIVAS, 2009). In a similar vein, Global Risk Network tries to establish the same aim through trying to identify the changing dynamics of the relationships between different risk events with the ideas of experts from different fields. These risk interconnection maps not only shows the strength of the relationship between different risks but also show the similarity of correlations, accordingly similar risks.

In a nutshell, in our century where the interconnectivity between risks increases in parallel to globalization, it's not feasible to manage the global risks with an "individual" and "silo" approach (World Economic Forum, 2007: 25). Accordingly much like we don't have the luxury of dealing with the risk events independently of other individuals, organizations, or countries, we don't have the luxury of evaluating them on an individual basis either. And this necessity of a systemic risk approach is reflected in the risk assessment phase of risk management through a more holistic risk identification and prioritization process exemplified by the composite measures of risks which takes the strength and nature of relationships between various risks into consideration. The combination of risks affects all nations and organizations. Furthermore combined risks may result in consequences that outsize the initial risk assessments. However, since interconnections between risks complicate prioritization with regards to cost-benefit analysis interconnections require an extended cooperation between different stakeholders, a systematic approach to risk assessment may be challenging for risk mitigation.

Discussion and Conclusions

Most risks, although pertaining to a specific locality or nation, directly or indirectly affect other regions and countries as well. With the systemic nature of the risk, the cooperation between governments, public and private sectors, and individuals becomes a prerequisite for a common framework for risk assessment and risk management.

However, as indicated earlier in the literature section and exemplified by the studies included in the method section, despite the increased interest in developing ways in which threats can be monitored and their consequences reduced, it is apparent that there is no real common methodology in the ways in which agencies across threat areas develop and conduct risk assessments. Unsurprisingly, against the challenges of the fast-paced, globalized world and with the lack of a common methodology

- the ample and efficient risk management at a local level;
- the efficient and unstrained communication of risk data across different levels of the same organization and different organizations;
- a "common consent" risk identification and prioritization;
- (existence of) parties eager to take responsibility for risk;
- long-term dedications for risk management; and
- an "all satisfying" risk response

all seem more like a fantasy than science.

A nation's, organization's, or individual's susceptibility to risk is affected by various natural and/or socially constructed factors. Among the socially constructed factors the limited access to information, resources, power, and structures constitute the root causes of vulnerability to any risks. Accordingly the resilience to any type of risk necessitates a systematic approach in which an ongoing and iterative risk assessment strategy is established through the cooperation of different stakeholders such as international standardizing bodies (e.g., the WHO), national governments, private and public institutions, and local communities. Although, in general, the localities affected by the risks are perceived as the main responsible parties of risk management, when the systemic nature of risks are taken into consideration the general frame of directions and resources should be allocated by the organizations in the higher levels of risk management (such as governments, national organizations, or international organizations). It is at this level of initial directions that standardizes the assessment; planning and development strategies will ease the communication of information between different levels of the risk management hierarchy. Accordingly, the risk plans and strategies at the supranational, national, or regional levels should be in relation to the local decision-making processes. When this cooperation is analyzed according to different levels of risk management (UN/ISDR, 2004: 397):

- at the international level, international organizations, non-governmental organizations, and the private sector should be a source of guidance and support for the lower levels of risk managements;
- at the national and local level, adopting the guidance provided on the international level, each country and locality should develop their own risk strategies. According to World Economic Forum (World Economic Forum, 2007: 25) this process may evolve in two innovative directions depending on the concern of flexibility and quick response to risks. In the first scenario a Country Risk Officer may be managing a portfolio of risk across different interests. In the second

scenario some relevant governments and organizations may create the "coalitions of willing," which may be considered as gradually expanding alliances when compared to permanent consensus;
- at the regional level, countries should constantly exchange information and resources.

Whether we are talking of centralization or decentralization of power in risk management, either approach should guarantee the uniform analysis and assessment of the risk. And the possibility of an integrated assessment shouldn't force stakeholders to comply with centralized implementation (World Economic Forum, 2008: 37).

Notes

1. The source of this information has a footnote that clarifies these terms: "The states who are parties to the IHR [International Health Regulations] include of 193 Member States, as well as the Holy See. While these terms are referred to as "States Parties" in the IHR, for ease of reference, this document generally uses the terms States or Member States."
2. Another example for this type of a correlation analysis may be the classical Bayesian network utilized in supply chain risk management that represents a set of variables and their probabilistic interdependencies.

References

Baker, T. and Simon, J., eds. 2002. *Embracing Risk: The Changing Culture of Insurance and Responsibility*. Chicago: University of Chicago Press.

Beck, U. 1992. *Risk Society: Toward a New Modernity*. London: Sage.

Bradley, B. S. and Morss, J. R. 2002. "Social Construction in a World at Risk: Toward a Psychology of Experience." *Theory & Psychology* 12:509–532.

Cummins, J. D. and Lewis, C. M. 2003. "Catastrophic Events, Parameter Uncertainty and the Breakdown of Implicit Long-Term Contracting: The Case of Terrorism Insurance." *The Journal of Risk and Uncertainty* 26:153–178.

Dimitrov, R. S. 2002. "Water, Conflict, and Security: A Conceptual Minefield." *Society & Natural Resources* 15:677–691.

Douglas, M. and Wildavsky, A. 1982. *Risk and Culture*. Berkeley, CA: University of California Press.

Foreign Policy. 2009. The Failed States Index 2009. Retrieved from http://www.foreignpolicy.com/articles/2009/06/22/2009_failed_states_index_faq_methodology.

GDACS (Global Disaster Alert and Coordination System). 2009. Retrieved from http://www.gdacs.org/about.asp.

GIVAS (Global Impact and Vulnerability Alert System). 2009. Retrieved from http://voicesofthevulnerable.org/about-givas.

Glass, T. A. and Schoch-Spana, M. 2002. "Bioterrorism and the People: How to Vaccinate a City against Panic." *Clinical Infectious Disease* 34:217–223.

Global Terrorism Database. N.d. Retrieved from http://www.start.umd.edu/gtd/.

Home Office, Office for Security and Counter Terrorism. N.d. Retrieved from http://security.homeoffice.gov.uk/counter-terrorism-strategy/about-the-strategy1/four-ps/.

Jackson, B. A. 2008. *Marrying Prevention and Resiliency: Balancing Approaches to an Uncertain Terrorist Threat.* Santa Monica, CA: RAND Corporation (www.rand.org).

McDaniels, T. and Small, M. J. 2004. *Risk Analysis and Society: An Interdisciplinary Characterization of the Field.* New York: Cambridge University Press.

Manuj, I. and Mentzer, J. T. 2008. "Global Supply Chain Risk Management." *Journal of Business Logistics,* 29(1):133–155.

Masse, T., O'Neil, S., and Rollins, J. 2007. *The Department of Homeland Security's Risk Assessment Methodology: Evolution, Issues, and Options for Congress.* Congressional Research Service Report RL33858. Retrieved from www.fas.org/sgp/crs/homesec/RL33858.pdf.

O'Malley, Pat. 1999. "Governmentality and the Risk Society." *Economy and Society* 28:138–148.

Patt, A. 2006. "Dealing with Uncertainty: How Do You Assess the Impossible?" Pp. 119–134 in *Assessments of Regional and Global Environmental Risks,* edited by A. E. Farrell and J. Jäger. Washington, DC: Resources for the Future.

Simon, J. 1988. "The Ideological Effects of Actuarial Practices." *Law and Society Review* 22:772–800.

UN/ISDR (United Nations Inter-Agency Secretariat of the International Strategy for Disaster Reduction). 2004. *Living with Risk (Vol. 1).* Geneva: UN Publications.

——. 2009. "UNISDR Terminology on Disaster Risk Reduction 2009." Retrieved from http://www.unisdr.org/eng/terminology/terminology-2009-eng.html.

UN OCHA (United Nations Office for the Coordination of Humanitarian Affairs). N.d. "Flood." Retrieved from http://ocha.unog.ch/drptoolkit/HFlood.html.

US Department of Health and Human Services. (2005). "HHS Pandemic Influenza Plan." Retrieved from www.hhs.gov/pandemicflu/plan/.

Van Brunschot, E. and Kennedy, L.W. 2008. *Risk Balance and Security.* Thousand Oaks, CA: Sage.

Viscusi, W. K. and Zeckhauser, R. J. 2003. "Sacrificing Civil Liberties to Reduce Terrorism Risks." *Journal of Risk and Uncertainty* 26:99–120.

WHO (World Health Organization). 2005. *International Health Regulations.* Retrieved from http://www.who.int/ihr/en/.

——. 2009. *Pandemic Influenza Preparedness and Response.* Retrieved from http://www.who.int/csr/disease/influenza/PIPGuidance09.pdf.

——. N.d. "Flooding and Communicable Diseases Fact Sheet." Retrieved from http://www.who.int/hac/techguidance/ems/flood_cds/en/index.html.

Wooldridge, M. 2001. "Risk Assessment and Risk Management in Policymaking." Pp. 81–95 in *Globalization and the Environment,* edited by D. Robertson and A. Kellow. Massachusetts: Edward Elgar.

World Economic Forum. 2006. *Global Risks 2006: A Global Risk Network Report.* Retrieved from http://www.weforum.org/pdf/CSI/Global_Risk_Report.pdf.

——. 2007. *Global Risks 2007: A Global Risk Network Report.* Retrieved from http://www.weforum.org/pdf/CSI/Global_Risks_2007.pdf.

——. 2008. *Global Risks 2008: A Global Risk Network Report.* Retrieved from http://www.weforum.org/pdf/globalrisk/report2008.pdf.

——. 2010. *Global Risks 2010: A Global Risk Network Report.* Retrieved from http://www.weforum.org/pdf/globalrisk/globalrisks2010.pdf.

Questions

1. When developing practical responses to dangers, how useful is it to develop a scenario that outlines how a disaster transpires and the effectiveness of responses that agencies can make to this type of event?
2. Can we learn about the risks involved in a terrorist attack by using models that focus on system failures in such things as aircraft crashes?

4.
THREATS, VULNERABILITY, AND EXPOSURE TO TERRORISM

Risk assessment must be considered in terms of *events* that occur over time rather than focusing solely on incidents that might happen. The precursors might include the hazards that we face but also the exposure and vulnerability that we experience relative to these hazards. Turning our attention away from single incidents and on to the broad evaluation of factors that precede and follow incidents allows us to incorporate a more contextual view into our assessment of risk. The incident itself has a significant impact on its victims and the area surrounding the attack. But, this incident cannot be understood outside of the more general assessment of the factors that led up to the attack and the monumental changes that it spawned in its aftermath. The retrospective analysis of the precursors to the incident shows up not only a failure to prevent the breach of safe environments by determined terrorists but also illustrated the failures of imagination in the thinking of security agencies that were tasked to assess the risk of attack and plan ways of averting these outcomes.

Mark Anarumo's chapter follows this theme further and addresses the extent to what law enforcement knows (or doesn't know) about the threat of terrorism and discusses the steps that they have taken to address this in face of uncertainty. In looking at the extent to which we can know about the real threat of terrorism, Anarumo focuses on three factors—local actors, local targets, and local agency. These, he says, comprise a trilogy that is a cornerstone of opportunity-based theories of crime. There is a logical partnering of these three elements and the elements are tested via a survey to discover "why" chiefs of police feared certain groups over others.

To measure the concepts set forth above, Anarumo describes a unique data collection tool that overcame the limitations in studying the threat of terrorists, due to the infrequency of incidents in this country, especially relative to other crimes. The several databases of terrorist events that exist are either in their infancy or list only international events, limitations that reduce their worth to this study and its hypotheses. Traditional measures in criminal justice—such as call for service, court records, and direct subject interviews, are similarly restrictive in that they are so scarce. The infrequency of calls for service and the very small number of court cases involving terrorism eliminates these as viable tools for the study. As Anarumo points out, a survey of key actors leaves us with the survey as the most appropriate tool.

From the survey results, Anarumo concludes that terrorism threats are based on local knowledge and are enhanced by the concerns that law enforcement see to their jurisdictions rather than being seen as deriving from national or global actions. He concludes that identification of the local terrorist threat is the foundation of effective anti-terrorism strategy development, whether at the local, regional, or national level.

THE PRACTITIONER'S VIEW OF THE TERRORIST THREAT

Mark Anarumo

Introduction

Although money and time are pouring into anti-terrorism and homeland security, comprehension of current terrorist threats within the United States remains generally weak. There is a perception of a "national" threat from Islamists; however, other groups, such as anti-abortionists and ethnic supremacists, were identified in pre-9/11 literature as greater threats.

Anti-terrorism is fast becoming an integral part of policing in the United States. Professional organizations such as the International Association of Chiefs of Police and sections of the Department of Justice have attempted to aid this process, but the law enforcement community still requires guidance via sound empirical research if they are to succeed in meeting this threat to public safety. After all, the threat of terrorism in the United States is not new; it has only gained prominence due to world events.

Police executives are being forced to ask the most basic of questions to assess the threat of terrorism in their jurisdictions. Simply put, they need to know of whom they should be afraid. The wide range of unique terrorist groups operating in the United States today, all with very specific goals and tactics, makes the correct identification of the threat the key component to developing an effective anti-terrorism strategy. This question of "who" can be addressed through the employment of police skills already possessed by professional police organization, such as surveillance and the use of informants. Group identification is only the first, albeit most crucial, step in evaluating local threat. The identification of which groups are predicted to act is also key to developing a true understanding of terrorism.

To develop an accurate and current assessment of terrorism in the United States, police executives were surveyed, resulting in 1,883 responses nationwide. The over 200 survey variables included terrorist group presence and the respondents' opinions of which groups were likely to attempt violence. Respondent telephone area codes allowed for spatial analyses.

Background

Although this study focused on police executives as key experts in the study of terrorism, there are still many respectable arguments that the proper field to study terrorism is not criminal justice, and further that the proper field to counter terrorist violence is not law enforcement. The argument against the utility of law enforcement in anti-terrorism is partially due to the lack of attention given to the unique identity, goals, and tactics of the distinctive terrorist groups operating in the United States today. The majority of existing research addresses terrorism in the frame that there is a common goal and purpose of all terrorist groups. This is a false assumption. The distinctions among terrorist groups are critical when addressing the threats to public safety posed by terrorism. Group distinctions are thoroughly addressed in the study's survey instrument and the portions of the analysis that are included in this reader.

The approach of this study argues strongly that the study of terrorism should reside primarily within criminal justice, primarily because civilian law enforcement agencies will ultimately be responsible for the prevention of and recovery from terrorist acts. Law enforcement executives are quickly realizing that terrorism is a phenomenon worthy of their concern and resource allocation, and that it can be defeated through an informed focus of their professional organizations. Reliance on federal agencies and the military has never been well founded. While organizations such as the Federal Bureau of Investigation (FBI) have done and continue to do much to support local agencies, they ultimately are too small and their missions too broad to meet the threat conclusively.

Beyond the challenges associated with federal law enforcement organizations, there are numerous legal and ideological reasons why the role of the armed forces has been so limited. A cornerstone of Western democracy is the trust in civilian law enforcement to handle even the most serious domestic threats without involving the military. In the United States, the removal of the military from domestic security is formalized in the Posse Comitatus Act. Although this once-sacred tenet is being reconsidered post-9/11, the American public (and courts) would not stand for any military augmentation of civilian police other than technological and logistical support. There are clearly skills sets that reside within the Department of Defense that would be of great benefit to public law enforcement in dealing with terrorism. The fact that at the time of this writing, more than eight years after 9/11, we still have not updated Posse Comitatus to allow for this support speaks volumes about the skittishness of the American legislature and judiciary in allowing the military to support public police in all but extreme, active emergencies. Given this lack of progression, it is clear that civilian police will have to meet the challenges of terrorism within the borders of the United States.

The literature, especially post-9/11, suggests that civilian police agencies possess unique capabilities to deal with terrorism. These capabilities have been identified in several practitioner publications, most notably by the International Association of Chiefs of Police (IACP) and the Department of Justice Office of Community Oriented Policing Services (COPS). Two recent publications, the IACP's Project Response—*Leading from the Front: Law Enforcement's Role in Combating and Preparing for Domestic Terrorism*, and the 2002 issue of *COPS Innovations*, identify examples of local agency capabilities, guidelines for first responders, suggested enhancements, and available resources. These publications also cite the views of several chiefs of police regarding their capabilities to deal with terrorism. One view that perfectly captures their value follows: "We're the ones out in the street. We talk to everyone in the community. We know what's happening in every block in every city. People need to reach out and get that information" (Norris 2002).

The acknowledgment of local law enforcement organizations as viable entities to counter terrorism is very recent. There are two main reasons for the delay in this recognition. First, terrorist events occur very infrequently, so much so that they have rarely been "on the scope" of law enforcement leaders who were focused on traditional crime and allocated their thin resources accordingly. The second reason is tied to the necessity to view terrorism in the United States in a historical context. A large-scale terrorist attack on US soil was inconceivable 15 years ago. Even the limited engagements of the 1960s had no lasting effects. The first World Trade Center bombing in 1993 was strategically a failure. It should have served as a wake-up call to the country, but in many circles it instead only reinforced the idea that no mass casualty event could be successfully perpetrated. The most significant pre-9/11 event was the 1995 bombing of the Murrah federal building in Oklahoma City. Lost in the atrocity of the act was the fact an Oklahoma highway patrolman, not the FBI or other federal agency, actually caught the perpetrator 80 minutes after the bombing. Even after this event, efforts by law enforcement to deal with terrorism were under-funded and relegated to far less importance than traditional police tasks.

The most significant terrorist events of the late 1990s—the 1996 bombing of US military barracks in Saudi Arabia, the 1998 bombings of US embassies in East Africa, and the 2000 bombing of the *USS Cole* off the coast of Yemen—all contributed to the perception of terrorism once again as a threat "somewhere else." Of course, the events of 9/11 changed this perception again, but for how long is impossible to gauge.

Group Variations: Goals, Tactics, and Target Sets

These issues of non-traditional policing and non-domestic threats have also contributed to the under-development of academic applications in terrorism, notably the development and exploration of typologies. The use of typologies is critical because the disparities between terrorist groups and goals are key elements in the construction of effective anti-terrorism policy and strategy. Unfortunately, the body of research, both pre- and post-9/11, attempts to explain terrorism in terms that do not account for these distinctions. For example, consider anti-abortionists and right-wing extremists, two group types that were active in the United States in the latter half of the twentieth century. Their goals were quite different, and goals defined their tactics. More specifically, anti-abortionists have targeted individuals who work in abortion clinics or those who support abortion as a legal right. Right-wing groups have sought to directly attack agents and symbols of the United States government. The scope of their respective operations is also much different. Anti-abortionists limit their operations to clinics and individuals, while anti-government extremists have no set parameter for their targets' size or the possibility that innocent bystanders may be harmed. They have conducted every conceivable type of operation from the assassination of police officers to large-scale attacks on government facilities (i.e., Timothy McVeigh's infamous attack in Oklahoma City), with resulting mass casualties.

A main theme of this portion of the reader is that terrorism, like politics, is local. Existing literature and the data gained from the supporting study clearly shows that there are distinctive types of terrorism, and argues that local anti-terrorism or first responder programs must take these distinctions into account when assessing and countering local risk.

The primary worth of the data in this study is to provide law enforcement professionals with an empirical study that could broaden their understanding of terrorism generally and their individual processes of designating their local threat. This is especially important to police leaders who have to decide how to formulate an anti-terrorism strategy and allocate resources to meet their designated local threat. Problems with resource allocation are exacerbated by the fact many police agencies have not received the manpower or technology needed to execute their anti-terrorism tasks. These new tasks are simply another drain on already stretched resources.

Given these issues, it is not difficult to envision how an agency can be distracted from homeland security tasks. Consider a jurisdiction with an active anti-government extremist group that is considered a threat by the local police chief. If the department is reactive to a national event and allocates its resources toward the surveillance and protection of abortion clinics, the anti-terrorism strategy will have little chance for success. Success will be similarly elusive if the goals and identity of the local group are treated as "universal," in other words, if the unique goals of the local threat are not accounted for in the local anti-terrorism plan.

Previous Research

This research is in many ways an update of two empirical studies undertaken in the mid-1990s. The first was accomplished in 1995 by Joseph Carlson. He surveyed police chiefs in all 140 US cities with populations above 100,000, requesting they choose from a list of ten potential terrorist groups and rank the four most likely to commit a terrorist act in the next two years. Eighty-six of the 140 chiefs responded. Anti-abortionists received 43 votes (50 percent of all responses), identifying them as the number one threat. The group with the second most votes as the greatest threat was White Supremacists. Middle East Groups and Black Militants were third and fourth, respectively. This means, according to this survey, that three of the top four organizations perceived by chiefs of police to present the greatest terrorist threat for "the next two years" were not international and not classified by the FBI as terrorist organizations. Only Middle East Groups satisfied the conspiratorial element required by the FBI (Carlson 1995: 83).

RAND published similar findings two years later. Through a national survey and ten case studies completed over 24 months, they analyzed the main issues confronting state and local agencies in employing anti-terrorism measures. In addition to finding that a "sizable majority" of state and municipal agencies considered terrorism a major problem, a much wider definition of terrorism was used than employed by the FBI. Specifically, "states and municipalities are equally adamant in identifying right-wing (Neo-Nazi, anti-Semitic, anti-federalist) and issue-specific (anti-abortion, animal rights, environmentalist) organizations as the most threatening actual and potential terrorist sources" (RAND 1995: 41). Again, the vast majority of the groups seen as the most dangerous threats were not considered terrorist organizations under the FBI definition.

Terrorist Group Typologies

In the paradigm of practitioner perspectives and threat projections a workable typology becomes critical. There are many existing typologies of terrorism, all built for specific research questions or agendas, just as the typology built for this study was constructed to best address practitioner concerns.

The key concept behind a workable typology is the context in which it must be employed. For example, studies of terrorism aiming to explain cultural factors leading to terrorist violence should develop or employ a typology along cultural-geographic lines. Depending on the specific research question, the typology may employ national boundaries, geographic features, or religious identity concentrations to distinguish among groups in a way that leads to reliable and valid research findings. Unfortunately, the macro-level views of such findings are worth little to the agencies that are responsible for the detection and defeat of terrorist groups operating within their jurisdictions. In this context, a typology must be developed that would be useful to the practitioner.

Considering the needs of law enforcement agencies, a workable typology must distinguish groups in a way that allows for specific deterrence and infiltration strategies. In other words, the typology must be based on group tactics. In this model, terrorist groups can be disaggregated into seven distinct categories: left wing, right wing, Islamists, eco-terrorists, anti-abortionists, ethnic supremacist, and foreign nationalists.

The identification of anti-abortionists as a distinct group may seem too specific for a typology, but previous research (most notably Carlson 1995 and RAND 1995) has shown this group to previously be the most feared among all types by United States law enforcement agencies. This translates to the identification of anti-abortionists as the group most likely to perpetrate an act of terrorist violence in the United States. Of course, as events occur fear of other groups may change based on who perpetrates terrorist acts. For example, the 1996 McVeigh bombing in Oklahoma City may affect the ranking of right-wing groups, just as the 9/11 attacks may change the ranking of Islamists. Regardless, the previous ranking of anti-abortionists so consistently high demands their status as a unique category. Specific research hypotheses thoroughly tested these statements.

Survey and Dataset

To build a suitable database, a survey was conducted of police executives in the 48 contiguous United States and the District of Columbia. The survey population included police chiefs, sheriffs, superintendents, and other executive positions in local police agencies, federal agents working in local field offices, and administrative agency heads of state counter-terrorism offices. The survey was internet-based, with automated coding of responses. There were no mailings, personal interviews, or manual data input.

The primary means of solicitation was through an online search of professional law enforcement organizations and agency websites. For example, a link to the study was included on the website for the IACP, and summary information and website address were included in the May 2004 State Associations of Chiefs of Police newsletter. The National Sheriff's Association (NSA) was likewise contacted and assistance was requested in advertising the study. The NSA was very supportive in advertising the study, informing 100 percent of their membership via email and through an advertisement in the September 2003 issue of their monthly membership magazine.

Of course, IACP and NSA members do not represent all members of the target respondent population, and efforts were made to ensure non-IACP/NSA members also knew of the study. To accomplish this, individual state associations of chiefs of police were contacted via email and telephone, and in some cases were provided with flyers announcing the study. Some state associations distributed these flyers at annual meetings, while others either emailed the study announcement to all members or provided a link to the study on their state association homepage.

Finally, the distinctive populations of tribal police agencies and college and university police executives were specifically targeted. To accomplish as wide an announcement as possible, the IACP Tribal Police representative faxed a copy of the announcement flyer to members of its tribal police section. To inform the college and university police populations, the president of the International Association of Campus Law Enforcement Administrators (IACLEA) email-forwarded the study solicitation to his membership population.

As mentioned, the survey for the study was web-based. There were no hard copies, mailings, or interviews. The target population was notified via the different methods of advertisement listed above, and directed to the survey's internet address. Once on the homepage, the initial survey screen provided a very brief introduction to the study, including a summary of information that was made available at the conclusion of the survey (generally a complete presentation of all responses up to that point so respondents could immediately compare their information with the rest of the population). A consent statement explaining that survey responses were anonymous followed the introduction screen.

Telephone area codes were chosen as the primary respondent identifier because they are not specific. Zip codes were rejected because, through their specific nature, they could potentially identify a specific department and thus a respondent. Since there are likely to be several or even several dozen departments within a given area code, there is no specificity.

Identification by state was rejected because the dissertation's central concept of locality would have been lost. Area codes are also valuable because they reflect regional dimensions of a state generally (for example the diverse regions of New Jersey are accurately reflected by the various area codes). Area codes also help disaggregate intra-state regions by identifying urban areas from suburban and rural. Finally, through geographic mapping software applications cross-boundary diffusion effects can be shown.

Since area codes are non-specific, and in almost every case there were multiple responses from each area code, the privacy of respondents was protected. There was no threat to privacy or confidentiality.

For analysis and the purposes of discussion, regions and states are listed as much by "identity" as they are geographically, though geographical boundaries are generally adhered to. The primary departure from geographical boundaries in favor of identity is the listing of Washington DC as in the Northeast while Maryland and Virginia are listed in the South. Arguments can be made for and against these designations, but it is the author's opinion that the US capital city has an identity much closer to that of New York City, Boston, and Philadelphia, than Atlanta, Charleston, and Richmond.

For reference, the survey listed regions and state affiliations as follows:

- Northeast (CT, DC, DE, MA, ME, NH, NJ, NY, PA, RI, VT).
- South (AL, AR, FL, GA, KY, LA, MD, MS, NC, SC, TN, VA, WV).
- Midwest (IO, IL, IN, KS, MO, NE, OH).
- Central (MI, MN, ND, SD, WI).
- Southwest (AZ, NM, OK, TX).
- West (CA, CO, ID, MT, NV, OR, UT, WA, WY).

The survey instrument was accessible on the study's internet homepage from June 6, 2004 through November 22, 2004. There were a total of 1,883 responses during this time frame. While some of these responses resulted from passive recruitment measures such as website links and flyer distribution, the vast majority of responses were received immediately after mass emailings inviting law enforcement agency heads to participate.

Overall, from June 6, 2004 through November 22, 2004, the study's homepage was visited 3,782 times. There is no way to distinguish unique "hits" from repeat visitors. Of these 3,782 visits, 1,883 surveys were submitted, for a rate of 50 percent.

To ensure there were no attempts to deliberately sabotage the data, the internet protocol (IP) addresses of respondents were collected and checked for features that may have threatened the study's reliability. IP addresses indicate specifically where a computer physically accesses the internet. To protect the anonymity of respondents in this study, IP addresses were only checked to ensure the respondents were accessing the internet from a location in the continental United States.

Finally, problems that could not be remedied resulted in the deletion of cases. For example, 135 respondents terminated their participation early in the survey process, resulting in the exclusion of the majority of data that was useful for analysis. All 135 of these cases were deleted.

Thus, of the original 1,883 responses, 1,744 were retained for analysis. These 1,744 cases served as the population that was analyzed and is discussed in the following sections.

Findings

Group Presence

To develop the overall picture of terrorist group presence, simple frequency distributions were run on the entire national sample of respondents and chi-square tests of independence were conducted. These tests were important to the models because they helped determine whether observed frequencies of reported terrorist group presence differed significantly from expected frequencies. These tests also indicated the strength of relationships, with large chi-squares indicating strong differences between observed and expected values and smaller values indicating weaker relationships.

Simple frequency tests were run for each group to obtain a snapshot of terrorist group presence at the national level, meaning an aggregate picture of group presence from all respondents. 1,732 respondents (99.3 percent of the 1,744 total) answered the survey questions on local terrorist group presence. All data included in Table 4.1, and in the discussion that follows, were gleaned from survey questions that asked respondents to list which terrorist groups were present in their local areas. Respondents were able to choose as many responses as were appropriate; thus, the list of percentages will not equal 100 percent, since respondents were able to report multiple groups as present.

As shown in Table 4.1, the national results of terrorist group presence ranked from most frequently reported to least were ($n = 1{,}732$): Ethnic supremacist, 1,028 (or 59.4 percent of all respondents); anti-abortionists, 1,004 (58 percent); right wing, 839 (48.4 percent); eco-terrorists, 675 (39 percent); Islamists, 486 (28.1 percent); left wing, 288 (16.6 percent); and foreign groups, 199 (11.5 percent). Other potential responses included a miscellaneous group presence, meaning the agency wanted to report a group presence but the group type was not listed as a choice in the survey. Eighty-eight respondents, or 5.1 percent, said this was the case. An additional 57, or 3.3 percent, said that a group was present but they were not sure what type.

Note that the presence of two groups—ethnic supremacists and anti-abortionists—was reported by a fairly sizable majority of respondents (59.3 percent and 57.9 percent respectively). The other five groups were not reported by a majority of respondents. Another interesting overview data point is that a large majority, 1,444 (84.4 percent of the 1,732 that answered these questions) reported the presence of a terrorist group, regardless of type.

Table 4.2 provides an overview of regional-level statistics on terrorist group presence, including rank ordering and percentage of respondents reporting presence per region.

Table 4.1 National Reporting of Group Presence

	No. of total respondents	*% of total study population*
Ethnic supremacist	1,028	59.4%
Anti-abortionists	1,004	58%
Right wing	839	48.4%
Eco-terrorists	675	39%
Islamists	486	28.1%
Left wing	288	16.6%
Foreign groups	199	11.5%

Note: $n = 1{,}732$.

Table 4.2 Group Presence by Region: Rank and Percentage Reporting

	Northeast	*South*	*Midwest*	*Central*	*Southwest*	*West*
Ethnic supremacist	2 56%	1 63%	2 59%	3 50%	1 66%	2 62%
Anti-abortionists	1 58%	2 58%	1 61%	1 65%	3 58%	4 53%
Right wing	3 37%	3 45%	3 51%	2 56%	2 62%	3 54%
Eco-terrorists	4 32%	4 31%	5 25%	4 37%	5 33%	1 70%
Islamists	5 28%	5 28%	4 27%	5 24%	4 38%	5 27%
Left wing	6 17%	7 13%	6 13%	6 10%	6 18%	6 24%
Foreign groups	7 15%	6 15%	7 10%	7 3%	7 8%	7 11%

There are several points of this summary data that are worthy of discussion. First, with one exception, ethnic supremacists were consistently reported as present either first or second most often in each region. The exception to this statement was the central region, which reported them as present third most often behind anti-abortionists and right-wing groups. Despite this relatively low ranking in the central region, ethnic supremacists were ranked first as the group most frequently reported as present in the national aggregate dataset.

Anti-abortionists were the second most often reported in regards to national presence. Though reported as locally present nearly as often as ethnic supremacists (only 1.4 percent less often nationally), the rate at which anti-abortionists were reported as present was consistent across regions. Even in the west, where they were ranked fourth of seven, anti-abortionists were reported as present by 53 percent of respondents. From these numbers it could be said that anti-abortionists were the most consistently reported group, as they did not experience the same fluctuation between regions as other groups.

Islamists experienced similarly consistent ranking across regions. Ranked as either fourth or fifth most present in every region, the percentage of respondents reporting an Islamist presence was between 24 percent and 28 percent, with a notable spike of 38 percent in the southwest region. Even with this spike, Islamists still placed only fourth of seven in the southwest.

Finally, the consistency of the two groups least frequently reported was notable. Left-wing and foreign groups were reported as present substantially less often than all other groups with such consistency that they were always ranked as the last two groups. The south region was the only population to deviate from the consistent ranking of left-wing sixth and foreign groups seventh, instead reversing the two lowest places.

The frequency reports suggested that there was not a consistent, national picture of terrorist group presence, but rather there were regional characteristics of the data. To discover relationships between regions and groups, crosstabulations and chi-squared analyses were conducted. In each crosstab, a single group's presence served as the dependent variable. Each group presence variable was dichotomous, meaning the answer to the question of group presence was either yes or no. Region variables were then recoded to build a new variable that allowed for the construction of a simple 2 × 2 table. Significance tests were set at a minimum value of $p < 0.05$, thought many models returned at $p < 0.01$ or $p < 0.001$. If any of the crosstabulations returned significance scores above $p < 0.05$, they were considered statistically invalid and thus not analyzed further.

Table 4.3 provides the critical measurement for discovering association in the crosstabulation models, Pearson's phi statistic. Phi was used to display a standardized measure of association between 0 and 1, with 0 indicating complete independence between variables and 1 indicating strong association. Though there are many ways to measure strengths of association, the phi statistic is the accepted measure of a 2 × 2 table, which was the primary model dimension used. When deciphering the phi statistic, the following range was applied to the level of association:

> Determinate; > 0.7 very strong; 0.5–0.7 strong; 0.3–0.5 medium; 0.2–0.3 weak; 0.1–0.2 very weak; < 0.1 extremely weak.

Table 4.3 Regional Crosstabulation Summary: Significant Group Presence Associations by Phi

	Eco-terrorists	*Islamists*	*Anti-abortion*	*Ethnic supremacist*	*Right wing*	*Left wing*	*Foreign groups*
Northeast	–0.079	—	—	—	–0.130	—	0.061
South	–0.081	—	—	—	—	–0.053	0.052
Midwest	–0.127	—	—	—	—	—	—
Central	—	—	—	–0.060	—	–0.054	–0.087
Southwest	—	0.074	—	—	0.091	—	—
West	0.321	—	—	—	0.055	0.106	—

When applying this scale, however, it must be noted that in social science even a weak association may be statistically valid and important for consideration. It is widely assumed that higher levels of association, such as 0.7 or larger, are impossible to achieve in all but the natural sciences.

Table 4.3 is presented as a summary for all regional crosstabulation models with the phi score indicating the strength of the relationship between the two variables by displaying the phi statistic for each significant region–group association. From these findings it can be said that there are clear regional dynamics to terrorist group presence.

Since it provides a quick visual reference for how terrorist groups are over- or under-represented in certain regions, Table 4.3 is also useful in providing an overview of how reported group presence was distributed across the respondent population. For example, it is easy to see that the negative associations between eco-terrorist group presence and the northeast, south, and midwest regions were more than offset by the tremendous over-representation of eco-terrorist group presence in the west region. It is therefore possible from Table 4.3 to appreciate where groups were concentrated and where they were sparse, though the table is obviously somewhat lacking in ease of interpretation.

Fortunately, through crime mapping software a more easily viewed spatial presentation of group presence can be created. Beyond entire regions, the resulting maps (Figures 4.1–4.7) show with much greater specificity where groups were reported to be, as associated with telephone area codes.

Projected Terrorist Activity

Though group presence was an important starting point, it is arguably more critical, especially in the context of a reader addressing the concept of risk, to forecast which groups are likely to attempt terrorist violence. To gain this forecasting data, frequency reports and crosstabulation procedures were applied to the dataset to see which groups were believed likely to act, which for the remainder of this study will be referred to as "group fear." First, simple frequency reports generated overview information of group fear and allowed for simple comparisons with group presence data. The familiar crosstabulation procedure then created models that highlighted differences in group fear both regionally and nationally.

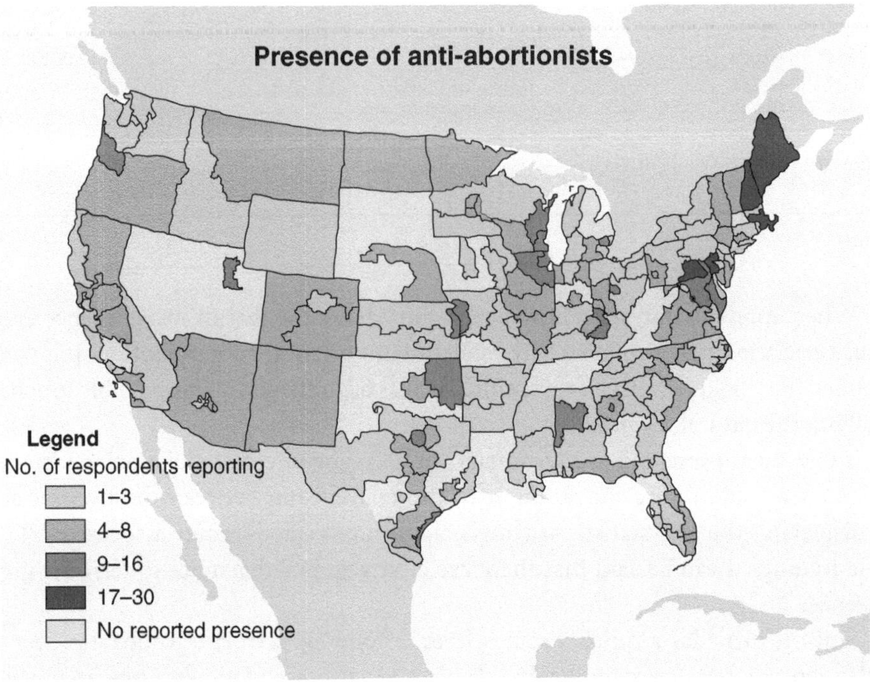
Presence of anti-abortionists
Legend
No. of respondents reporting
1–3
4–8
9–16
17–30
No reported presence

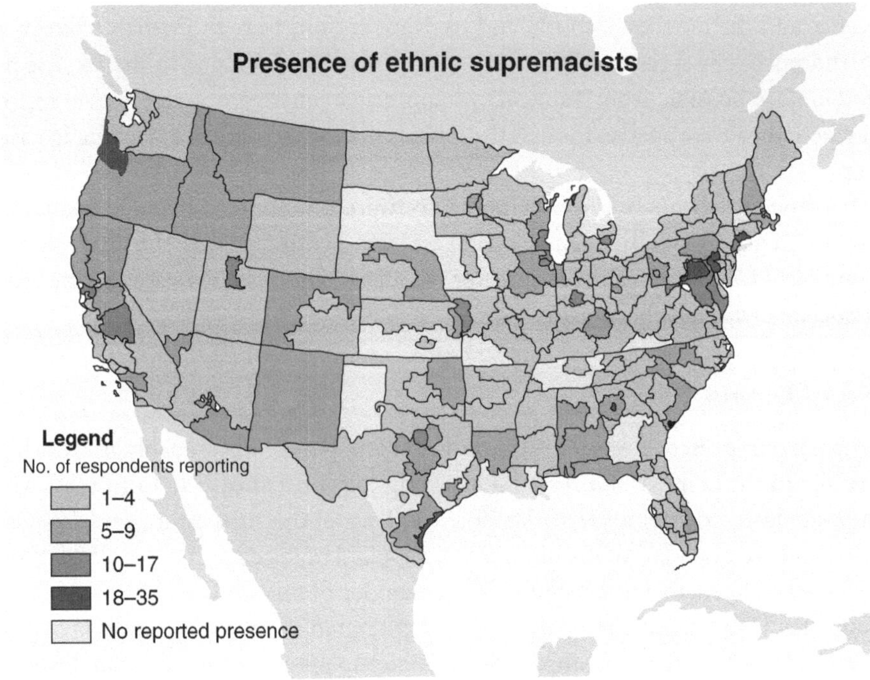
Presence of ethnic supremacists
Legend
No. of respondents reporting
1–4
5–9
10–17
18–35
No reported presence

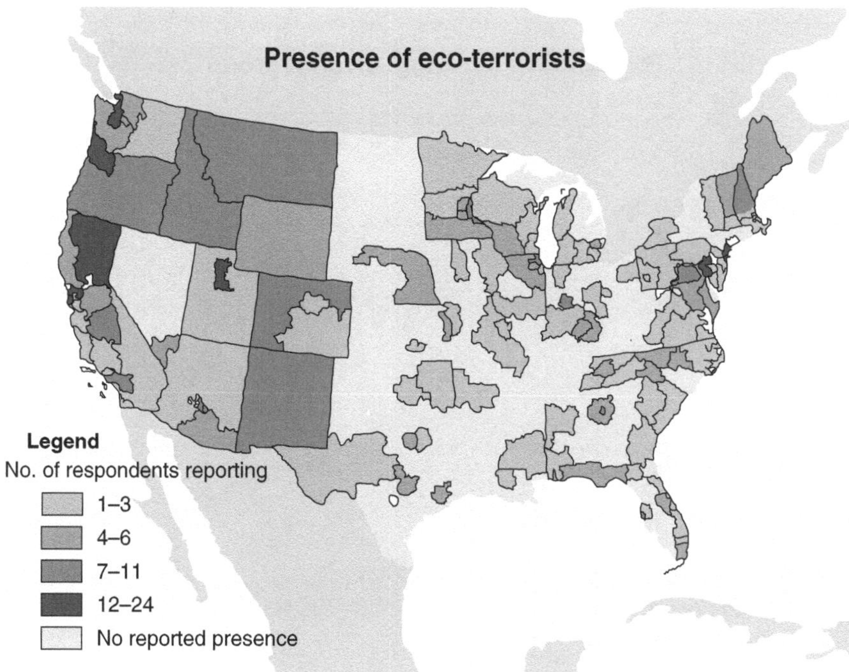
Presence of eco-terrorists
Legend
No. of respondents reporting
1–3
4–6
7–11
12–24
No reported presence

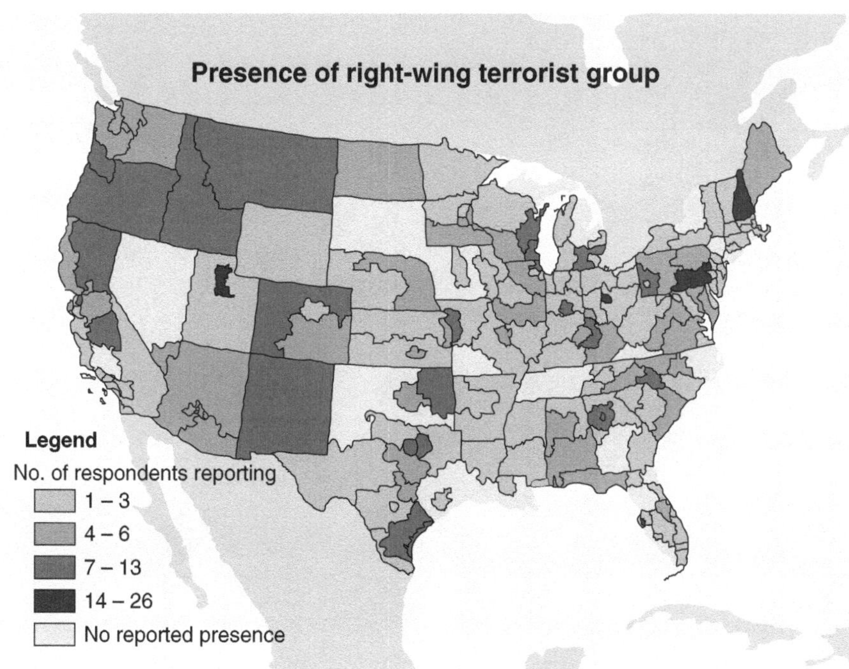
Presence of right-wing terrorist group
Legend
No. of respondents reporting
1 – 3
4 – 6
7 – 13
14 – 26
No reported presence

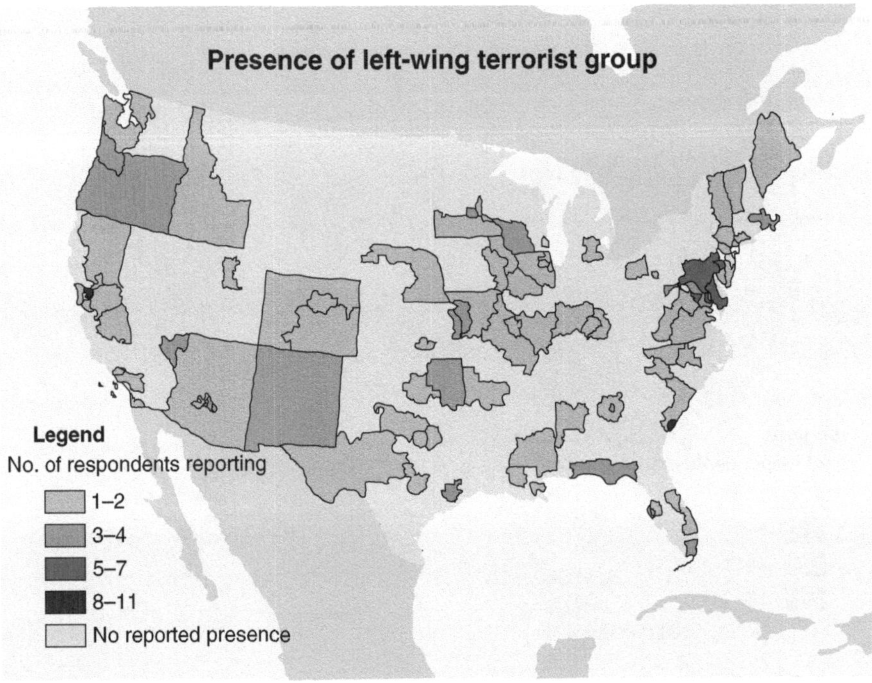
Presence of left-wing terrorist group
Legend
No. of respondents reporting
1–2
3–4
5–7
8–11
No reported presence

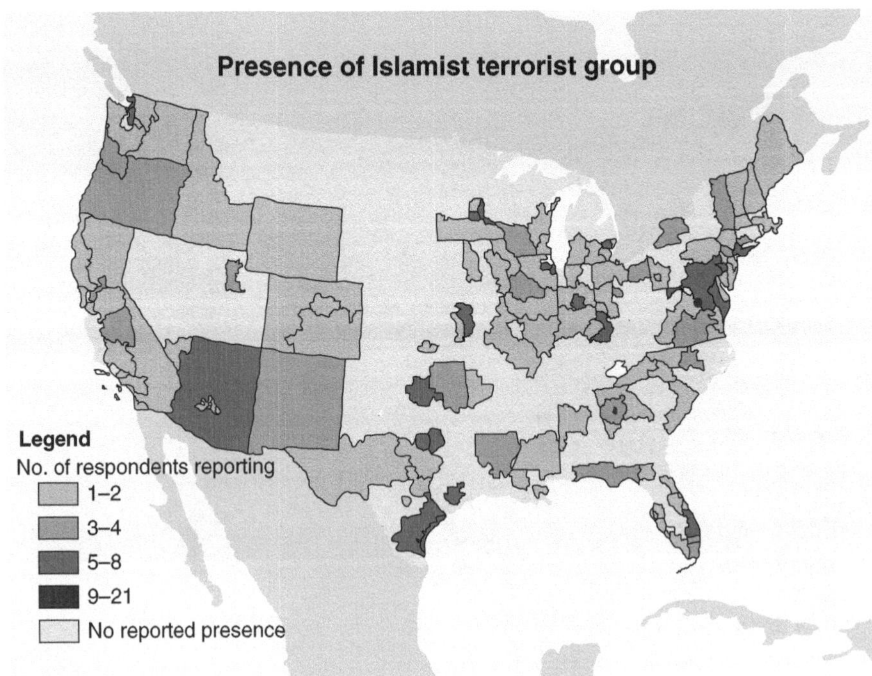
Presence of Islamist terrorist group
Legend
No. of respondents reporting
1–2
3–4
5–8
9–21
No reported presence

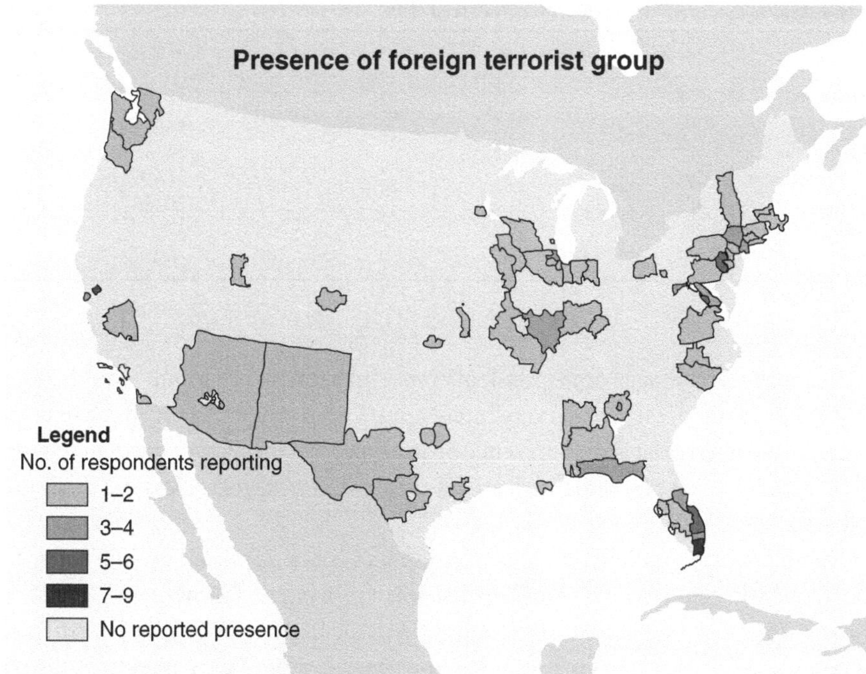

Figures 4.1–4.7 Reported presence of terrorist groups

National Frequencies

Before reviewing the frequency report results, the measurements should be clearly understood. First, the percentages and raw numbers of group fear represent the entire sample population. For the national section, the total number of cases was 1,732. All percentages in the national data section are based on this number. Likewise, each region's frequency data was based on the entire region's population.

It must also be noted that the survey instrument did not restrict respondents on how many groups they could report as feared to act. Thus, the list of percentages will not equal 100 percent, since respondents could choose to report fear of none, several, or all of the groups listed. The percentage of respondents that reported fear of a specific group simply represents how many of the total sample feared that a particular group would act, regardless of their fear of other groups (Table 4.4).

The national frequency reports generated two unexpected findings. First, as reflected in Table 2.1, no group was predicted to act by a majority of respondents, the highest percentage being eco-terrorists at 43.8 percent. Second, and most surprisingly, was that the rank order list of groups feared to act was entirely inconsistent with group presence reported in the previous section. Eco-terrorists (43.8 percent of the total study sample) and Islamists (40.4 percent) were the top two most feared groups, and the only two to have been reported by over 40 percent of respondents as likely to perpetrate an act of terrorist violence in the future. This was a very intriguing data point, especially when considering that these groups were ranked fourth and fifth out of seven in reported group presence.

Table 4.4 Fear a Terrorist Group will Act: National Data

	No. of total respondents	*% of total study population*
Eco-terrorists	758	43.8%
Islamists	700	40.4%
Anti-abortionists	629	36.3%
Ethnic supremacist	591	34.1%
Right wing	487	28.1%
Left wing	207	12%
Foreign groups	112	6.5%

Note: n = 1,732.

The lower percentages of respondents reporting fear that a group would attack as compared to group presence was very interesting. Recall that 84.4 percent of all respondents reported a terrorist group presence of any type. It was noteworthy then to see that only 78 percent of all respondents feared a terrorist attack, regardless of which group was feared.

Overall, the national distribution of group fear is best reported in rank order from most frequently reported to least. This list was reported as follows (n = 1,732): eco-terrorists, 758 (or 43.8 percent of all respondents); Islamists, 700 (40.4 percent); anti-abortionists, 629 (36.3 percent); ethnic supremacist, 591 (34.1 percent); right wing, 487 (28.1 percent); left wing, 207 (12 percent); and foreign groups, 112 (6.5 percent). A surprisingly high 228, or 13.2 percent, said that a group would attempt an attack in the future, but the respondent was not sure what type. Finally, there was also an option on the survey to forecast an attack by a group type that was not listed as a choice in the survey. This miscellaneous category was chosen by 29 respondents, or 1.7 percent of the total response population.

Summarized in Table 4.5, analyses of group fear data yielded clear regional characteristics. Some aggregate national consistencies were suggested, but none without regional exceptions as discussed below.

In regard to overview data, the apparent lack of consistency between overall rates of group presence and overall group fear was notable. For example, national data showed a fairly wide gap of 6.4 percent (with 78 percent reporting group fear and 84.4 percent reporting group presence). Across regions, the difference between the two measurements varied widely, from almost no difference (0.1 percent in the west) to a wide gap (10.9 percent in the southwest).

Table 4.5 Most Feared Group by Region: Rank and Percentage Reporting

	Northeast	*South*	*Midwest*	*Central*	*Southwest*	*West*
Eco-terrorist	3 38%	4 31%	4 32%	1 44%	3 40%	1 75%
Islamist	1 45%	1 41%	2 36%	5 30%	1 45%	2 41%
Anti-abortionist	2 38%	2 35%	1 37%	2 42%	5 33%	4 35%
Ethnic supremacist	4 32%	3 32%	3 34%	4 32%	2 42%	3 38%
Right wing	5 19%	5 24%	5 29%	3 40%	4 37%	5 33%
Left wing	6 11%	6 11%	6 8%	6 8%	6 14%	6 19%
Foreign group	7 8%	7 9%	7 4%	7 2%	7 9%	7 7%

The most compelling result of group fear analysis was the surge of Islamist groups in regional and national reporting. Islamists were ranked as the first or second most-feared group in every region except the central, which had them ranked very low in the fifth of seven rank order positions. Even in spite of the very low reporting in the central region, Islamist groups were ranked as the second most-feared group nationally. The only group ranked ahead of Islamists nationally was eco-terrorists, which is ranked in the top two in only two regions, central and west. If not for the west region's extraordinary reporting of fear of eco-terrorist groups, Islamists would be by far the most feared group nationally. There is clearly something about Islamist groups that causes respondents of this study to fear they will act, regardless of whether they actually exist in the respondent's local area.

As mentioned previously, eco-terrorists were considered most likely to act by two regions, the central and west, and second most likely by respondents from the southwest (behind Islamists). This finding was particularly surprising when considering the apparent lack of similarity between the central and west regions.

Building a Predictive Model

The disparity between group presence and group fear demanded further exploration. To discover the value of group presence in predicting group fear, seven regression models were developed (one for each region) that used fear that a certain group would act as the dependent variable. Local presence of terrorist groups was used as the independent, or predictor, variables. In each regression model the independent variables were comprised of seven named terrorist groups and two general representations of group presence—the "miscellaneous" and "any group" categories. Since each model included all nine types of group presence as predictor variables, interaction among the variables was controlled for and resulted in statistically stringent models.

Each model was considered significant, and returned high Nagelkerke R-square statistics. These R-square statistics provided the percentage of the variance in the group fear dependent variable explained by the group presence predictor variables. While all R-square statistics could be considered high, they varied according to which group served as the dependent variable. All R-square statistics and odds ratios are summarized in Table 4.6.

The value of group presence in predicting fear of that group was clear in each model, though the odds ratios varied widely depending on the group listed as the dependent variable. From this data it is apparent that though fear of a group is predicted by the group's presence, the strength of the prediction depends on which group is being discussed. Stated differently, how much relationship there is between group presence and group fear depends on the group type. Despite the distinctions among group types, however, the models presented above show a clear and universally strong relationship between local actor presence and fear that the same actors will attempt a terrorist attack.

As with reported group presence, data from group fear variables were analyzed with crime mapping software. Figures 4.8–4.14 show the concentration of reported

Table 4.6 Summary Statistics for Fear/Presence Regression Models: R^2 and Odds Ratios

Group type	*Nagelkerke R^2*	*Odds ratio*
Left wing	0.38	17
Right wing	**0.32**	**12**
Islamist	0.27	6
Eco-terrorist	0.45	19
Anti-abortionist	0.20	9
Ethnic supremacist	0.24	6
Foreign groups	0.30	14

Note: All models were significant at $p < 0.001$.

group fear by area code. The concentration of group fear is extraordinary even within regions, reinforcing the localized effects of fear.

Summary of Findings

Terrorism is a local phenomenon. Analyses performed in this study demonstrated this fact without exception. This cannot be overstated when discussing how to most effectively assess risk. Among explanations for the presence of terrorist organizations and fear of experts that they will act, the most useful rationales were found at the local level.

Statistical procedures and computer mapping applications generated conclusive responses to the research questions posed earlier in the study. First, the presence of terrorist groups and fear that they would act showed highly localized characteristics, a

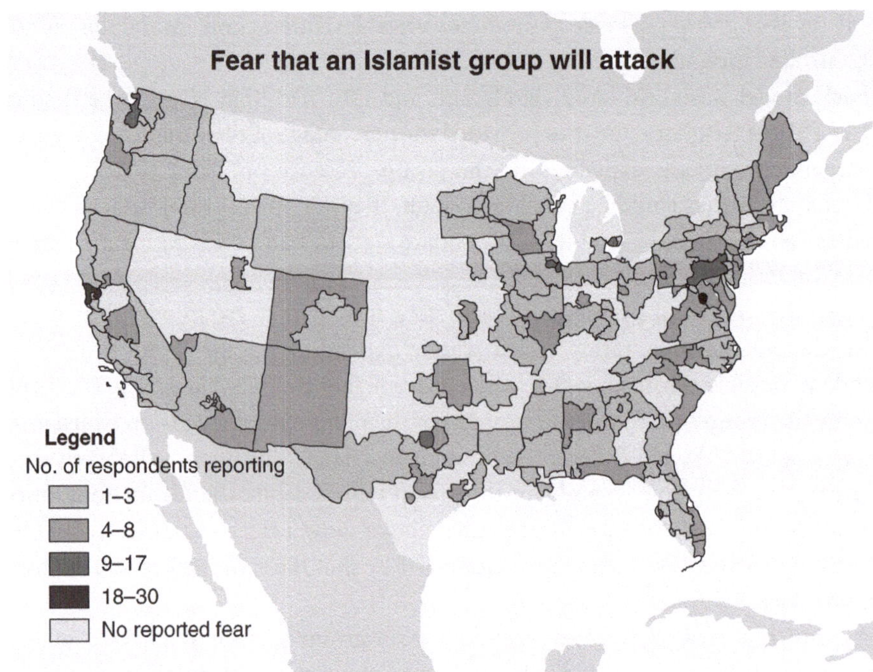

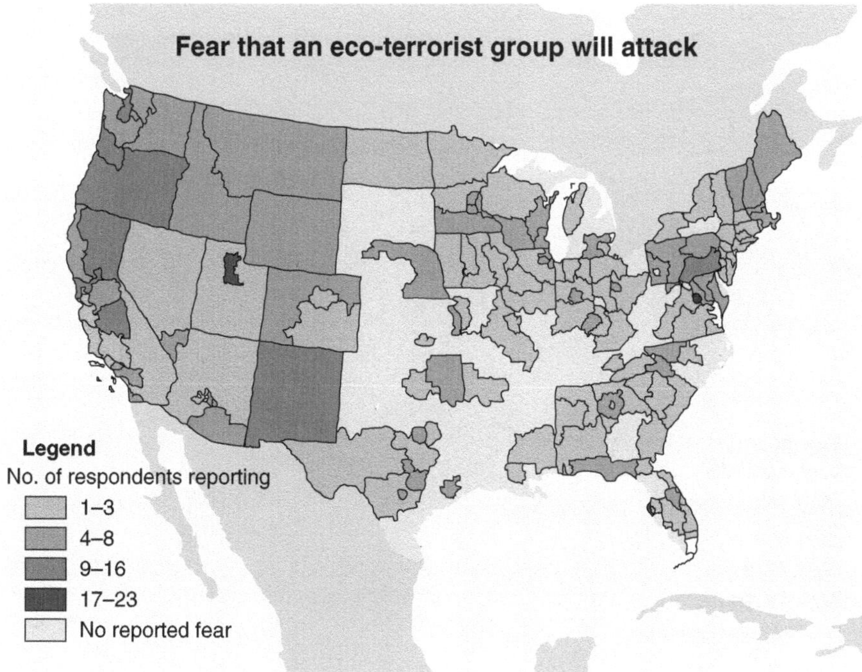
Fear that an eco-terrorist group will attack
Legend
No. of respondents reporting
1–3
4–8
9–16
17–23
No reported fear

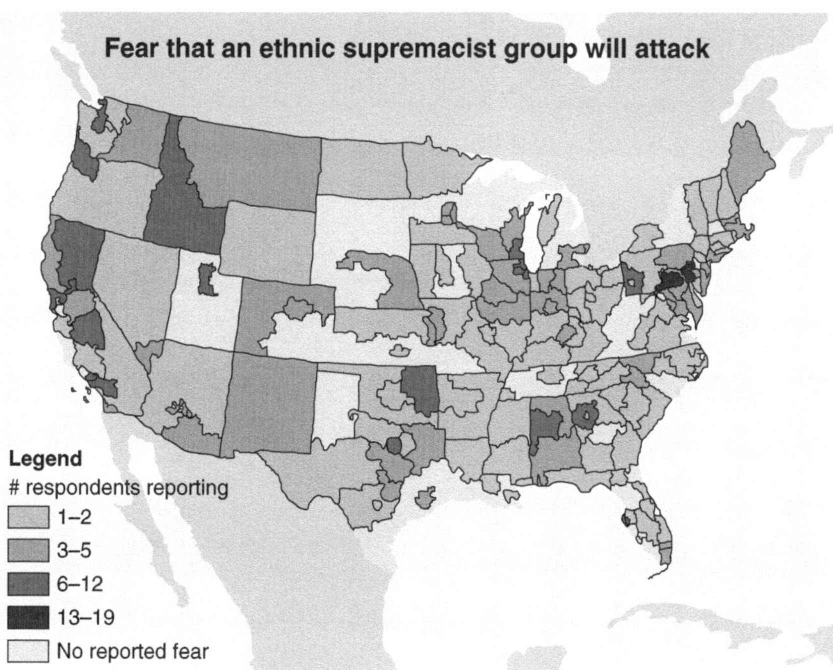
Fear that an ethnic supremacist group will attack
Legend
respondents reporting
1–2
3–5
6–12
13–19
No reported fear

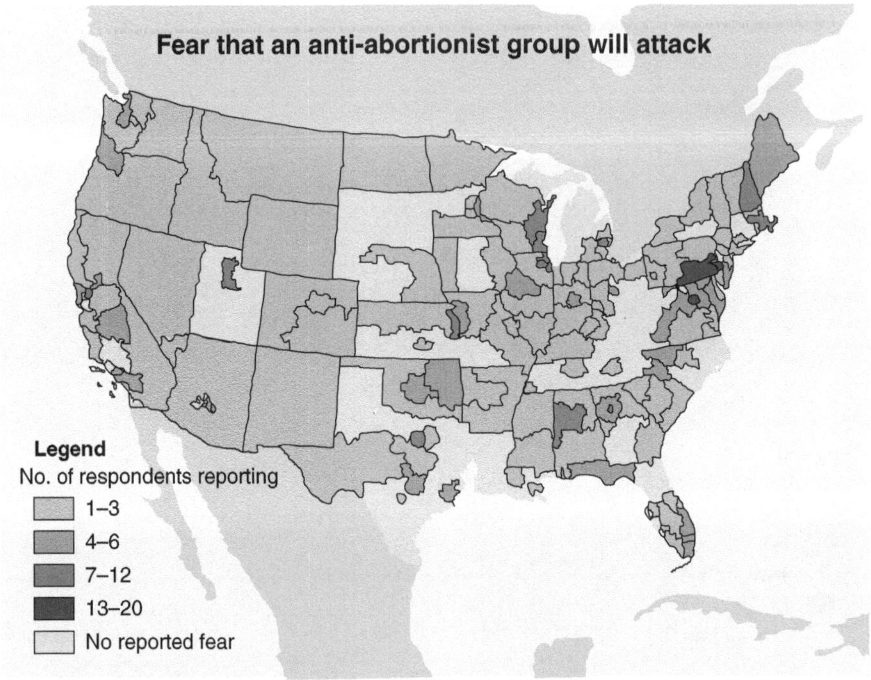
Fear that an anti-abortionist group will attack
Legend
No. of respondents reporting
1–3
4–6
7–12
13–20
No reported fear

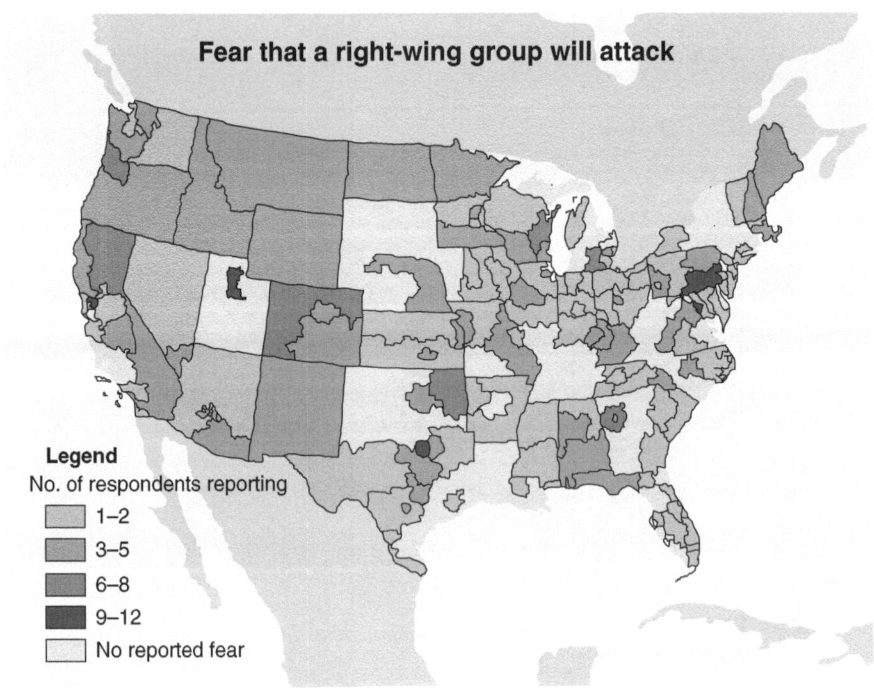
Fear that a right-wing group will attack
Legend
No. of respondents reporting
1–2
3–5
6–8
9–12
No reported fear

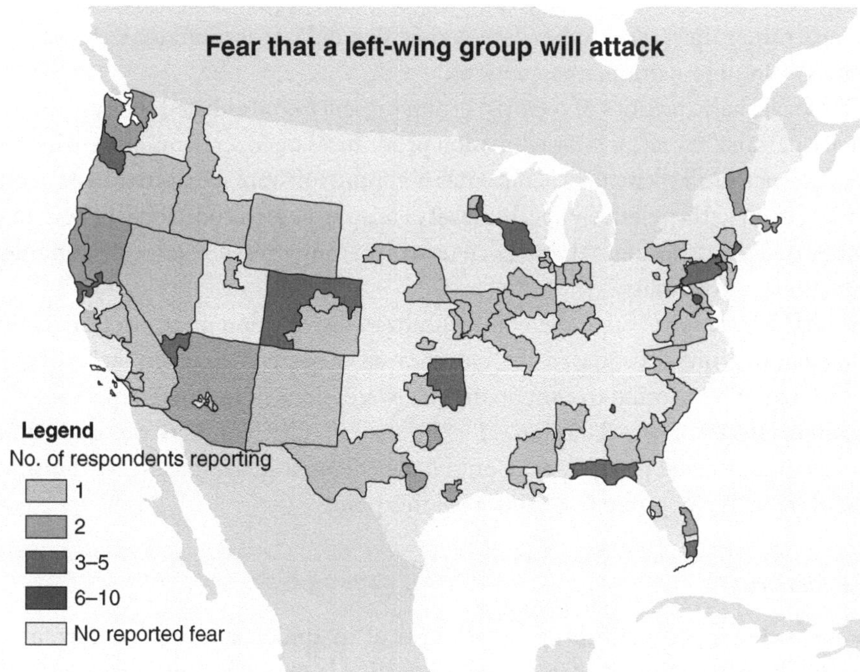

Figures 4.8–4.14 Reported fear that an attack will occur

significant departure from past research. These findings were partly due to the structure of this study, which put such emphasis on the uniqueness of group types and the individualized perceptions of geographic areas. This unique framework was a significant departure from earlier studies of terrorism that did not account for terrorist group identities or local practitioner perceptions.

The local characteristics of terrorist group presence contradicted the current paradigm of a national threat, as crosstabulation procedures uncovered unique attributes of group presence in all six regions. Computer mapping software illustrated an even more localized effect, to the relatively specific level of telephone area code zones. These analyses supported the hypotheses that viewed terrorist group presence as local phenomena, not a national universality.

Similar tests also revealed the highly localized distribution of group fear. Though crosstabulations and chi-squared analyses showed cross-region consistencies of group fear, each region was found to contain unique perceptions of fear that a specific terrorist group would act. These data showed without ambiguity that there was no national-level fear of any single group type. Computer-generated maps that clearly illustrated localized reporting of group fear reinforced this point.

Why it Matters

This study developed concepts that are critical in understanding the current terrorist threat in the United States. This criticality is based on the recognition that identification of the local terrorist threat is the foundation of effective anti-terrorism strategy development, whether at the local, regional, or national level. Integral in this statement is the recognition that local strategy will likely be very different than regional strategy, which will in turn exhibit considerable differences from the national strategy. The central issue at all levels is to understand the threat against the population being served.

Regarding the issue of distinctive threats at the various levels of jurisdiction, the study showed without a doubt that terrorist group presence and forecasted activity varies widely across geographic areas. Frequency reports and crosstabulations showed that each region had a unique view of its terrorist threat. Through mapping software, these threats were shown to be even more area-specific, concentrated not in entire regions but in specific areas of regions. These analyses indicated that there were clear and persistent local effects on terrorist group presence and fear of future attack.

The importance of these findings to policy development is illustrated through documents published to guide anti-terrorism strategy. As stated in the opening page of the Government Accounting Office's (GAO) *Combating Terrorism: Linking Threats to Strategies and Resources*: "The first step in developing sound programs to combat terrorism is to develop a sound understanding of the terrorist threat" (GAO 2000: 1). Among the pitfalls of anti-terrorism strategy development listed in the report is the use of improbable scenarios to develop programs. In other words, the GAO cautioned against the ineffectiveness of an anti-terrorism strategy based on a threat that is not present or likely to act.

The GAO also recommended that policymakers consider the local threat's "qualifications" during strategy development, including the local group's intentions weighed against capabilities, the targets they are likely to hit, and the resources available for groups to carry out an attack (GAO 2000: 3). Data from the current study provides policymakers with much of this information at a level as detailed as a practitioner's telephone area code. Of course, the data is self-reported by law enforcement professionals that are assumed to know the criminal actors in their jurisdictions. Hence, information gained from this study is only as good as the respondents' level of knowledge.

Through a comparison with earlier studies, the current project indicated interesting trends in terrorist attack forecasting. Consider, for example, a 1993 mail survey of police chiefs from 140 cities with a population over 100,000, in which respondents were asked to rank terrorist groups as the greatest threat to the United States for the next two years. The study hypothesized that Middle Eastern groups and anti-abortionists would be heavily reported, since the first World Trade Center bombing had taken place soon before the survey and there were many acts by fringe anti-abortion groups that year. Not surprisingly, the two groups were ranked high, but only second and third of ten groups likely to be active. The unexpected most-feared group was white supremacists. Also noteworthy was the lumping of eco-terrorists into an "other" category that was reported as likely to act by just 13 percent of respondents (Carlson 1995).

The RAND study administered during the same year surveyed a sample of law enforcement agencies in January 1993 (one month before the first World Trade Center bombing). Respondents from this study reported right-wing groups and issue-specific organizations as the most prominent threats (RAND 1995: 24). The issue-specific group aggregation included eco-terrorists and anti-abortionists, two groups that were disaggregated in the current survey. It is important to note that even individually these two groups were reported at rates as high or higher than when grouped together in the earlier study.

The Carlson and RAND studies were similar methodologically to the current project, but the findings contrast sharply. First, the almost dismissive importance of eco-terrorists in the other studies indicates that they were not considered much of a threat at the time. This is noteworthy because respondents in the current study identified eco-terrorists as the most feared group nationally, with some regions reporting their presence and likelihood to act at rates approximating 70 percent. Likewise, Islamists (identified in the previous two studies as "Middle Eastern" or "Foreign terrorist groups") were reported as likely to act in the current study at rates far beyond those in earlier surveys.

Through evaluating the evolution of perception of these two groups alone, it must be acknowledged that the presence of terrorist groups and perceived likelihood that they will act has evolved significantly in just ten intervening years. This is true not just of eco-terrorists and Islamists, but of all groups used in this study's typology. Clearly, the threat shifted, making the data available through this study critical to policymakers who are formulating anti-terrorism plans at all levels of responsibility. The shift in local threats suggests that care must be taken when using data that are not current. Findings from this study are important now, but will require new data for comparison in the intermediate future if the information is to be used for policy development. Like all forms of crime, the threat of terrorism is fluid. It demands current data if it is to be understood and controlled.

The importance of hard, current data to anti-terrorism policy development is even more valuable when financial shortfalls in law enforcement are taken into account. Fiscal challenges are well illustrated through analysis of the three primary federal funding sources for local law enforcement; the Local Law Enforcement Block Grant, Edward Byrne Memorial Grant Program, and the COPS program. Since 2002, just as law enforcement agencies were taking on new homeland security tasks in the post-9/11 environment, these programs have been cut by $1.24 billion, representing 50 percent of available funds. Evaluation of the one-year cuts in the individual programs suggests an even more harrowing picture; for example, the COPS program was allocated $150 million less in 2005 than in 2004, a one-year decrease of 19.8 percent, and a 47 percent decrease since 2003 (Boyter 2004: 8).

To counter the financial strains on local law enforcement, the availability of data from studies like the current one is increasingly important. Practitioners must have access to up-to-date information to better formulate their anti-terrorism strategy and disaster response plans. When tackling the issue of local terrorist threat identification, reliance on outdated data would result in a strategy built on false information. The stakes are simply too high to have to rely on old data.

Closing Comments

Developing sound data for field use should be a priority in the field of risk management, never more so than when discussing terrorism. There are few places to find better data for field use than in the field itself, in the case of this study from the practitioners tasked with countering and responding to the terrorist threat in their jurisdictions.

Police cannot be everywhere, and, despite political pressure otherwise, they cannot do everything. In the current environment of increased tasks and decreased funding for equipment and training, identifying the local threat is the critical step in formulating a lean, competent strategy for dealing with this relatively new phenomenon.

Finally, the importance of using current data from the field is absolutely critical. Note that the findings from this study vary significantly from studies conducted just ten years prior. This can be due to world events, police effectiveness in eliminating earlier threats, or other mitigating factors. Yet there is no question that the threat will continue to change. Consider some seminal events of just the past four short years: US military involvement in two geographically distinct locations; the increasing pervasiveness of the internet to recruit and indoctrinate the willing into hate groups of various affiliations; the ongoing abortion debate and all the passion it evokes on both sides; religious recruitment and conversion programs in our prisons; and the increased militarization and politicization of gangs, especially from Central America. Each of these will have obvious effects on the changing landscape of political violence in the United States and abroad.

Threat is local, it is fluid, and to be properly mitigated the responding risk management programs must also be local and current. Not doing so with traditional crimes can result in loss of life and property that, while tragic, pales in comparison to the next mass casualty event that is no doubt being considered in the mind of a threat that has yet to be identified.

References

Boyter, Jennifer. 2004. "Congress Passes Omnibus Appropriations Bill; State and Local Law Enforcement Assistance Funds Cut by 24 Percent." *Police Chief* LXXI(12):8.

Carlson, Joseph R. 1995. "The Future Terrorists in America." *American Journal of Police* 14:71–91.

GAO (US General Accounting Office). 2000. *Combating Terrorism: Linking Threats to Strategies and Resources.* GAO/T-NSIAD-00-218.

Norris, Ed (Baltimore City Police Commissioner). 2002. Introduction. *COPS Innovations.*

RAND Corporation. 1995. *Domestic Terrorism: A National Assessment of State and Local Preparedness.* Santa Monica, CA: RAND.

Questions

1. When considering risk, how important are threats when there is no vulnerability to these threats?
2. From an all hazards perspective, does it make sense to consider threats to public security as having the same impact? How do we adjust for the difference in these threats (e.g., the exposure of individuals to these threats or their vulnerability)?

5.
TERROR GROUPS AND RADICALIZATION

Although the study of terrorism and terrorist groups certainly pre-dates the events of 9/11, and it is also the case that since 9/11 there has been a significant increase in such research, it is also clear that there remains a lack of systematic empirical research upon which risk assessment and management can proceed. Indeed, Lum, Kennedy, and Sherley's (2006) review found that only 3 percent of the terrorism studies involved could be classified as empirical research. This lack of basic research on participation in terrorist groups and in the radicalization process represents a significant shortcoming in the knowledge base for assessing differential risk. The following chapter by Suttmoeller, Chermak, Freilich, and Fitzgerald helps to fill this void through the systematic analysis of white supremacist groups. The authors compare these groups by their involvement in violence. This, in turn, provides a window to the processes in which these groups radicalize their members. Continuing to build this type of database, on white supremacist groups and other extremist groups, is critical to effective risk assessment.

RADICALIZATION AND RISK ASSESSMENT[1]

Michael Suttmoeller, Steven Chermak, Joshua D. Freilich, and Sarah Fitzgerald

Introduction

The risk of attack by terrorists is a major concern. Managing this risk was primarily the responsibility of federal law enforcement agencies before the September 11, 2001 (9/11) attacks changed perceptions about public vulnerabilities to terrorist attacks. Subsequently, concern about additional attacks have increased, expectations about developing effective counter-measures have become more urgent, and shortcomings in our understanding of the nature of the risks posed by terrorist and terrorist groups have become more visible. Indeed, one of the most troubling conclusions that emerged from scholarly reviews of the empirical literature on understanding the risks of terrorism was how little is actually known. There are excellent critiques of the terrorism literature that highlight several concerns (LaFree & Dugan 2004; Ross 1993; Silke 2001), but the use of limited and questionable data was clear (Hamm 2005; Ross 1993). Statistical analysis is rarely used in terrorism research (Merari 1991). For example, Silke's (2001) review indicates that only 3 percent of articles in terrorism journals used inferential analysis compared to 86 percent in forensic psychology and 60 percent in criminology. Lum, Kennedy, and Sherley's (2006: 8) important Campbell systematic review of over 14,000 terrorism articles published between 1971 and 2003, similarly found that only 3 percent were empirical.

Law enforcement has also had to adapt in response to new expectations about preventing and/or responding to acts of terrorism. The Federal Bureau of Investigation's website (FBI 2010), for example, states:

> Following the attacks of 9/11, we were called on once again to change. What ensued has been one of the most far-reaching transformations in our history, as we've realigned our structure, created new capabilities, hired different talents, and shifted more personnel into intelligence and terrorism operations—all to meet our now overriding priority to protect the U.S. from terrorist attacks.

Another change to law enforcement has been increased expectations related to the collection, processing, analysis, and sharing of critical intelligence about the activities of terrorists and terrorist groups. Law enforcement agencies are now encouraged to share information so that decision makers can make timely strategic and tactical decisions to prevent terrorism. In an effort to promote data and intelligence sharing, States have created and staffed intelligence fusion centers. These centers are designed to process local security information and pass along critical intelligence to national networks. These changes have increased the burden on local law enforcement as they are now responsible for better understanding terrorism risks in their communities and collecting and sharing information with federal authorities and/or fusion centers without receiving additional resources.

There have clearly been some successes, as law enforcement agencies have identified the potential risks of certain individuals and/or groups that posed a threat, investigated, and then apprehended the suspects prior to the planned action occurring. For example, one post-9/11 thwarted terrorist act was the planned attack of the Fort Dix Military Base by a group of home-grown, Al-Qaeda-inspired terrorists. In 2007, the FBI arrested six men involved in a plan to "invade" the military base with a large cache of weapons, open fire, and kill as many soldiers as possible. There are several interesting elements to the case, but two are relevant to understanding and assessing risk related to terrorism. First, the act was prevented because of effective sharing of information between law enforcement agencies. Local law enforcement officials received a tip from a video retail clerk who was asked by the suspects to convert video to DVD. The clerk observed the suspects participating in "disturbing activities"—firing assault weapons and calling for jihad and screaming "Allah Ackbar" ("God is Great") (see *United States v. Mohamad Ibrahim Shnewer*). The clerk reported this information to local law enforcement, who shared it with a terrorism coordinator, who provided the information to the FBI. After viewing the DVD, the FBI and the Joint Terrorism Task Force conducted the successful investigation. Second, the individuals who plotted the massacre came together as an extreme group, were not directly connected to international terrorism networks, studied and embraced radical Islam, and independently radicalized toward violence. Five of the six were convicted of conspiracy to commit murder in late 2008, receiving long prison sentences. The sixth member was sentenced to a five-year term after being convicted of weapons offenses.

Little empirical work exists that allows us to effectively understand the processes that result in extremists or extremists groups radicalizing to commit violent terrorist activities as what occurred in the Fort Dix case. This is a critical gap as law enforcement has limited resources for monitoring the extremist groups in their jurisdiction involved in criminal activities. Law enforcement has to make informed risk assessments: law enforcement agencies must decide of the X number of radical left, right, and Al-Qaeda-inspired groups and/or individuals in their jurisdiction, which pose the greatest risk and warrant close law enforcement attention. It is thus critical to understand the existing schools of thought related to the radicalization process. Moreover, it is important to identify basic differences between groups that have radicalized to commit violence compared to ideologically extreme, but non-violent groups.

This chapter engages both of these issues. First, we provide a synopsis of the radicalization literature, highlighting the key steps that contribute to groups becoming radicalized. What is radicalization? Is radicalization necessary for a group to partake in violent activity? Second, we discuss some basic findings related to identifying the differences between white supremacist groups involved in violence compared to those groups that were not known to be involved in violence. Here we review some important differences that highlight the process groups use to radicalize.

Group and Individual Radicalization Issues

Radicalization has been conceptualized in several different ways. Many scholars, for example, describe radicalization as the process of adopting extreme beliefs and being willing to utilize violence in support of that ideology. For example, the Dutch Intelligence Service (AIVD) defined radicalization as the "use of undemocratic methods including violence" (Pressman 2008: 3). Others, however, argue that the use of violence should not be a requirement of radicalization. For example, the Royal Canadian Mounted Police excludes a predisposition or adoption of violence as a necessary result of radicalization. Similarly, Mandel (2010: 111) argued that "radicalization refers to an increase in and/or reinforcing of extremism in the thinking, sentiments, and/or behavior of individuals and/or groups of individuals."

It is our view that the operationalization of radicalization should include both attitudinal and behavioral attributes. Radicalization generally begins as an attitudinal change, which then may result in a behavioral change. For example, at the individual level, Silber and Bhatt (2007) discussed how during the early stages of radicalization, individuals explore radical ideologies, and their attitudes begin to change. Then as the individual moves toward the later stages of the radicalization process, the attitudinal changes are manifested in behavioral changes. Similarly, Sprinzak's (1995) discussion of group radicalization included both attitudinal and behavioral characteristics. During the earlier stages of Sprinzak's (1995) radicalization model, groups begin participating in activities such as protests, demonstrations, and symbolic resistance (see also Bjorgo and Horgan 2009; Horgan 2009). Even though the behavioral component of radicalization consists of increasingly radical behaviors, they are not necessarily violent behaviors. From a law enforcement perspective, although counter-measures may include attempting to "de-radicalize" or "disengage" individuals, their primary concern has been to respond to illegal behaviors that result from the radicalization process.

Researchers have also discussed non-violent behaviors that may occur after radicalization (Horgan 2009; Sprinzak 1995). During the early stages of this radicalization model, groups begin participating in activities such as protests, demonstrations, and symbolic resistance. While these activities have the potential to become violent, they are not inherently violent. However, as the radicalization process unfolds the potential for violent or terrorist behaviors increases. In the section examining differences between violent and non-violent hate crimes, we consider a group's behavioral involvement in violent criminal activities as a key distinguishing feature. All of the groups in the sample have extreme beliefs, thus we are much more concerned with the behavioral indicators of radicalization than attitudinal indicators.

While scholars have not agreed on a definition of radicalization, they also do not agree on how an individual or a group becomes radicalized. Some scholars have proposed that radicalization is a process of specified steps that an individual or group moves through sequentially (or not) as they become more extreme. However, scholars do not necessarily agree on the number of steps in the process, or what is involved at each step.

Radicalization Process

Scholars have generally discussed the radicalization process as pertaining to individuals (Kirby 2007; Sageman 2007, 2008; Silber and Bhatt 2007; Wintrobe 2006; Zaman 1998) or groups (Ezekiel 1996; Hamm 2002; Sprinzak 1995). Silber and Bhatt (2007) examined individual-level radicalization and developed a model that consisted of a distinct series of steps: pre-radicalization, self-identification, indoctrination, and jihadization. In their model, they acknowledge that not all individuals who begin the radicalization process necessarily progress through all the stages—individuals may stop or quit at any time throughout the process. They believe, however, that those who progress through and complete the four steps are more likely to participate in terrorist activities. While their conclusions were applied to understanding Islamic radicalization in the West, with slight modification, the argument could be extended to other types of extremists (for example, see Aho 1990, 1994; Hamm 2002).

During the pre-radicalization stage, a person is living an ordinary life and has not accepted the radical ideology that will provide the basis for future stages (Mandel 2010), however, during this stage, certain characteristics such as the demographic composition of the town or area where the person lives may provide the background for radicalization (Silber and Bhatt 2007). In the self-identification stage, the individual becomes more personally aware and begins to explore the ideology, which may move them away from their previous life and identity (Silber and Bhatt 2007). Individuals who have experienced some type of crisis are particularly vulnerable during this stage. The crisis could be economic, social, political, or personal. These crises sometimes cause a person to seek out others who are similarly suffering and could lead to an increased exposure to radical ideology (Silber and Bhatt 2007; see also Kirby 2007; Zaman 1998).

During the indoctrination stage, the radical views are intensified, and the individual unquestioningly adopts the radical Islamic ideology that justifies, encourages, and supports the use of violence against anything or anyone in conflict with these beliefs (Silber and Bhatt 2007). These ideas are usually supported by a group of like-minded individuals or a leader. In the final stage, jihadization, members accept their duties to commit acts of terrorism or martyrdom (Mandel 2010). Also during this stage, group dynamics begin to play a much larger role than in previous stages. During this stage, the group becomes a cohesive entity (see also Della Porta 1998; Kirby 2007; Wintrobe 2006) and group loyalty surpasses all previous relationships (Silber and Bhatt 2007). Importantly, the individual is increasingly isolated from other influences (Hamm 2002; Kirby 2007; Silber and Bhatt 2007).

While some believe the radicalization process to be a distinct series of steps, others, such as Sageman (2007, 2008) do not believe that radicalization is a specified linear process, but rather consists of four prongs that can occur in any order and at any time. Additionally, not only may the four prongs occur concurrently, but they may also interact with each other. These four prongs are: moral outrage, interpretation, resonance with personal experience, and mobilization through networks (Sageman 2007, 2008). Similar to the application of Silber and Bhatt's (2007) conclusion, although Sageman focuses on Islamic groups, his four prongs could probably be applied to other contexts. These prongs are discussed briefly below.

Moral outrage is caused by a specific event or events (e.g., a crisis) that cause extreme anger and may incite a need to retaliate. In order for the outrage to reach this level, there must be enormous suffering caused by human beings. Even though events such as tornadoes, tsunamis, or earthquakes, can cause great suffering, they do not fit this criteria since they are naturally occurring and cannot be directly attributed to anyone in particular (Sageman 2008). Examples of events that would cause the moral outrage as described by Sageman (2007, 2008) are murders, killings, rapes, or arrests. Sometimes the media portrayal of an event can inflame some segments of the population and cause moral outrage (Sageman 2008). Events such as discrimination and corruption, when independent from other factors, do not cause moral outrage of this nature.

Interpretation (Sageman 2007, 2008) is how the moral outrage is interpreted. Depending on how it is interpreted will determine whether or not the moral outrage contributes to the process of radicalization or not. Sageman (2007) surmised that too much emphasis has been placed on a terrorist's ideology as an explanation of radicalization. He further believed that radicalization is more related to how a person feels, and because of this focus on a person's feelings, rather than ideology, Sageman (2008) explained that for the moral outrage to contribute to radicalization, the person must interpret the event and place it into a context based on their personal experiences.

During resonance, people examine the current event through the lens of the social, economic, political, and religious factors that comprise their personal life. The example in Sageman (2007, 2008) is that European Muslims are generally affected differently by the perceived War on Islam that is occurring due to the differences in the personal experiences Muslims have had in Europe versus the United States (see also Ezeikeil 1996; Zaman 1998).

The final prong is that radicalization is mobilized by networks (Sageman 2007, 2008; see also Della Porta 1998). These networks can take different forms, but within this context, individuals are radicalized through group dynamics. Individuals are radicalized through their interactions with other people and groups. Through these interactions, potential extremists are radicalized through discussions with other potential extremists or extremists. As a result of these discussions, fledgling extremists gain moral support and validation from the other members. In small groups of potential extremists, the choice to radicalize becomes a group or collective decision rather than simply an individual decision. Traditionally, these dynamics played out in a face-to-face setting, but since the advent of the internet, group forums, and discussion pages have contributed to this process (Sageman 2007, 2008).

Even though the majority of the radicalization literature has focused on the individual, some scholars such as Sprinzak (1995) have studied group radicalization. Sprinzak (1995) argued that the group radicalization process consisted of three distinct stages: crisis of confidence, conflict of legitimacy, and crisis of legitimacy. A crisis of confidence is the first stage of the process, which is the initial stage of a group forming and developing its beliefs. It is generally moderate and does not consist of any violent behavior, but is rather where a group begins to develop its beliefs and uses more socially accepted ways to showcase them. This stage is further characterized by decreased confidence in the established government and its policies. The group, in this stage, will articulate their complaints with ideological statements, protests, symbolic demonstrations and other direct, non-violent action (Sprinzak 1995).

The second stage identified by Sprinzak (1995: 19), conflict of legitimacy, is characterized by increased anger and demonstration from the group, or the emergence of "an alternative ideological and cultural system." This new system attempts to decrease the legitimacy of the established system. At this stage the group will develop anger and frustration causing the group's internal cohesion to increase, isolation from the larger community, and cause more questioning of the legitimacy of the government or regime, which does not seem to be taking action (see Aho 1990, 1994; Ezekiel 1996; Zaman 1998). During this stage, the group may participate in intense political actions that take the form of angry protests, demonstrations, confrontations, and vandalism. These new activities may include low-level violence (Sprinzak 1995).

The final stage identified by Sprinzak (1995) is the crisis of legitimacy. This stage is the culmination of the two previous stages. During this stage, the previous attacks on the legitimacy of the government are extended to all that are associated with that entity. These individuals are dehumanized in an effort to allow the individuals involved to morally disengage from the situation and prepare them to commit violent terrorist acts (Sprinzak 1995). While the previous process is part of the "process of delegitimization" (1995; 18), Sprinzak (1995: 20) argued that the process can sometimes be different, a process of "split delegitimization."

During this process, the previously discussed stages are present, but not necessarily occurring in the same order as before. Consistent with the name "split delegitimization," the focus of the extremist group is split: an intense delegitimization focused on a non-governmental entity and a less intense effort of delegitimization aimed at the ruling regime or government. This form of delegitimization is especially applicable to right-wing groups. Within this context, the groups begin by believing that a particular non-governmental group is not legitimate or that they are inferior to group members as human beings (conflict of legitimacy) (Sprinzak 1995). However, they may eventually begin targeting the ruling regime or government. The group may perceive that the government is not protecting their "legitimate" group, but rather is supporting the inferior elements of society. At this point the group is suffering a crisis of confidence. However, sometimes the "split" disappears and the group begins to believe that the regime or government and the illegitimate group have become one entity, with the illegitimate group having taken over the government (Sprinzak 1995). An example of this would be the Zionist Occupation Government that many right-wing groups espouse to be in conflict with them.

Moskalenko and McCauley (2009) and McCauley and Moskalenko (2008) also discussed the process of group radicalization. Instead of a step-by-step process of radicalization, they proposed a pyramid model. The different levels of the pyramid represent varying levels of commitment to a cause. The lowest stage or base of the pyramid represents those that may be sympathetic to the cause, but are not involved. Each level of the pyramid then represents an increased level of commitment, with the apex representing those involved in violent or terroristic behaviors. While some may pass through all the levels of the pyramid, it is not necessary in order to reach the apex. Outside influences may cause levels of commitment to change dramatically, which in turn causes levels of the pyramid to be skipped. Additionally, not all those who begin the radicalization process at a lower level necessarily reach the apex and commit violent or terrorist acts (Moskalenko and McCauley 2009).

While scholars may not agree on the exact steps that are involved in the radicalization process, or even whether it is a process at all, one commonality is the escalation of the types of behavior. Even though Silber and Bhatt (2007) examined individuals and Sprinzak (1995) and Moskalenko and McCauley (2009) examined groups, a common theme through each of their models was the escalation of behavior toward violence. In the earlier stages of Silber and Bhatt's (2007) model, the individual changes their behavior to accommodate their new ideology, whether it is attending a Salafi mosque, or by becoming more involved in social activism and community issues. Similarly, in Sprinzak's (1995) model of group radicalization, as the group radicalizes, it will begin to participate in more radical behaviors such as protests or symbolic actions. Generally, these types of actions in both contexts are legal; however, based on their models, the individual or group behavior may escalate into illegal violent behaviors (Silber and Bhatt 2007; Sprinzak 1995).

Prior to this point, the focus of this discussion has centered on how an individual or group becomes more engaged in the radicalization process. However, a discussion of disengagement should also be included. Horgan (2009) defined disengagement as the process through which an individual's role or purpose within the radical movement changes usually resulting in a decrease in their participation in violent activities. When an individual disengages, they may not necessarily abandon the movement or group, but rather take on a different role. Often disengagement and the changing of the individual's role is the result of some psychological change such as disillusionment (Horgan 2009). Because an individual who is disengaged from the movement or group does not necessarily abandon those entities, they are not necessarily deradicalized.

Desistance from terrorist activities is somewhat similar to desistance from crime (Horgan 2009). Desistance is not characterized by a distinct event that marks the end of a terrorist or criminal career, but rather is a process of continually avoiding and refraining from participating in those activities. Even though some comparisons may be made between desistance from crime and desistance from terrorism, those who participate in those activities often characterize their actions very differently. Since terrorists generally perceive their activities as part of a larger ideological framework and do not perceive themselves as criminals, a comparison to traditional criminals is difficult (Horgan 2009). Further research into the process of disengagement from terrorism and deradicalization is warranted.

Groups may decline due to individual desistance, or as Oots (1989) discussed, group-level reasons for a group's decline and desistance of terrorist activities. Issues such as ineffective leadership, poor recruitment, the inability to form coalitions, a lack of outside support, increased competition with rival groups, and a lack of internal cohesiveness can all contribute to a group's demise.

While some scholars have examined the process of radicalization in terms of the steps toward radicalization, others have identified other reasons, social, political, cultural, historical, etc. why a group may radicalize. Although a complete discussion of all factors are beyond the scope of this chapter, it is important to briefly discuss how groups may radicalize as a process of group dynamics—as a rational decision to accomplish goals or to gain supporters and sympathizers.

Group Dynamics

As groups radicalize, group dynamics may play an important role in their progression. One aspect of group dynamics that is prevalent in the radicalization process is the concept of "groupthink." (Janis 1972: 9) According to Janis (1972: 9), groupthink is a "mode of thinking that people engage in when they are deeply involved in a cohesive in-group, when the member's strivings for unanimity override their motivation to realistically appraise alternative courses of action." As a result of this mode of thinking, groupthink may result in an erosion of "mental efficiency, reality testing, and moral judgment" (1972: 9) that resulted from the internal group pressures.

Some of the characteristics identified by Janis (1971, 1972, 1986) that are associated with groupthink are invulnerability, rationalizations, morality, stereotypes, pressure, self-censorship, and unanimity. Groups involved in groupthink may irrationally believe that they will not suffer any harsh consequences for their actions, believing that the group dynamic will protect them from outside threats. As threats or issues arise, group decisions are made. These decisions affect the cohesiveness of the group. As the group becomes more cohesive pressure is applied to those members who may have dissenting opinions concerning the group's goals or actions. This pressure to conform may result in unanimity, which is the belief that the group's goals must be valid because everyone in the group believes in them (Janis 1971, 1972, 1986).

Cohesiveness of a group is a particularly important aspect necessary for the formation of a groupthink mentality. As cohesiveness grows it often causes common sense thinking to be reduced as well as mental efficacy to deteriorate (Janis 1971, 1972). As Janis (1986: 98) describes, "the more cohesiveness in the group, the greater the inner compulsion on the part of each member to avoid creating disunity, which includes him to believe in the soundness of whatever proposals are promoted by the leader or by the majority of the group's members." The creation of cohesiveness in extremist groups is quite similar to the creation of it in religious groups or cults, where it is seen through mutual trust as well as interdependence (Hudson 1999).

An aspect that contributes to group cohesiveness is leadership. Because of their position of power, the views or ideology of the leader are often the views and ideology that are adopted by the group (Aho 1990; Ezeikiel 1996; Hamm 2002; Long 1990; Oots 1989). To maintain authority within the group and keep its cohesiveness it is common to exclude those from the group who are doubters or question authority, as they may harm the internal power structure. The conflict that can arise between members of a same group through questioning of authority can be rather damaging to the group, since secrecy, security, and maintenance of the group are highly important (Long 1990;

Ozer 2006). Therefore, the high levels of cohesion in the group create an extreme pressure for there to be agreement and no conflict within the group (McCauley and Moskalenko 2008; Post 1998). This may lead to what is referred to as group extremity shift where there is an increase in agreement among the members resulting in more radical beliefs (McCauley and Moskalenko 2008; McCauley and Segal 1987). As this process is repeated, the group becomes more radical due to those members with less radical views being expelled from the group.

As the group grows, its norms are set, and members avoid deviating from those norms and causing tension within the group. The pressure to conform is even stronger after many members have isolated themselves from their previous lives and made the group their family. The development of this mindset within the group, as it becomes radicalized can have dramatic effects on the group moving toward more violent and terroristic behaviors.

In addition to the role of leadership in maintaining group cohesion, the type of leader can also contribute to a group's risk for radicalization and violent behavior. According to Post, Ruby, and Shaw (2002a, 2002b), the characteristics of the group's leader may help to determine the possible risk associated with that group for violent radicalization and behavior. The authors believed that past experiences with violent behavior are a good indicator of future violent behavior. If a group is headed by a leader who has previously taken part in violent campaigns or been a member of a violent organization, that group is more prone to participating in violent activities (Post et al. 2002a, 2002b). In addition to the past behaviors of leaders, leadership style also may be an indicator of a propensity toward violence.

A charismatic leader can be very influential over the members of the group. In a situation where an organization is headed by a charismatic leader, the group members believe the leader to be superhuman or god-like. Because the charismatic leader is bestowed these types of characteristics, the followers or group members will be unquestioningly loyal, believe what he says and follow his directions. The group members have relinquished their personal judgment to the leader and to the group direction identified by the leader (Post et al. 2002a, 2002b).

Authoritarian leadership also influences a group's risk for violent behavior. Groups with this type of leadership are more vulnerable to violent action through groupthink processes. In authoritarian groups, all decision-making powers are held by those in power positions—often a decision-making body consisting of the leader and their appointed or chosen group members (Post et al. 2002a, 2002b). These types of leaders demand loyalty and obedience. If the leaders are prone to violent activity, then the group is also at risk for violent activities.

Additional Reasons for Radicalization

While group dynamics and leadership play an important role in group radicalization, other reasons also contribute to a group's radicalization. Groups may also radicalize due to competition with other similar groups, to gain supporters and sympathizers, or due to relationships with other radical, violent groups. As groups compete for a base of supporters, resources, and power they may increase their violent tactics in an attempt to win over supporters and eliminate competition from other groups. The competition that occurs between groups, which includes competing for publicity, prestige, benefits, support, and recruits, can help to create a more radical environment within the group (McCauley & Moskalenko 2008; Post et al. 2002a, 2002b).

Not only will groups increase their violent acts in response to competition with another similar group, they also may increase their propensity for violence, absent of competition, in an effort to gain supporters and sympathizers. However, groups must be very cautious not to become too radical or violent and alienate the very base of support they are hoping to win. A group that becomes too radical and surpasses the threshold of acceptance by their supporters may actually lose support, rather than win support (McCauley & Moskalenko 2008; Post et al. 2002a, 2002b).

Just as a group may radicalize and become more violent due to competition with another similar group, a group may radicalize and become violent due to a supportive relationship with a more violent or radical group. When a group maintains active, supportive relationships with other violent or radical groups, this may be an indicator of their intention to increase their level of radicalization or their propensity for violence. Additionally, when groups form these types of relationships, their choice of targets may also align with those chosen by the other group (Post et al. 2002a, 2002b).

As is apparent from the previous discussions, the process of radicalization is not a consistent, known entity. It seems to depend on the situation and type of group. The majority of the literature that exists on radicalization is based on a European context. The limited literature focused on the United States is mainly focused on Islamic radicalization. This is not surprising due to the lasting effects that the terrorist attacks of 9/11 have had on the United States. Because of this focus on Islamic radicalization, very little information concerning the radicalization of United States hate groups exists. We will now attempt to address this gap in the literature and will be the focus of the rest of this chapter.

The Characteristics of Violent and Non-violent Groups

Below we present some preliminary findings from a study that compares group-level differences between violent and non-violent white supremacist groups. Although there is a rich body of literature that defines issues related to radicalization and the radicalization process, there has been very little empirical work with quantitative data. The rest of the chapter makes an attempt to begin to fill this gap by providing empirical results related to assessing the differences between violent and non-violent hate groups. Based on the discussion of the literature from above, there are at least three ways we would expect violent groups to be different than non-violent groups. First, we would suspect them to be more actively involved in recruiting new members, using various strategies. Second, we would suspect that these groups were more actively linked with other extremist networks in both positive and negative ways. Finally, we would suspect the leadership structures of violent and non-violent groups to be quite different.

Before addressing these issues and discussing the results, we wanted to provide a brief overview of how we have developed the sample, collected information about the groups, and coded characteristics about them.

We identified all hate groups existing in the United States between 1990 and 2008 according to the Southern Poverty Law Center's (SPLC) intelligence report. In these reports they provide a list of groups annually. We identified over 4,500 hate groups, but eliminated any group that was not reported for at least three consecutive years. Our primary concern was with groups that had demonstrated at least some survival capacity. In addition, we focused on the activities of the umbrella chapters of specific groups. For example, the SPLC intelligence report noted 18–20 different chapters of the World Church of the Creator in existence, but we only included the umbrella organization. After applying these two inclusion criteria, there were approximately 280 hate groups. We then stratified the groups into violent and non-violent categories, and then randomly picked 30 groups from each category for analysis. These 60 groups are used for the analysis below.

Each group was then assigned to a specific researcher who systematically searched for open source information through 26 search engines, including Google, Copernic, News Library, Infotrac, Lexis-Nexis, and All the Web. Our searches uncovered all public source materials on that group, including court documents such as indictments and appellate decisions when individuals were involved in criminal cases. The name of the group was used as the initial search term, but as individuals and locations were identified in the initial search materials, follow-up searches were conducted to collect additional information. Finally, targeted searches were conducted when specific information was lacking using existing search materials.

Finally, the information about each group was read by a coder who then inputted the materials into an ACCESS file using a group-level codebook. Although only a small portion of the variables are presented below, we identified over 200 characteristics of each group. The variables included information about the size, location, and demographic make-up of the group, its history, ideology, and structure, how it interacted with other extremist groups, member's involvement in violent activity, recruitment strategies, and communication strategies.

Recruitment Strategies

The first issue that we wanted to examine was whether violent groups more aggressively recruited new members. In addition, we wanted to assess the types of strategies used to recruit new members. Table 5.1 presents these results.

The results indicate that violent groups are more likely to use a diverse range of strategies to recruit individuals into their group, and, in general, a higher percentage of these groups use these strategies. Some of the violent groups attempted to recruit individuals through all avenues considered: in prison, in schools, by newsletter, through family and friends, in church, through the Internet, personal visits, at protests, through the mail, in the military, and at concerts. Although a somewhat higher percentage of non-violent groups recruited through the Internet and at church, a higher percentage of violent groups recruited new members using every other type of strategy.

Table 5.1 Recruitment Strategies of Violent and Non-Violent Groups

Recruitment strategy	*Violent groups*	*Non-violent groups*
In prison	20.7%	9.7%
In schools	17.2%	6.5%
By newsletter	24.1%	16.1%
Through family/friends	3.4%	0.0%
In church	10.3%	16.1%
Internet	27.9%	38.7%
Personal visits	6.9%	6.5%
Protests	17.2%	3.2%
Through the mail	6.9%	3.2%
In the military	3.4%	0.0%
At concerts	24.1%	6.4%

It is interesting that such a higher percentage of violent groups recruited at locations often thought of as "hot beds" of radicalization. Thus, it appears that they invest resources to recruit in those areas with the greatest potential of attracting disenfranchised individuals. Many law enforcement officials and policy makers have raised concerns that prisons are breeding grounds for terrorism. Over 20 percent of the violent groups recruited actively in prison, but less than 10 percent of the non-violent groups recruited here. Violent groups also actively recruited at concerts, such as hate rock shows, and protests, and the differences in using these recruitment strategies compared to non-violent groups are quite large. Over 17 percent of the violent groups recruited at protests and over 24 percent recruited at concerts.

Positive and Negative Links with Other Hate Groups

Research indicates that groups may also radicalize toward violence due to their relationships with other hate groups or because of competition with these groups. We attempted to capture these relationships in two ways. First, we documented whether the violent and non-violent groups in the sample had any positive or negative links with international or national domestic far-right groups. When the linkages existed, we also collected information on the groups that they were linked with and the type of linkages. Second, we documented whether these groups had specific relationships with far-right leaders. Table 5.2 presents these results.

The results presented in Table 5.2 indicate that the violent groups were more likely to have positive links with both international and domestic far-right groups. Over 24

Table 5.2 Linkages with Other Hate Groups

Type of linkage	*Violent groups*	*Non-violent groups*
Positive links with international right-wing groups	24.1%	9.7%
Positive links with domestic right-wing groups	72.4%	61.3%
Positive links with international far-right figures	6.9%	9.7%
Positive links with domestic far-right figures	31.0%	29.0%
Negative links with international far-right groups	0.0%	0.0%
Negative links with domestic far-right groups	13.8%	6.5%

percent of the violent groups were positively linked with international far right groups and over 72 percent were positively linked with domestic far-right groups. In comparison, less than 10 percent of the non-violent groups were positively linked to international groups and just over 61 percent were linked to domestic groups. It is interesting as well that the nature of the linkages between the groups in the sample and domestic far-right groups was quite different. For example, the primary type of linkage between domestic groups and the non-violent groups in the sample was the sharing and dissemination of radical literature or music. In contrast, the violent groups were linked to the far-right groups through friendships, recruitment initiatives, and for collaborating to accomplish criminal objectives.

Although the linkages between violent and non-violent groups with far-right figures was quite similar, violent groups were more likely to also have negative linkages with domestic far-right groups. Approximately 14 percent of the violent groups were negatively linked with domestic far-right groups compared to 6.5 percent of the non-violent groups. These linkages were also somewhat different. The nature of the negative links for the non-violent groups was that groups either stole their name or hacked their website. In contrast, the negative linkages between violent groups and the domestic far-right groups were concerned with organizational rivalries and internal splits that occurred resulting in competition between groups and the end result was multiple new groups being formed.

Leadership Structures

As previous research has shown, leadership plays an important role in the radicalization of groups. We wanted to explore whether violent and non-violent groups have different leadership structures by exploring three issues. First, we examine whether the leadership structure of the violent and non-violent groups was different. Second, we examine whether how a leader achieves their position in the organization was different. Finally, we discuss whether the leader specifically advocates violence. These results are presented in Table 5.3.

Similar to the results presented in Tables 5.1 and 5.2, the results in Table 5.3 indicate that violent groups are quite different than non-violent groups. Although there was a considerable amount of missing data for the non-violent groups, the findings

Table 5.3 Leadership Issues in Violent and Non-Violent Groups

Leadership variable	*Violent groups*	*Non-violent groups*
Leadership structure		
1. Single command	78.9%	51.6%
2. Ruling council	5.3%	0.0%
3. Competing leaders	5.3%	0.0%
4. Leaderless resistance	10.5%	0.0%
Leaders chosen		
1. Self-appointed	56.3%	41.9%
2. Appointed with election	18.8%	0.0%
3. Election	25.0%	0.0%
Leader advocates violence	10.3%	0.0%

show that the dominant structure for violent groups was a single command structure. Nearly 79 of the violent groups had a single command structure compared to only 52 percent of the non-violent groups. Most of the leaders in the violent groups were self-appointed with over 56 percent of the groups. Over 18 percent of the groups were appointed but confirmed with election and 25 percent chose leaders by democratic election. Approximately 10 percent of the violent group leaders advocated that their members use violence.

Conclusion

This chapter accomplished two objectives. First, it reviewed the literature on the radicalization of individuals and groups. On the one hand, this literature is valuable for assessing risk, as it highlights the many potential factors that might impact the radicalization of individuals and/or groups. It highlights the social, political, and cultural factors, and discusses how issues might push an individual or a group toward a deeper commitment to extreme ideologies and ultimately to violence. There appears to be some potential for reducing risk, as certainly not all and maybe most individuals who may possess extreme views never commit any violent acts. Moreover, some researchers discuss how individuals may be less engaged over time. However, the literature is limited because there has been little empirical work that has directly assessed the relative importance of these radicalization factors, how they might vary by different movements (far-right v. Al-Qaeda, for example), and what types of triggers might push individuals or groups toward or away from violence. This is a particularly important gap from a risk perspective as it is difficult to prioritize targets without a more refined understanding of these issues.

The second objective of this chapter was to provide a preliminary assessment of some of the ways that violent groups are structurally and behaviorally different than non-violent hate groups. Although the results are only based on a small sample of groups, the results show that violent groups are very different in terms of how they recruit new members, the nature of their linkages with other international and national far-right groups, and the leadership structures of their organization. Such findings are a good first step in determining risk priorities—it is perhaps most important that the results clearly indicate that although both violent and non-violent groups share ideological and radical viewpoints, they were quite different in terms of behavior. They were different in whether they committed violence, and also, importantly, the type of violence the group committed. It would make sense to examine these and other factors in more detail to better assess and predict which groups pose the greatest threats for committing serious violence and terrorist acts.

In conclusion, it is important to acknowledge that the results are preliminary in nature. The results are based on a portion of the groups identified for the analysis, thus, it will be critical to see what impact increasing the sample size will have on the findings. Moreover, there are other factors that need to be considered when comparing violent to non-violent groups, including the type of ideological position, geographic locations, and membership demographics. Finally, it is critical to acknowledge that organizational change is a moving target, and thus the groups "radicalization level" is not static, and is susceptible to changes in membership, leadership, and critical events. It would be helpful to understand how such events impact the attitudes and behavior of extremist organizations and the risk of violence in future research.

Note

1. This research was supported by the United States Department of Homeland Security through the National Consortium for the Study of Terrorism and Responses to Terrorism. However, any opinions, findings, and conclusions or recommendations in this document are those of the authors and do not necessarily reflect views of the US Department of Homeland Security.

References

Aho, James A. 1990. *The Politics of Righteousness: Idaho Christian Patriotism.* Seattle: University of Washington Press.

——. 1994. *This Thing of Darkness: A Sociology of the Enemy.* Seattle: University of Washington Press.

Bjorgo, T. and J. Horgan, eds. 2009. *Leaving Terrorism Behind: Individual and Collective Disengagement.* New York: Routledge.

Della Porta, D. (1998). "Recruitment Process in Clandestine Political Organizations: Italian Left-wing Terrorism." *International Social Movement Research* 1:155–169.

Ezekiel, R. (1996). *Racist Mind: Portraits of American Neo-Nazis and Klansmen.* New York: Penguin Books.

FBI. 2010. Retrieved from: http://www.fbi.gov/aboutus/transformation.htm.

Hamm, M. S. 2002. *In Bad Company: America's Terrorist Underground.* Boston: Northeastern University Press.

——. 2005. "Crimes Committed by Terrorist Groups: Theory, Research and Prevention." Final Report to the National Institute of Justice, September 2005. Retrieved from: www.ncjrs.gov/pdffiles1/nij/grants/211203.pdf.

Horgan, J. 2009. *Walking Away from Terrorism.* New York: Routledge.

Hudson, R. A. 1999. *The Sociology and Psychology of Terrorism: Who Becomes a Terrorist and Why?* Washington, DC: Library of Congress, Federal Research Division.

Janis, I. L. 1971. "Groupthink among Policy Makers." Pp. 71–89 in *Sanctions for Evil,* edited by N. Sanford and C. Comstock. San Francisco: Jossey-Bass.

——. 1972. *Victims of Groupthink.* Boston: Houghton Mifflin.

——. 1986. "Groupthink." Pp. 97–104 in *Decision Making: An Organizational Behavior Approach,* edited by J. M. Pennings. Amsterdam: IOS Press.

Kirby, A. 2007. "London Bombers as 'Self Starters': A Case Study in Indigenous Radicalization and the Emergence of Autonomous Cliques." *Studies in Conflict and Terrorism* 30:415–428.

LaFree, G. and L. Dugan, 2004. "How Does Studying Terrorism Compare to Studying Crime?" *Sociology of Crime, Law and Deviance* 5:53–74.

Long, D. E. 1990. *The Anatomy of Terrorism.* New York: Free Press.

Lum, C., Kennedy, L.W., and Sherley, A. J. 2006. *The Effectiveness of Counter-Terrorism Strategies: A Campbell Systematic Review.* Retrieved from: www.campbellcollaboration.org.

McCauley, C. and S. Moskalenko 2008. "Mechanisms of Political Radicalization: Pathways toward Terrorism." *Terrorism and Political Violence* 20(3):415–433.

McCauley, C. R. and M. E. Segal 1987. "Social Psychology of Terrorist Groups." Pp. 231–256 in *Group Processes and Intergroup Relations,* edited by C. Hendrick. Newbury Park: Sage.

Mandel, D. R. (2010). "Radicalization: What Does it Mean?" Pp. 101–113 in *Home-grown Terrorism: Understanding and Addressing the Root Causes of Radicalization among Groups with an Immigrant Heritage in Europe*, edited by T. M. Pick, A. Speckhard, and B. Jacuch. Netherlands: IOS Press (NATO Science for Peace and Security Series E: Human and Societal Dynamics: Vol. 60).

Merari, A. (1991). Academic research and government policy on terrorism. *Terrorism and Political Violence, 3*(1), 88-102.

Moskalenko, S. and C. McCauley 2009. "Measuring Political Mobilization: The Distinction between Activism and Radicalism." *Terrorism and Political Violence* 17(2):239–260.

Oots, K. L. 1989. "Organizational Perspectives on the Formation and Disintegration of Terrorist Groups." *Terrorism* 12(3):139–152.

Ozer, M. 2006. "The impact of group dynamics on terrorist decision making." Pp. 63–75 in *Understanding Terrorism: Analysis of Sociological and Psychological Aspects*, edited by S. Ozeren, I. D. Gunes, and D. M. Al-Badayneh. Amsterdam: IOS Press.

Post, J. L. 1998. "Terrorist psycho-logic: Terrorist behavior as a product of psychological forces." Pp. 25–40 in *Origins of Terrorism: Psychologies, Ideologies, Theologies, States of Mind*, edited by W. Reich. Washington, DC: Woodrow Wilson Center Press.

Post, J., K. Ruby, and E. Shaw 2002a. "The Radical Group in Context: 1. An Integrated Framework for the Analysis of Group Risk for Terrorism." *Studies in Conflict & Terrorism* 25:73–100.

Post, J., K. Ruby, and E. Shaw 2002b. "The Radical Group in Context: 2. Identification of Critical Elements in the Analysis of Risk for Terrorism by Radical Group Type." *Studies in Conflict & Terrorism* 25:101–126.

Pressman, D. E. 2008. "Exploring the Sources of Radicalization and Violent Radicalization: Some Transatlantic Perspectives." *Journal of Security Issues* 2(1):1–20.

Ross, J. I. 1993. "Research on Contemporary Oppositional Political Terrorism in the United States: Merits, Drawbacks and Suggestions for Improvement." Pp. 101–116 in *Political Crime in Contemporary America: A Critical Approach*, edited by K. D. Tunnel. New York: Garland Publishing

Sageman, M. 2007, June 27. *Radicalization of Global Islamist Terrorists.* Testimony presented to US Senate Committee on Homeland Security and Governmental Affairs. Retrieved from: hsgac.senate.gov/public/index.cfm?FuseAction=Hearings.Hearing&Hearing_ID=9c8ef805-75c8-48c2-810d-d778af31cca6.

——. 2008. *Leaderless Jihad: Terror Networks in the Twenty-first Century.* Philadelphia: University of Pennsylvania Press.

Silber, M. and A. Bhatt 2007. "Radicalization in the West: the Homegrown Threat." *The New York City Police Department* 1:1–90.

Silke, A. 2001. "The Devil You Know: Continuing Problems with Research on Terrorism." *Terrorism and Political Violence* 13(4):1–14.

Sprinzak, E. 1995. "Right-wing Terrorism in a Comparative Perspective: The Case of Split Delegitimization." *Terrorism and Political Violence* 7(1):17–43.

United States v. Mohamad Ibrahim Schnewer, 1:07-cr-00459-RBK(2007).

Wintrobe, R. 2006. *Rational Extremism: The Political Economy of Radicalism.* New York: Cambridge University Press.

Zaman, M. Q. (1998). "Sectarianism in Pakistan: The Radicalization of Shi'i and Sunni Identities." *Modern Asian Studies* 32(3):689–716.

6.
PROSPECTIVE RATHER THAN REACTIVE RISK ASSESSMENT

Risk assessment is prospective rather than reactive. It forces us to anticipate the future and, while not necessarily providing firm predictions of what may occur, allows us to plan for and prevent the negative effects of hazards and threats. In looking forward to risk, individuals and agencies are encouraged to take into account what might happen and plan accordingly. As we move from what we don't know (under the heading of uncertainty) we determine what we do know about what hazards we face and our vulnerabilities in face of these hazards. In doing this, we must account for preparedness and prevention in light of the exposure that they have to certain types of hazards. While there obviously is still a reactive component to what we do in light of these challenges, this is based on firm plans and readiness to act in a strategic way. Plans outline what needs to be done, the resources that must be allocated, and the priorities that must be set.

Hot spot analysis is the identification of crime trends, an important new tool being used by law enforcement to manage risk and allocate resources. Caplan, Moreto, and Kennedy suggest that hot spot maps simply show where criminal activity has taken place but do not impart any real information of where events are likely to occur in the future, that is, the risk of future victimization. What this means is that the likelihood of future crime in a location is enhanced by the occurrence of a crime in a measureable fashion. In looking at their results from a study of international piracy, Caplan et al. report an increase in the predictive capability of this type of risk-based mapping over the more conventional forecasts offered by retrospective hot spot mapping, reinforcing the idea that these approaches provide a greater focus for hazard detection and response.

FORECASTING GLOBAL MARITIME PIRACY UTILIZING THE RISK TERRAIN MODELING APPROACH TO SPATIAL RISK ASSESSMENT

Joel M. Caplan, William D. Moreto, and Leslie W. Kennedy

Introduction

While maritime piracy is not a new phenomenon and can be traced as far back as ancient Greece (Gottschalk & Flanagan 2000; Ong-Webb 2007; Konstam 2008), the current state of contemporary maritime piracy[1] has garnered significant attention from international governments, the media, policy strategists, and academic researchers. Maritime piracy incidents have not only increased globally in recent years, but have increased exponentially in specific regions. Moreover, the tactics and strategies utilized by pirates have evolved in certain piracy-prone areas. For example, reports of maritime piracy in Southeast Asia have shown an increased prevalence in violence (Raymond 2009), while incidents off the coast in Somalia and within the Gulf of Aden (GOA) have displayed increased levels of organization and sophistication, including the use of "mother vessels" and advanced weaponry (Anyu & Moki 2009; Murphy 2009; United Nations Office on Drugs and Crime 2009; ICC–IMB–Piracy Reporting Center 2010). Maritime piracy can result in assaults, hostage situations, kidnappings, homicides, and extortions.

Due to the heavy reliance of global businesses and state economies on the maritime transportation of goods (Barrios 2005; Bateman 2007)—with approximately 80 percent of global commerce being transported by sea (Luft and Korin 2004)—maritime piracy also presents a problem beyond the direct harm of ship crews and passengers, as it affects a myriad of stakeholders including shipping companies, insurance agencies, and individual states. Among other things, it can stifle business transactions and increase the financial cost of doing business, affect the allocation of limited state resources, and affect the reputation of specific states and ports by condoning them as piracy-prone locations and invariably affecting trade relations (Middleton 2008; Murphy 2009). It has also been argued recently that the money gained from maritime piracy-related extortions is used to finance future incidents of maritime piracy and possibly even global terrorism (Murphy 2009).

While maritime piracy can be considered an international security problem, it should not be considered a global problem, but rather a problem that is localized and occurs in specific places (Murphy 2009). Arguably, all states with coastlines and ports are susceptible to different forms of maritime piracy, including minor forms of maritime mugging; however, there are several factors that either exacerbate or mitigate opportunities for maritime piracy and certain locations. Favorable geographic factors and the presence of potential targets, whether they be local pleasure crafts or international commercial vessels, are not sufficient to establish and sustain maritime piracy at a region. A fortuitous combination of non-existent or ineffective state laws, inadequate monitoring of territorial waters, and/or a tolerant (and possibly even accepting) cultural backdrop is also required. Indeed, maritime piracy has been acknowledged as primarily occurring in specific areas where such attributes and qualities of space are present, including "Maritime Asia" and more recently Somalia and the GOA (Lehr 2007).

Changes in hot spots and increases in the frequency and the severity of maritime piracy incidents in recent years warrant that this contemporary international security threat be addressed forthright by being analyzed in an appropriate way and responded to in an effective and efficient manner. An actuarial assessment of the potential for maritime piracy across all global regions based on aggravating or mitigating qualities of space is required. While aggregating or enabling conditions (i.e., risk factors) may vary from region to region; measurements used to assess such qualities of places need to be similar in order to compare maritime regions to one another in a valid, reliable, and actionable manner. In other words, risk factors specific to a region (i.e., domestic and/or international coastal/naval support) can be incorporated in a general assessment of the full global extent to enable policymakers, counter-piracy strategists, and researchers to comparatively assess the risk of piracy at certain places relative to all other places. Within this analytical and empirical approach, maritime piracy can be analyzed from its fundamental core aspects while also addressing dynamic place-based factors that influence such core aspects.

Risk terrain modeling (RTM) is an approach to risk assessment that standardizes risk factors to common geographic units over a continuous surface. Separate map layers representing the presence, absence, or intensity of each risk factor at every place throughout an area are created in a geographic information system (GIS) and then all risk map layers are combined to produce a composite "risk terrain" map with attribute values that account for all risk factors at every location throughout the geography. RTM has been proven to aid in strategic decision-making and tactical action at municipal extents by showing where conditions are ripe for events to occur in the future (Caplan, Kennedy, & Miller 2010; Kennedy, Caplan, & Piza 2010). But, there is also reason to believe that RTM can be applied to spatial risk assessment at larger, even global, extents, such as where maritime piracy threatens international security.

Anecdotal evidence supports the contention that specific geospatial, temporal, seasonal, socio-political, and cultural factors facilitate an environment that is conducive to maritime piracy. However, to the best knowledge of the authors, there has been no empirical attempt to examine how such factors combine to create high-risk environments. There have been calls by some experts to identify the principal land-based factors that contribute to maritime piracy and create responses to alleviate such enabling conditions (Chalk 2009). While such an assessment is necessary, a focus on only land-based factors without operationalizing how they proximately influence the waterways and increase risks at certain places on the water could result in underestimating the interaction effects of several interdependent factors of maritime piracy. Current sea-based risk assessments rely solely on hot spot locations of past incidents to predict locations of future similar events.[2] This univariate analysis assumes a static environment—that the locations of pirate attacks do not change over time, which is an unreasonable assumption given the dynamic nature of the social, natural, and built environments in our world today. The inability to acknowledge such dynamic environments leads to an inaccurate representation of the incident–environment nexus that occurs, resulting in potentially erroneous assumptions and inappropriate and/or inadequate responses. This chapter applies RTM to the international security problem of maritime piracy and tests the predictive validity of using this approach for risk assessment, strategic planning, and tactical responsive action. We continue in the next section with a more in-depth discussion of the security dilemmas posed by contemporary maritime piracy and the ways in which RTM can be uniquely useful for addressing them. We then discuss the causes and correlates of contemporary maritime piracy before explaining the specific steps taken to develop and test a risk terrain model. We conclude that RTM can significantly identify the locations where future maritime piracy incidents will occur and, therefore, supports the push toward a proactive and more efficient allocation of tactical resources to combat piracy in the twenty-first century.

Security Dilemmas, Maritime Piracy, and RTM

Booth and Wheeler (2008) explained that security dilemmas have two parts. The first consists of the dilemma of interpretation about the motives, intentions, and capabilities of others. Here the decision-makers are faced with trying to overcome the uncertainty present in reading their opponents' actions to determine whether they are intent on a purely defensive and protective stance or are more likely to result in an offensive position. The second part relates to the dilemma of response, whereby decision-makers faced with security threats must determine how best to react. Pirates tend to be offensive actors, so the first "dilemma" tends to be a less onerous one since experts can assume at the most basic level what most pirates' motives and intentions are. They are motivated and driven by economic objectives (Chalk 2009; Murphy 2009). This particular motivation principally distinguishes "maritime piracy" from "maritime terrorism," with the latter driven by political, religious, and/or ideological motives (Murphy 2009). Albeit the proposition that pirates are economically driven does not suffice in providing a useful explanation to address the piratical problems of a specific place or region. While experts and practitioners are becoming more aware of the capabilities of pirates in certain regions of the world, specifically tailored responses can only be created and implemented successfully by addressing most of the underlying factors that promote and enable maritime piracy at these locations.

Consistent with suggestions by Booth and Wheeler (2008), we propose viewing the security dilemma posed by maritime piracy in the context of a multi-level, multistate perspective because it is in understanding the dynamic interactions that occur at certain places that international maritime security can best be ensured. As Booth and Wheeler (2008: 147) said,

> security should be understood as the consequence of a wider range of threats (e.g. poverty, the environment, the global economy) than that of military violence, and, second,

contemplating a wider range of referents (e.g. individuals, regions, common humanity) than sovereign states.

Several steps are required to enhance maritime security including understanding the motives and capabilities of pirates so that their intended targets can be hardened and potential victims can be strengthened. Another important but often overlooked step is to understand and routinely assess the environmental conditions that enable pirate attacks to occur successfully. Until very recently, this has been made difficult by problems in obtaining accurate data. It has also been confounded by analytical challenges that develop around the correlates of these behaviors, where hypotheses about causes are more often tested through descriptive analysis—citing examples of how things work rather than using a systematic approach to the study of the patterns of risk and their connection to piratical outcomes. Fortunately this is precisely the type of analytical problem that RTM helps to overcome, as it allows for the articulation of everything we know about maritime piracy causes at units of analysis that are meaningful to shipping stakeholders—maritime places.

RTM[3] assumes a step that is basic to the development of GISs in assuming that certain places can acquire attributes that, when combined in prescribed ways, create contexts in which certain outcomes are made more likely. A risk terrain map provides a composite picture of the underlying micro-level conditions throughout a neighborhood, city, region, or waterway. The ways in which these conditions combine is an important aspect for setting up the "meaning" that a risk terrain model will carry. The backdrop for pirate attacks is not static but it is recognizable and sustainable. Its characteristics emerge over time and connect the structure and qualities of the maritime environment to the activities that occur at these locations. For example, the combined attributes of open space, presence of children, and proximity to schools may indicate a playground. These attributes combined can be used to anticipate the types of behavior that we would expect in a playground, reducing the uncertainty in our forecasts about what would transpire there. In this way, environmental attributes are used as a means of assigning risk (or likelihood) that certain events will happen at a particular place.

It is not really enough to say that risk of maritime piracy increases when the numbers of pirates increase. What is more likely is that the risk of piracy at places that have enabling attributes is higher than other places because these locations attract motivated offenders (or more likely concentrate them in close locations) and are conducive to allowing certain events to occur. This is different from saying that piracy concentrates at highly dense hot spots. It suggests, instead, that individuals most likely to commit piracy will congregate at locations that are best suited for perpetrating it. Oddly enough, this is not to say that there are more or better targets of piracy, but rather that the conditions for piratical behavior (e.g., lower risk of apprehension or retaliation) are better at these places than in others.

RTM is consistent with the approaches discussed in the early parts of this book because it sees maritime piracy as more than just an opportunity crime, but one that is influenced by the conditions that exist in certain areas of the world. Pursuant to environmental criminology, the RTM approach acknowledges that the surrounding "environment is not just a passive backdrop for criminal behavior; rather, it plays a fundamental role in initiating the crime and shaping its course" (Wortley and Mazerolle 2008: 2). In other words, the surrounding environment is very much an active feature in any criminal event (Brantingham and Brantingham 1991). RTM also supports and is supported by the two other propositions of environmental criminology: (1) that crime does not occur randomly through time and space and is dependent upon situational factors and criminogenic environments; and (2) that by properly identifying such characteristics and environments and the subsequent patterns that occur within and around such environments, investigators may be able to analyze and examine specific types of pirate attacks and implement appropriate preventative responses. Within this scope, RTM can aid in strategic decision-making and tactical action by showing where conditions are ripe for events to occur in the future and provide important insights into the nature of these events that permits the prioritization of ways in which to mitigate one or more risk factors and prevent piracy altogether.

Causes and Correlates of Contemporary Maritime Piracy

There is no universal definition of maritime piracy which implicitly limits a proper assessment of the "real extent of piracy" (House of Commons Transport Committee 2006: 11; see also Beckman 2002) and invariably results in disconnect between actual incidents and documented incidents (Keyuan 2005). For the purposes of this study we use the definition suggested by the International Maritime Bureau (ICC–IMB–Piracy Reporting Center 2010): "An act of boarding or attempting to board any ship with the apparent intent to commit theft or any other crime and within the apparent intent or capability to use force in the furtherance of that act." Though not typically an act that is promoted by state action,[4] maritime piracy often thrives in the absence of state government control and requires military response on the part of nations whose nationals and flag-bearing ships are attacked. This means that vessels that operate in international waters may find themselves vulnerable to attack with no real protection from the pirates. For example, the case of Somalia—a country that has had no functioning government since 1991 and was put under the control of the Transnational Federal Government in 2004 (Lehr and Lehmann 2007)—exemplifies how this type of chaotic situation, rife with turmoil, poverty and corruption, enables piratical activity. Sitting right next to the Bab el-Mandeb strait, separating the GOA and the Red Sea, the sea lanes near Somalia (and Yemen) are filled with ships from numerous international states transporting goods into and out of the Middle East and East Africa. As a result, pirates have recently thrived and proliferated in this region.

Responses to maritime piracy in piracy-prone regions have varied. Such variation is attributed to state differences manifested through cultural and socio-political factors, including legal and administrative disparities, capacity of accessible and expendable internal resources, and regional and international external state support. Decreases in maritime piracy incidents in the Southeast Asia region, for instance, have been attributed to the regional cooperation among littoral states in direct proximity of the Malacca straits (e.g., Indonesia and Malaysia) and other surrounding nations. Initiatives such as Operation MALSINDO, EiS ("Eyes in the Sky"), and ReCAAP (Regional

Cooperation Agreement on Combating Piracy and Armed Robbery Against Ships in Asia) are believed to have minimized the maritime piracy problem in the region due to appropriate information sharing, cooperative jurisdictional agreements, effective shared resource allocation, and strategic monitoring and deployment (Raymond 2009).

The highly touted collaborative international response to the maritime piracy problem off the coast of Somalia and within the GOA further exemplifies how responses can vary in scope. The inability of Somalia to fully govern its territorial waters has led to its reliance on the international communities for support. In 2003, the Combined Joint Task Force Horn of Africa (CJTF-HOA) was created in order to "establish long-term regional stability [while also] protecting U.S. and Coalition interests (Sampson 2009: 4). Within the GOA, there are also several other maritime task forces operating, including the Combined Task Force 150 (CTF-150), which is responsible for the monitoring the Red Sea, GOA, Indian Ocean, Gulf of Man, and the Arabian Sea, and the Combined Task Force 151 (CTF-151), which is directly responsible for the conducting of counter-piracy operations around Somalia and the GOA. Other anti-piracy initiatives are also being implemented by NATO, the European Union, and other international navies (e.g., Sampson 2009).

The dynamic interplay of motivated offenders, potential victims, intervening naval forces, and vast operating spaces makes monitoring and responding to piracy difficult as domestic and international policing resources are limited and pirates fear little or no retribution from their own government (or nation states) and have yet to be completely deterred. There has been some evidence of fewer pirate attacks happening in hot spots targeted for intervention, but (perhaps as a direct effect) pirates have increasingly moved further away from the coastlines and have targeted commercial vessels at greater distances from their home countries. As maritime pirate attacks shift downward or expand to new areas, so does the margin of error for naval/policing agencies. Stakeholders' desires for piracy reduction are ever-present, as is the political motivation for maximizing maritime security. For these reasons, a technique such as RTM that accurately forecasts the locations of future incidents (not just future totals) given what is known about the risk factors for piracy can be a valuable tool for navies and other organizations looking to build upon their recent successes. For instance, we know that maritime piracy is not widely dispersed at the macro level, but generally concentrated at specific regions and nearby failed or failing nation states; that the global transportation of goods to certain ports is mostly limited to specific sea lines of communication (SLOC), or maritime shipping routes; and that many SLOCs go through maritime chokepoints to help alleviate travel distance and related costs.

State Status

Maritime piracy has "its origins on land, its bases and markets are on land, and many of its pernicious effects are felt on land" (Murphy 2009: 130), so it can be seen as partly a manifestation of land-based issues. Directly related to the association between land-based issues and maritime piracy is the fact that a significant number of maritime piracy incidents occur within or close to the territorial waters of specific states, such as Somalia and Yemen. The interests of nation states are now intricately interwoven in the era of globalization (Sorensen 2005), and nations in which governments are either dysfunctional or operate according to rules outside of generally accepted practice or protocol pose serious threats to global order. The Fund for Peace and Foreign Policy (www.fundforpeace.org) has created what they call the Failed State Index that ranks countries on a dozen social, economic, and political indicators including: the movement of refugees and internally displaced persons, evidence of demographic pressures and serious ethnic or sectarian grievances, gaps between rich and poor, economic growth or recession, performance of public services, corruption, the human rights situation and rule of law, and the intervention of other states or foreign non-state actors. "For each indicator, countries are given a numerical score of between one (for best performance) and 10 (for worst)" (Lobe 2008). This Index is intended to provide an early warning of the potential failure or collapse of countries. In 2009, at the top of the list was Somalia followed closely by Sudan and Chad.

In a connected world, statewide economic disadvantage, civil and human rights infringements, corruption, and interventions by other states can have deleterious and far-reaching effects in a number ways. Notably, pirates can find safe haven on the land and territorial waters of such nation states where outlaws can prosper, alternative militias can develop, and predatory behavior goes undeterred. It should be no surprise that the countries that top the Failed States Index are also near to many maritime pirate attacks and that failed states will be one risk factor included in our risk terrain model of maritime piracy. Specifically, places near littoral "failed" states will be at the greatest risk for maritime piracy than places near "sustainable" states.

Shipping Routes

Commercial shipping routes are a critical component of maritime piracy because they represent areas where potential targets and pirates can intersect in time and space. As described by Toth and cited by Peele (1997), SLOCs are critical for the economic development and security of states. Pursuant to the routine activities approach (Cohen and Felson 1979; Worrall 2000), shipping routes, which are a component within important SLOCs, enable vulnerable vessels to become attacked by motivated pirates usually in the absence of capable guardians (as represented by naval forces and/or coast guards), whose abilities to protect such vessels are restricted due to limited resources and the vast operating area that the ocean provides for pirates. Due to the routinized use of SLOCs by shipping companies, pirates may be able to synchronize and schedule attacks based on approximations of the movement of vessels. Shipping routes, then, are another risk factor for piracy. Specifically, places where there is a high concentration of shipping routes will be at greater risk than places with lower concentrations or no shipping routes.

Maritime Chokepoints

A maritime chokepoint is another critical component of maritime piracy due to its purpose and physical geography. Whether natural (i.e., Bab el-Mandab) or artificial (i.e., Suez Canal), maritime chokepoints limit the speed and maneuverability of passing vessels as these areas condense traffic passing through (Rodrigue, 2004; Dela Pena, 2009). As defined by Rodrigue (2004: 359), a chokepoint "is a location that forces [maritime] traffic to converge and, by virtue of its physical characteristics, namely, depth, width or navigability, limits movement." Such limitations can begin at a significant distance away from a chokepoint and, therefore, increase the likelihood of a vessel being targeted by pirates in areas immediately surrounding the chokepoint and up to a certain distance away. Nevertheless, chokepoints are strategically utilized by shipping companies to alleviate costs related to traversing much longer alternate routes. For example, 20,000 vessels travel through the Gulf of Aden (GOA) annually (Raymond 2009; Davies 2009) instead of sailing around the southern tip of Africa, which provides a substantial amount of potential targets for pirates within a condensed area. Chokepoints, then, are another risk factor for piracy. Specifically, maritime places within a certain distance of chokepoints will be at greater risk for piracy than other places.

Operationalizing Risk Factors of Maritime Piracy According to their Influence on Places

Once the risk factors for maritime piracy have been identified—state status, shipping routes, and chokepoints—their proximal influences to maritime places must be operationalized in order to articulate the intensity of risk for pirate attacks at those places. To do this, we used a GIS (ArcGIS by ESRI) to create seven risk map layers of the Earth's oceans based on the proximal influence of the three risk factors. These risk map layers were actually grids of equally-sized 42 × 42 nautical mile cells that represented all non-land places on Earth. Each cell (i.e., maritime place) within the risk map layers, respectively, was assigned a numerical ordinal value signifying that place as being a certain level of risk according to how it was deemed to be influenced by the respective risk factor. These seven risk map layers were then summed together to produce a composite risk terrain map in which each cell's composite value signified the combined intensity of risk factors at that place[5] during 2008. Maritime piracy incident data were then obtained from the 2009 Annual Report on Piracy and Armed Robbery against Ships from the International Maritime Bureau. Coordinates from the 2009 report was geocoded using ArcGIS and superimposed on the 2008 maritime piracy risk terrain map for further visual and statistical analyses of predictive validity.

Operationalizing State Status

Data on state status were obtained from the 2008 Failed States Index in which states were categorized as alert ("failed"), warning ("weak"), moderate, and sustainable. While there has been some prior acknowledgments of the correlations between failed and/or weak states and maritime piracy (e.g., Chalk 2009; Murphy 2009; Hastings 2009), there has been less focus on distinguishing these aggravating factors from the mitigating effects of "moderate" and "sustainable" states. We accounted for all four types of states by classifying risks within the exclusive economic zones (EEZ) of each state accordingly. One risk map layer designated maritime places within the EEZ of littoral alert ("failed") states with a risk value of +2 (i.e., highest risk); all other places outside of the EEZ of alert states in that risk map layer were given a value of zero (i.e., not highest risk). Another risk map layer designated maritime places within the EEZ of littoral warning ("weak") states with a risk value of +1 (i.e. highest risk); all other places outside the EEZ of warning states in that risk map layer were given a value zero (i.e., not highest risk). The positive risk values were based on the assumption that the influence of such states would provide aggravating environments from which maritime piracy may flourish (i.e., lack of maritime and land-based resources to alleviate maritime piracy threats; accepting cultural backdrop, etc); "alert" states were weighted as more aggregating than "warning" states. Figures 6.1 and 6.2 represent "alert" and "warning" states along with their accompanying place-based risk levels.

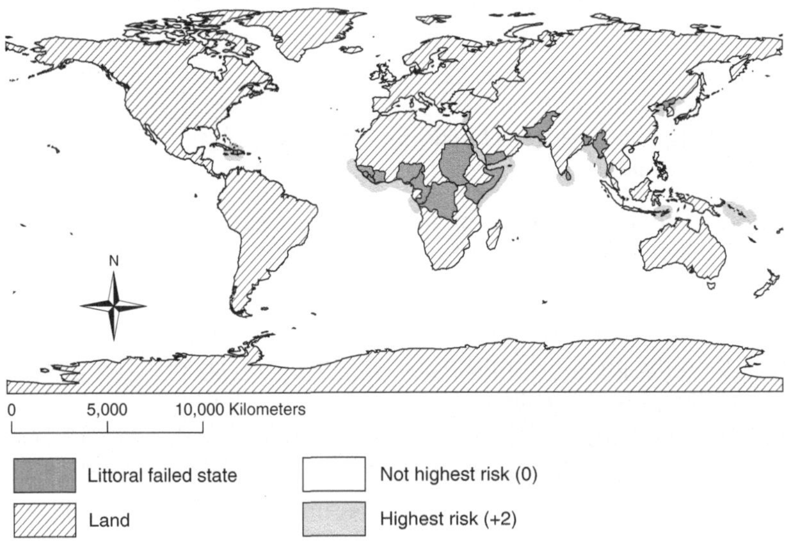

Figure 6.1 Risk level of littoral failed states within EEZ (200nm), 2008
Source: Data from 2008 Failed States Index, Fund for Peace; map created by William Moreto and Joel Caplan.

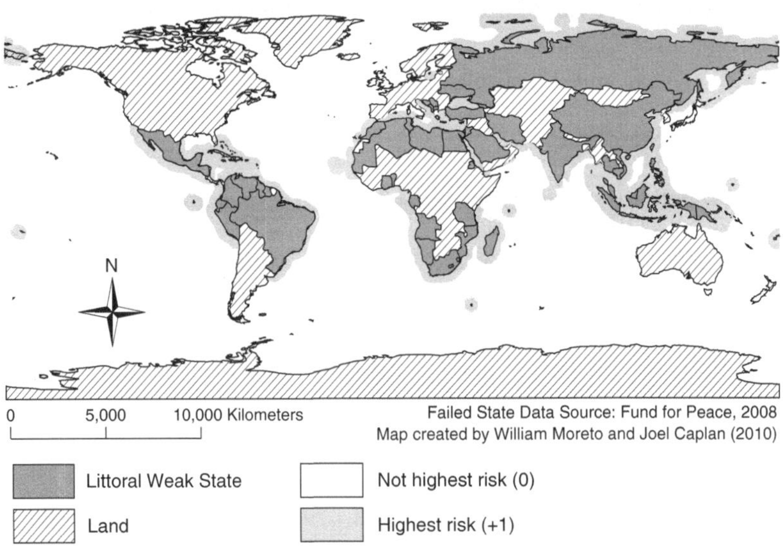

Figure 6.2 Risk level of littoral weak states within EEZ (200nm), 2008
Source: Data from 2008 Failed States Index, Fund for Peace; map created by William Moreto and Joel Caplan.

A third risk map layer designated maritime places within the EEZ of littoral moderate states with a risk value of –1 (i.e., lowest risk); all other places outside the EEZ of moderate states in that risk map layer were given a value of zero (i.e., not lowest risk). A fourth risk map layer designated maritime places within the EEZ of littoral sustainable states with a risk value of –2 (i.e. lowest risk); all other places outside the EEZ of sustainable states in that risk map layer were given a value of zero (i.e., not lowest risk). The negative values were based on the assumption that the influence of such states would provide mitigating environments that may hinder piratical activity (i.e., presence of a domestic coast guard); "sustainable" states were weighted as more mitigating than "moderate" states. Figures 6.3 and 6.4 represent "sustainable" and "moderate" states along with their accompanying place-based risk levels.

A fifth risk map layer was created to account for the recent trend of maritime piracy incidents occurring as far as 900 nautical miles from the coastline of failed states (ICC–IMB–Piracy Reporting Center 2010). All places between 200 nautical miles (end-point of a state's EEZ) and 900 nautical miles from a failed state were assigned a value of +1 (highest risk for this particular risk factor). All other places within that risk map layer were assigned a value of zero (not highest risk). The increased level of risk between 200 and 900 nautical miles from a failed state is represented in Figure 6.5.

Operationalizing Shipping Routes

The influence of shipping routes on maritime piracy was operationalized as places with a high concentration of shipping routes will be at greater risk because these places will be known by pirates to have the most ships and, thus, the most potential targets for attack. Concentration of shipping routes was based on data collected by the National Center for Ecological Analysis and Synthesis on commercial shipping activity from

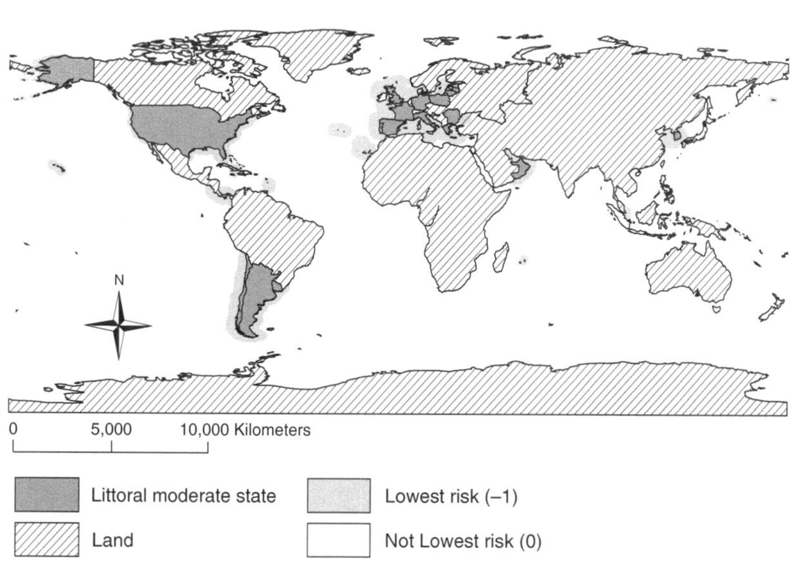

Figure 6.3 Risk level of littoral moderate states within EEZ (200nm), 2008
Source: Data from 2008 Failed States Index, Fund for Peace; map created by William Moreto and Joel Caplan.

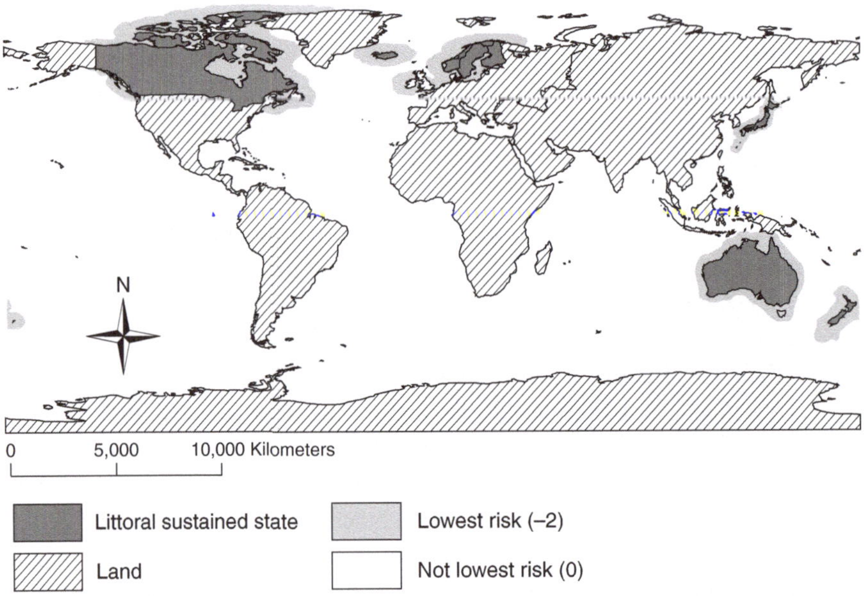

Figure 6.4 Risk level of littoral sustained states within EEZ (200nm), 2008
Source: Data from 2008 Failed States Index, Fund for Peace; map created by William Moreto and Joel Caplan.

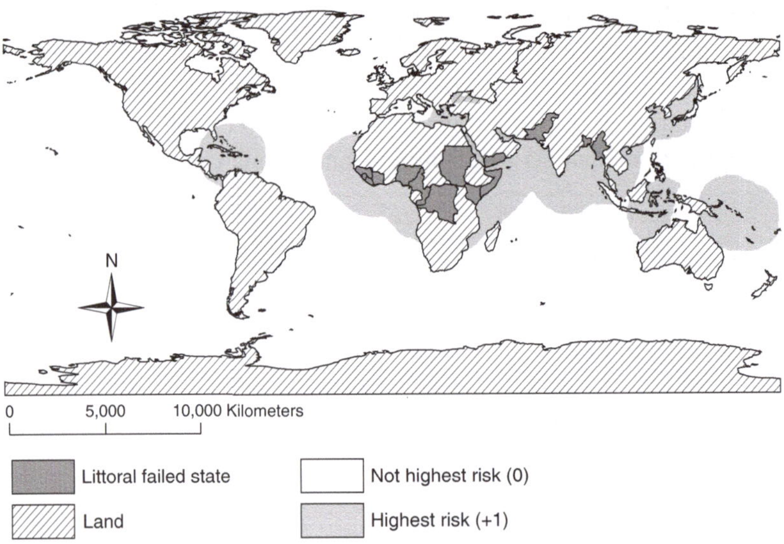

Figure 6.5 Risk level up to 900nm from a littoral failed state, 2008
Source: Data from 2008 Failed States Index, Fund for Peace; map created by William Moreto and Joel Caplan.

October 2004 through September 2005 (Rodrigue 2004).[6] A sixth risk map layer was produced from a density raster map of these commercial shipping data. All places with a density value that was greater than positive two standard deviations from the mean density value were assigned a risk value of +1 (i.e., highest risk). All other places in this risk map layer were valued at zero (i.e., not highest risk) (Figure 6.6).

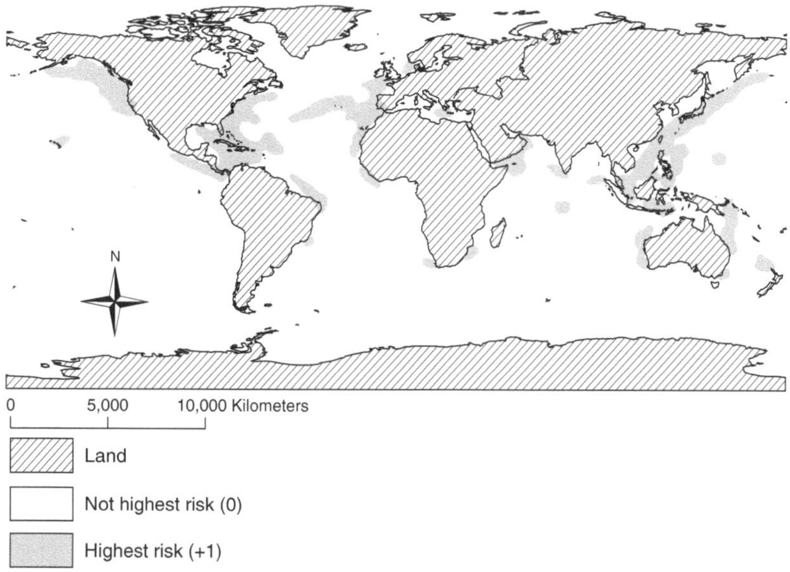

Figure 6.6 Risk level for maritime shipping routes, 2008

Source: Personal communication from Jean-Paul Rodrigue, 2010; map created by William Moreto and Joel Caplan.

Note: The authors would like to thank Dr Jean-Paul Rodrigue for codifying and providing data on international commercial shipping routes and important maritime chokepoints.

Operationalizing Chokepoints

Maritime chokepoints were considered aggravating factors for maritime piracy because their physical geographies limit speeds and maneuverability of (target) vessels, yet they are logistically vital locations for shipping companies despite the risks. Although there is no known empirical evidence outlining risky distances from maritime chokepoints, we utilized the same distance of EEZ, 200 nautical miles, which seemed reasonable given that chokepoints are (by their geographic nature) generally within exclusive economic zones of one or more nation states. The seventh and final "chokepoint" risk map layer was operationalized as all places within 200 nautical miles of a chokepoint were the highest risk (+1); all places beyond 200 nautical miles were not highest risk (0) (Figure 6.7).

Results

The seven aforementioned risk map layers, all with data representative of calendar year 2008, were combined in a GIS using map algebra (Tomlin 1994; "Raster Calculator" in the Spatial Analyst Extension) to produce a risk terrain map (see Figure 6.8) of places on the Earth's waterways where maritime piracy incidents were more or less likely to occur during 2009 because of the nexus of risk factors at those locations. As shown in Table 6.1, a logistic regression analysis reveals that for every unit increase in risk at a 42 × 42 nautical mile area, the likelihood of a pirate attack occurring there increases by

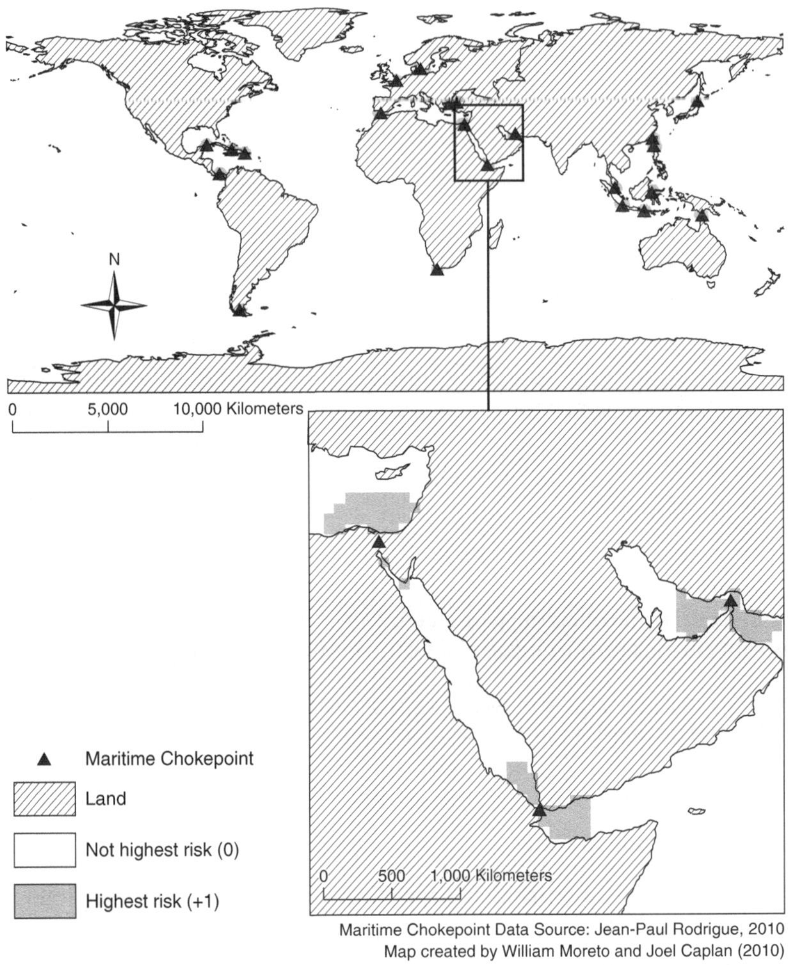

Figure 6.7 Risk level for maritime chokepoints, 2008

Source: Personal communication from Jean-Paul Rodrigue, 2010; map created by William Moreto and Joel Caplan.

184 percent ($p < 0.001$). According to a Pearson chi-squared test, the top 10 percent of high-risk cells in the 2008 risk terrain map correctly predicted the locations of 61 percent of the maritime piracy incidents that occurred during 2009 ($N = 88{,}093$; value = 502.29; $p < 0.01$). This is quite a feat considering that pirate attacks were located in approximately 1.3 perecent of all the cells, globally. So out of the vast ocean, this risk terrain model was able to identify piracy incident locations in advance with more than 60 percent accuracy. In this example, 1,764nm^2 cells are by no means small areas, but they are arguably meaningful and more manageable for targeted interventions and naval patrols compared to the information that dictates existing security activities. As a forecasting tool, RTM provides immediate empirical evidence for locations of future behavior and, importantly, does not rely on the occurrence of pirate attacks as a foundation for predicting future incidents.

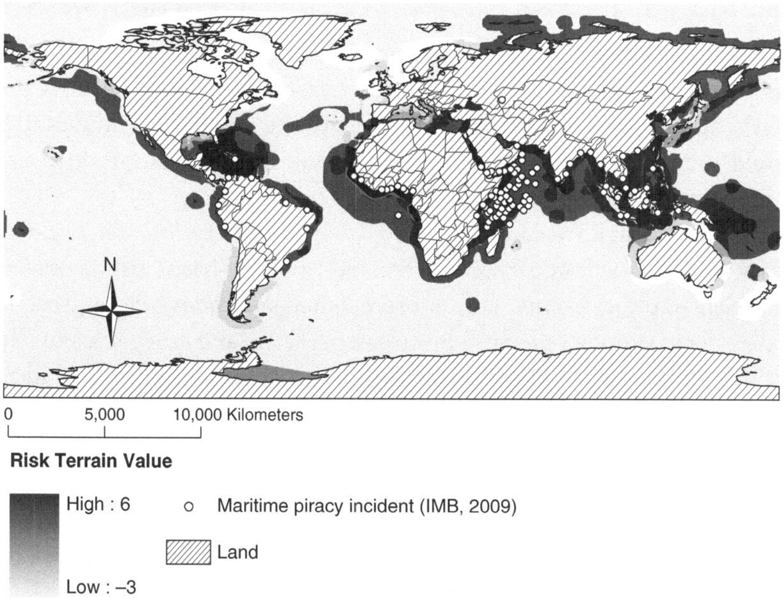

Figure 6.8 2008 maritime piracy risk terrain composite map with 2009 maritime piracy incidents
Source: Data from ICC–IMB–Piracy Reporting Center 2009; map created by William Moreto and Joel Caplan.

Table 6.1 Logistic Regression for Risk Value on Pirate Attack or Attempt

Period 1 Risk Terrain* Forecasting Period 2 Incidents

	B	*S.E.*	*Wald*	*df*	*Sig.*	*Exp(B)*	*95% C.I. for Exp(B)*	
							Lower	*Upper*
Risk value	1.04	.047	500.00	1	< .001	2.84	2.59	3.11

* –2Log likelihood = 2141.47; Nagelkerke R Square = .15; n = 88093 cells

Discussion

Commonsense suggests that mapping hot spots of past pirate attacks is a reasonable method, and research confirms its predictive power (Berk 2009; Chainey, Tompson, and Uhlig 2008). However, research also indicates that the predictive efficacy of mapping can be improved over conventional approaches (Johnson et al. 2007), using multivariate methods (Gorr and Olligschlaeger 2002), raster data (Groff and La Vigne 2001), and diverse techniques such as RTM, which builds in part on the theoretical insights already reviewed that seek to explain the clustering of maritime piracy. The identification of risky areas should permit shipping companies, nations and their navies, and other stakeholders to intervene and allocate resources to reduce risk without analysts having to fear the positive impact of intelligence-led interventions. Paradoxically, too much success will avert pirate attacks to the point that piracy counts will be reduced and there are none or too few past attacks to be used in traditional inferential statistical forecasting models (e.g., see Gorr and Olligschlaeger 2002; Berk 2009). This is a problem that emerges from the past hot spot mapping approach. In real-world settings, as locations evolve and as stakeholders respond to problems in them, security problems and patterns also change. RTM does not suffer the fate of other techniques that could (and should) put themselves out-of-business if their information products are valid, reliable, and can be easily applied to intervention operations.

Results suggest that RTM can significantly identify the locations where global maritime piracy incidents will occur, given the land- and sea-based risk factors that are known to relate to piracy events. This unprecedented and innovative approach to spatial risk assessment supports the push towards a proactive and more efficient allocation of tactical resources to combat piracy through traditional deterrence- and incapacitation-based strategies, such as patrolling waterways and apprehending pirates, but also by simultaneously addressing one or more of the risk factors in the risk terrain model. By focusing on the qualities of maritime places that enable piracy to happen most easily, stakeholders at all levels may respond to this international security threat in a more effective manner. The flexibility of RTM to be applied to any extent—global, regional, or local—is also a noteworthy advantage, as it could be used to assess the risks of maritime piracy at certain regional (i.e., micro) levels and then be used to evaluate the impact of interventions at those areas through an iterative process or risk terrain map making and assessments of changes in risks over time.

RTM moved the study of maritime piracy beyond a univariate analysis of known past piracy incident locations and showed the meaningful and actionable value of synthesizing aggravating and mitigating place-based factors—to create a more accurate representation of future risks of maritime piracy on the Earth's waterways. This study only incorporated a few of many possible risk factors into a risk terrain model and it did not incorporate seasonal factors that change throughout the year. So while it serves as a proof of concept for the applicability of RTM at a global scale, the model developed here may also be considered a conservative estimate of RTM's applicability to maritime piracy. The risk terrain model presented here can be revised, weights can be applied, and/or more advanced techniques can be incorporated to improve the model and maps as needed.

Indeed, it can be argued that the capacity of pirates to successfully complete an attack depends heavily on facilitating opportunities and conditions, and such conditions can vary by the month, day, and week (Townsley & Pease 2002). Utilizing an RTM approach in resource allocation and strategic deployments of interventions may prove especially useful in proactively identifying when and where incidents will occur. But by incorporating temporal factors in subsequent risk terrain maps, preventative resources can be allocated not only to appropriate areas, but at appropriate times as well. Research is currently underway to develop multi-dimensional risk terrain models that incorporate temporal factors that not only identify regions of high risk but also when such locations are at their highest levels of risk. Though for now, we believe it safe to conclude that the RTM approach to spatial risk assessment enables stakeholders to be more effectively proactive and identify areas with the greatest probability of becoming hot spots for maritime piracy in the future. Not just because statistics show that incidents occurred there yesterday, but because the environmental conditions are ripe for incidents to occur there tomorrow. Tools and procedures for accurate risk assessment, such as RTM, that can help experts anticipate and respond to problem areas in the near- and long-terms are key to achieving real and comprehensive security on the Earth's waterways.

Notes

1. For an in-depth discussion on the different eras of maritime piracy, see Gottschalk and Flanagan (2000) and Konstam (2008).
2. See http://www.icc-ccs.org/index.php?option=com_fabrik&view=visualization&controller=visualization.googlemap&Itemid=219.
3. It is important to differentiate between risk terrain modeling and maritime security and safety risk management. While the latter is premised on the accurate identification of risky locations through the combination of a myriad of conducive factors specific to the problem being addressed, maritime security and safety risk management is "aimed at proposing appropriate and cost-effective solutions [in relation] to security and safety matters" (Psarros, Skjong, and Eide 2009: 151). Within this scope, risk terrain modeling would aid in strategic decision-making and tactical action by showing were conditions are ripe for events to occur in the future and provide important insights into the nature of these events that permits the identification and prioritization of ways in which to prevent them.
4. It should be noted that Murphy (2009: 54) also identifies another form of maritime piracy, *state piracy*, which is "the use of maritime depredation by a state for the purposes of policy." Murphy argues that the overt focus on maritime piracy as an act for *private* means as opposed to *public* means results in the neglect of this other form of piracy.
5. The authors acknowledge that an approach that utilizes a weighted distribution of risk may be more accurate and therefore, more beneficial; however, for the purposes of the current pilot study, the authors believe that the dichotomous categories utilized are sufficient as a proof of concept.
6. See the abstract section of the metadata information on commercial shipping activity from the National Center for Ecological Analysis and Synthesis: http://ebm.nceas.ucsb.edu/GlobalMarine/impacts/transformed/fgdc/shipping.xml; Two categories of shipping routes were created: primary routes and secondary routes, with primary routes receiving the majority of maritime commercial traffic. Special thanks to Dr Rodrigue for providing the shapefile of these data.

References

Anyu, J. N. and S. Moki. 2009. "Africa: The Piracy Hot Spot and Its Implications for Global Security." *Mediterranean Quarterly* 20(3):95–121.

Barrios, E. 2005. "Casting a Wider Net: Addressing the Maritime Piracy Problem in Southeast Asia." *Boston College International and Comparative Law Review* 28(1):149–164.

Bateman, S. 2007. "Outlook: The New Threat of Global Terrorism." Pp. 249–258 in *Violence at Sea: Piracy in the Age of Global Terrorism*, edited by P. Lehr. New York: Routledge.

Beckman, R. C. 2002. "Combatting Piracy and Armed Robbery Against Ships in Southeast Asia: The Way Forward." *Ocean Development & International Law* 33:317–341.

Berk, R. 2009. "Asymmetric Loss Functions for Forecasting in Criminal Justice Settings." Unpublished manuscript.

Booth, K. and N. J. Wheeler. 2008. *The Security Dilemma: Fear, Cooperation and Trust in World Politics.* New York: Palgrave MacMillan.

Brantingham, P.J., and P. L. Brantingham, eds. 1991. *Environmental Criminology.* Prospect Heights, IL: Waveland Press, Inc.

Caplan, J. M., L. W. Kennedy, and J. Miller. 2010. "Risk Terrain Modeling: Brokering Criminological Theory and GIS Methods for Crime Forecasting." *Justice Quarterly.*

Chainey, S., L. Tompson, and S. Uhlig. 2008. "The Utility of Hotspot Mapping for Predicting Spatial Patterns of Crime." *Security Journal* 21:4–28.

Chalk, P. 2009. *Maritime Piracy: Reasons, Dangers and Solutions.* Testimony presented before the House Transportation and Infrastructure Committee, Subcommittee on Coast Guard and Maritime Transportation on February 4, 2009. Santa Monica, CA: RAND Corporation.

Cohen, L. E. and M. Felson. 1979. "Social Change and Crime Rate Trends: A Routine Activity Approach." *American Sociological Review* 44(4):588–604.

Davies, P. M. 2009. "US House of Representatives Sub-Committee on Coastguard & Maritime Transportation." Written testimony (February 4). London: Oil Companies International Marine Forum.

Dela Pena, J. 2009. "Maritime Crime in the Strait of Malacca: Balancing Regional and Extra-Regional Concerns." *Stanford Journal of International Relations* 10(2):1–8.

Gorr, W. and A. Olligschlaeger. 2002. *Crime Hot Spot Forecasting: Modeling and Comparative Evaluation.* Washington, DC: National Institute of Justice.

Gottschalk, J. A. and B. P. Flanagan. 2000. *Jolly Roger with an Uzi.* Annapolis, MD: Naval Institute Press.

Groff, E. R. and N. G. La Vigne. 2001. "Mapping an Opportunity Surface of Residential Burglary." *Journal of Research in Crime and Delinquency* 38:257–278.

Hastings, J. V. 2009. "Geographies of State Failure and Sophistication in Maritime Piracy Hijackings." *Political Geography* 28(4):213–223.

House of Commons Transport Committee. 2006. *Piracy. Eighth Report Session 2005–06, HC 1026.* London: The Stationery Office Limited.

ICC–IMB–Piracy Reporting Center. 2009. *Piracy and Armed Robbery against Ships (1 January–31 December 2008).* Annual Report. London: International Chamber of Commerce–International Maritime Bureau.

——. 2010. *Piracy and Armed Robbery against Ships (1 January–31 December 2009).* Annual Report. London: International Chamber of Commerce–International Maritime Bureau.

Johnson, S., D. J. Birks, L. McLaughlin, K. J. Bowers, and K. Pease. 2007. *Prospective Crime Mapping in Operational Context. Final Report.* London: Home Office.

Kennedy, L. W., J. M. Caplan, and E. Piza. 2010. "Risk Clusters, Hotspots, and Spatial Intelligence: Risk Terrain Modeling as an Algorithm for Police Resource Allocation Strategies." *Journal of Quantitative Criminology.*

Keyuan, Z. 2005. "Seeking Effectiveness for the Crackdown of Piracy at Sea." *Journal of International Affairs* 59(1):117–134.

Konstam, A. (2008). *Piracy: The Complete History.* Westminster, MD: Osprey Publishing.

Lehr, P. (2007). "Introduction." Pp. vii–xii in *Violence at Sea: Piracy in the Age of Global Terrorism,* edited by P. Lehr. New York: Routledge.

Lehr, P. and H. Lehmann. 2007. "Somalia – Pirates' New Paradise." Pp. 1–21 in *Violence at Sea: Piracy in the Age of Global Terrorism,* edited by P. Lehr. New York: Routledge.

Lobe, J. 2008. 'Weak States Got Weaker in 2007." *Global Policy Forum,* June 24. Retrieved from the World Wide Web on October 23, 2010: http://www.antiwar.com/lobe/?articleid=13037.

Luft, G. and A. Korin. 2004. "Terrorism Goes to Sea. Foreign Affairs (November/December, 2004)." Retrieved from the World Wide Web on June 24, 2010: http://www.foreignaffairs.com/articles/60266/gal-luft-and-anne-korin/terrorism-goes-to-sea.

Middleton, R. 2008. "Piracy in Somalia: Threatening Global Trade, Feeding Local Wars." Briefing Paper. Africa Programme. London: Chatham House.

Murphy, M. N. 2009. *Small Boats, Weak States, Dirty Money: Piracy and Maritime Terrorism in the Modern World.* New York: Columbia University Press.

Ong-Webb, G. G. 2007. "Piracy in Maritime Asia: Current Trends." Pp. 37–93 in *Violence at Sea: Piracy in the Age of Global Terrorism*, edited by P. Lehr. New York: Routledge.

Peele, R. B. 1997. "Maritime Chokepoints: Key Sea Lines of Communication (SLOCs) and Strategy." USAWC Strategy Research Project. Carlisle Barracks, PA: US Army War College.

Psarros, G., R. Skjong, and M. S. Eide. 2009. "The Acceptability of Maritime Security Risk." *Journal of Transportation Security* 2(4):149–163.

Raymond, C. Z. 2009. "Piracy and Armed Robbery in the Malacca Strait: A Problem Solved?" *Naval War College Review* 62(3):32–42.

Rodrigue, J. P. 2004. "Straits, Passages and Chokepoints: A Maritime Geostrategy of Petroleum Distribution." *Cahiers de Geographie du Quebec* 48(135):357–374.

Sampson, D. W. 2009. "USAFRICOM's role in counter-piracy operations in the Horn of Africa." Final Report. Newport, RI: Joint Military Operations Department, Naval War College.

Sorrenson, G. 2005. "State Transformation and New Security Dilemmas." Pp. 81–150 in *Globalization, Security, and the Nation State: Paradigms in Transition*, edited by E. Aydinli and J. Rosenau. Albany, NY: SUNY Press.

Tomlin, C. D. 1994. "Map Algebra: One Perspective." *Landscape and Urban Planning* 30:3–12.

Townsley, M. and K. Pease. 2002. "Hot Spots and Cold Comfort: The Importance of Having a Working Thermometer." *Crime Prevention Studies* 13:59–69.

United Nations Office on Drugs and Crime. 2009. "UNODC and Piracy." Retrieved from the World Wide Web on December 11, 2009 from: http://www.unodc.org/unodc/en/piracy/index.html.

Worrall, J. L. 2000. "The Routine Activities of Maritime Piracy." *Security Journal* 13:35–52.

Wortley, R. and L. Mazerolle. 2008. "Environmental Criminology and Crime Analysis: Situating the Theory, Analytical Approach and Application." Pp. 1–15 in *Environmental Criminology and Crime Analysis*, edited by R. Wortley and L. Mazerolle. Portland, OR: Willan Publishing.

Questions

1. How could the data on the different types of recruitment processes, different linkages to other terrorist groups, and leadership structures (see the tables) for violent and non-violent terrorist groups be used in a risk assessment process?
2. If you were an intelligence analyst and you hoped to extend the analysis presented in this chapter to better understand the risk posed by far-right terrorist groups, what types of information would you try to collect and analyze?

7. RISK ASSESSMENT MODELS IN THE SECURITY SECTOR

Although there is evidence that product counterfeiting has been around for ages, globalization and technology have increased the risks significantly. Although the financial risks to businesses are apparent, the following chapter by John Spink documents that counterfeiting also poses significant risk to human health and that the range of products subject to counterfeiting covers virtually all sectors of the economy (see also Chapter 14 by Roy Fenoff and Jeremy Wilson, this volume). Given the global nature of this risk, it is very challenging for corporate security officials, public law enforcement, and policy makers to develop an accurate picture of the nature and scope of the risk. Consequently, in this context of uncertainty, the development of meaningful risk assessment models is critical. In this chapter, Spink presents an initial product counterfeiting model based on history, ability or ease, attractiveness/profit, hurdles, and the market profile. The risk assessment model was developed through a review of prior research and in consultation with an expert panel of industry officials. In addition to being a tool for corporate security officials to employ and refine, the approach utilized in developing this model may prove beneficial for addressing other security risks.

PRODUCT COUNTERFEITING

Developing a Risk Assessment Model

John Spink

Product counterfeiting does not lend itself to traditional criminology or risk analysis analytical methods due to the unique, expanding, and evolving nature of the crime and the criminals. This is precisely the type of risk that requires practical, flexible, interdisciplinary assessments to determine the most effective countermeasures. On June 23, 2010, the United States Government presented the first national Joint Strategic Plan for Intellectual Property through the Intellectual Property Enforcement Coordinator office. This chapter provides an overview of product counterfeiting, perspectives on assessing the risks, and methods to help select countermeasures that will reduce the vulnerability or likelihood of a threat. The US Food and Drug Administration (FDA) food industry has been focusing on this risk under the more general concept of Economically Motivated Adulteration (EMA), or the more common term Food Fraud.

This chapter will introduce the counterfeit threat, scale, and scope, including an overview of regulations and countermeasures. This chapter intentionally presents a very wide overview of the counterfeit threat and countermeasures, to present a broad view of the topic, and to better assist in developing the prediction of a risk model. Particular attention will be paid to the aspects of the counterfeit threat that apply to establishing the root factors in the Counterfeit Product Risk Model: Counterfeit-History, Counterfeit-Attractiveness, Counterfeit-Ability, Counterfeit-Hurdles, and Market Profile.

Scale of the Threat

There are many reasons for the increase in scope and scale of the counterfeit threat: globalization, the growth of brands, the growth of imaging and printing technology, and organized crime shifting its focus and resources in this direction (Acheson 2007; Arruñada and Vázquez 2006; Closs and McGarrell 2004; Hopkins, Kontnik, and Turnage 2003; Kelly 2005; Naim 2005; van Schendel and Abraham 2005; Yar 2005). The underlying driver of the counterfeit boom is the massive profit that can be made by unscrupulous manufacturers, in addition to relatively low non-deterrent enforcement and penalties for counterfeiting.

In 1997 and then reiterated in 2007, the Counterfeit Intelligence Bureau (CIB) of the International Chamber of Commerce (ICC) estimates the worldwide trade in fakes to be 5–7 percent of the world's trade (CIB 1997, 2007). That estimate, which equates to approximately US$600 billion annually in 2010, has been referenced by a wide range of other agencies and organizations (European Commission 2007; United States Department of Justice 2006; World Customs Organization 2006). The US Federal Bureau of

Investigation has called counterfeiting "the crime of the 21st century" (Fourtou 2006). Interpol has stated: "Counterfeiting is increasing relentlessly year on year" (Withington 2005), and "the losses due to transnational organized criminals to be 'hundreds of billions of dollars' annually" (INTERPOL 2007). What is especially surprising is that only 5–10 percent of all counterfeited goods are believed to be in what would be considered "luxury goods" (Alcock et al. 2003; Phillips 2005). "In 2006, luxury goods accounted for only one percent of all the items intercepted by the 27 Customs administrations of the European Union (EU)" (World Customs Organization 2007b). The counterfeit product market has an even faster growth rate; it is now the size of the worldwide trade in illegal drugs (e.g., heroin, cocaine, and marijuana), which is estimated at approximately US$500 billion (Budima 2006; Union des Fabricants 2004; United Nations International Narcotics Control Board 2006). Criminal organizations that expanded from illegal drugs to counterfeiting have essentially doubled their potential market size:

> Counterfeiting overlaps with other illicit trades, organized crime, and terrorist networks. A natural connection, given the similarities in product and format, is between the fake medicine and illegal drug trade, which have come to follow many of the same routes. (Naim 2005: 126)

The estimates for the impact of counterfeiting on the global marketplace are elusive and the CIB states, "Given the clandestine nature of the business, such estimates are no more than educated guesses" (CIB 1997). There are a number of reasons for this: the clandestine nature of the counterfeiters, the reduction in consumer detection of fakes due to increased quality, the low incentive for product companies to divulge counterfeiting, and the fact that the core data collection comes only from seized goods (Bialik 2007; Ivanova 2007; World Customs Organization 2007a). Unfortunately, determining the level of counterfeiting is made more complex because the dollar amounts are sometimes defined using the value of the genuine product (if sold at genuine retail price) and sometimes by the value of the sale of fake products (sold as fake product) (OECD 2007c). PIRA's David Collins (2004) also states that: "Statistics in this area are difficult to validate both because of the clandestine nature of the threats and some brand owners' reluctance to talk, and the array of different authorities collecting statistics for varying purposes." The estimates are given as a range, since the data are based on a combination of estimating what percentage of product gets caught, and educated guesses (heuristic perceptions) about what has passed through (Ivanova 2007; Los Angeles County Development Corporation 2007). Regarding the customs seizure results: "We wouldn't consider the seizure rate to be a random sample of the extent of counterfeit goods coming in," says Loren Yagar, of the Government Accountability Office study (as quoted in Bialik, 2007). To further compound the complexity of estimating a scale of counterfeiting, a large amount of the work in the counterfeit area is either classified by governments or considered confidential by companies.

The economic impact can also be measured in lost sales for companies and lost tax revenue for governments (OECD 2007b). The tax revenue loss can be in the billions, even for US cities. Los Angeles estimates its annual losses to be $483 million (Los

Table 7.1 Estimated Value of Counterfeit Goods Based on Seizure Success Rate of US Customs and Border Protection, 2004 (percentage rate and billions of dollars)

US Customs assumed interdiction rate (%)	10.0%	5.0%	1.0%	0.2%	0.1%
Implied value of counterfeit goods entering the US (billions)	$1.4	$2.8	$13.8	$69.1	$138

Angeles County Development Corporation 2007) and New York City estimates its annual losses to be $1 billion (New York City Comptroller's Office 2004).

Table 7.1 demonstrates how some counterfeit estimates are derived. For example, the US Customs and Border Protection Bureau of the Department of Homeland Security (US Customs) present actual data on the amount of product seized during a year—2004 in this example. If US Customs stated that $138 million was seized, then to figure the total estimated value of counterfeited goods, the analyst must estimate the percentage of product that was confiscated. This table then estimates the "implied value of counterfeit goods" based on an "assumed interdiction rate (%)." So if $138 million represents 1 percent of all counterfeit goods, then the implied value of counterfeit goods is $13.8 billion, as the table demonstrates (Los Angeles County Development Corporation 2007). Of course, the actual counterfeit value and the actual interdiction rate are unknown and unknowable.

The US Customs and Border Protection Bureau reported that 26 million commercial shipping containers were imported in 2008 (US Customs and Border Protection Bureau 2008). The US FDA estimates that it inspects around 1 percent of all imported food products, and that around 0.3 percent is actually lab tested (FDA 2002; Paradise 1999). The Rotterdam shipping port, one of the busiest in the world, estimates it "inspects" approximately 3 percent of the products that pass through its system (Phillips, 2005). "Inspects" could mean anything from simply reviewing documentation to opening and physical inspection of the shipping containers (European Union 2003). The World Customs Organization (WCO) has discussed the difficulty of determining fakes even when the product is inspected (World Customs Organization 2007a).

While quantitative determination of the counterfeit threat may be unknowable, the existing data points and anecdotal evidence emphasize that this is a serious threat (Bialik 2007; United States Government Accountability Office 2008). It is important to accept that the financial opportunity for the counterfeiters is astronomical, which leads to great incentive and motivation to grow the scope and scale of the crime.

Scope of the Threat

The scope of the counterfeit product threat is widespread. Product counterfeiting could be an economic threat, a public health threat, and often both. It is important to note that for the criminals doing the counterfeiting, the public health threat may be quite real but occurs through negligence and not intent (PriceWaterhouseCooper 2007; United States General Accounting Office 1995).

A general list of some counterfeited products is noted from the Global Business Leaders Alliance Against Counterfeiting (GBLAAC) in Table 7.2 (GBLAAC 2003).

There are even reports of foodstuffs such as waffles, apples, candy, and chewing gum in Table 7.3 (European Commission 2005). A WCO report lists some other counterfeited products (World Customs Organization 2007a).

Table 7.2 List of Counterfeit Products from GBLAAC

Adhesive strips	Agrochemicals	Baby food
Batteries	Canned meat and fish	Cigarettes
Clothing and footwear	Coffee	Cosmetics and skin creams
Credit cards	Diving masks	Dry food goods
Golf clubs	Household appliances	Household detergents and cleaning products
Inkjet and toner cartridges	Motor oils	Parts for automobiles, trucks, and other machinery
Perfume	Pharmaceuticals	Photographic film
Razor blades	Rubbish bags	Shampoo
Soft drinks	Software, audio, and video	Sportswear
Super glue	Tea	Toothpaste and toothbrushes
Toys	Wines and spirits	Zippers

Table 7.3 List of Counterfeit Products from WCO

Soup	Mineral water
Breast implants	Contact lenses
Pharmaceutical products for treating life-threatening conditions such as breast cancer and high blood pressure	Pacemakers
Weapons of war	Automobile brake disc pads
Even a WCO training course on CD!	

A random listing of the public health threats from the World Health Organization (WHO) shows the following (WHO 2007):

- Eighty-nine children died in Haiti in 1995 and 30 infants died in India in 1998 due to the consumption of paracetamol cough syrup prepared with diethylene glycol (a toxic chemical used in antifreeze).
- In 2001, in South-East Asia, a Wellcome Trust study revealed that 38 percent of 104 anti-malarial drugs on sale in pharmacies did not contain any active ingredients.
- In Cambodia, in 1999, at least 30 people died after taking counterfeit anti-malarials prepared with sulphadoxine-pyrimethamine (an older, less effective anti-malarial), which were sold as artesunate.

The US FDA specifically identified four examples of EMA, also called Food Fraud (FDA 2009). These are all examples of product counterfeiting:

- In March 2007, the FDA received reports of kidney failure among cats and dogs and a report that cats died during taste tests of certain brands of pet food. In the subsequent investigation, melamine and melamine-related compounds were found in products labeled as wheat gluten and rice protein concentrate that had been imported from China.
- In January 2008, the FDA received reports of adverse reactions in pediatric dialysis patients in the US. Initial investigations by the Centers for Disease Control and Prevention indicated that the adverse events appeared to be associated with heparin manufactured by Baxter Healthcare Corp. that was administered during the dialysis procedures.

- In September 2008, the FDA issued a Health Information Advisory in response to reports of melamine contaminated milk-based infant formula manufactured in China. Melamine was apparently added to diluted milk in order to increase measured nitrogen levels (indicators of protein content) and thereby inflate the apparent protein content found in the product.
- Adulteration of glycerin, an ingredient in cough syrup and other drugs, with diethylene glycol has resulted in several mass poisonings around the world in the past two decades.

All countries, even the biggest source economies, are adversely impacted by product counterfeiting. It is estimated that 250,000 Chinese die each year from counterfeit or "substandard" medicines though this does not account for all counterfeit products, so the actual numbers may be even more staggering (Morris and Stevens 2006; Schwarz and Wong 2006).

Summary

Product counterfeiting is *not* a victimless crime. Counterfeit products are a recognized and significant economic and public health threat. The counterfeit product threat is growing in scope and scale. An interdisciplinary, multi-layered approach is necessary to curb the growth in counterfeits and the risk to the public.

The quantitative data on the overall threat are very sparse and the threat has been described as "unknowable." Counterfeit product estimates are a combination of actual seizures and educated guesses.

The Counterfeit Product Risk Model

This chapter explores the root factors that address the counterfeit threat that were used in the development of the Counterfeit Product Risk Model. Agency reports—specifically the FDA Combating Counterfeit Drugs Report—stress the need for a counterfeit risk model that would help brand owners review aspects of the counterfeit risk and prioritize the highest risk products for deterrent actions (FDA 2004).

Five factors were selected for the Counterfeit Product Risk Model, based on literature and report reviews, and peer consultation:

- Counterfeit-History: evidence of past counterfeiting.
- Counterfeit-Ability: ease of mimicking genuine product.
- Counterfeit-Attractiveness: scale of fraud in relation to costs.
- Counterfeit-Hurdles: deterrent activities or systems.
- Market Profile: aspects of the consumer, the sales channel, the manufacturing system, etc.

Risk Factor Selection Analysis

Sodipo (1997) presented one of the most complete and detailed discussions of the attributes impacting counterfeiting, based on research related to the TRIPS (Trade Related

Aspects of Intellectual Property Rights) agreement. Sodipo's key counterfeit attributes were:

- profits;
- cheap to copy;
- easy to copy;
- unsatisfied market demands;
- difficulties of detection and proof;
- non-deterrent laws and lacunae in laws;
- poor government policies;
- location of countries (production).

From a counterfeit product supply-side perspective, there is an additional set of more detailed sub-factors to be considered from the driving factors of counterfeit supply.

The Organization for Economic Co-operation and Development (OECD) listed a *supply-side* perspective set of detailed factors that should be considered as potential counterfeit drivers (OECD 2007a):

- market characteristics;
  - high unit profitability;
  - large potential market size;
  - genuine brand power;
- product, distribution, and technology;
  - moderate need for investments;
  - moderate technology requirements;
  - unproblematic distribution and sales;
  - high ability to conceal operations;
  - easy to deceive consumers;
- institutional characteristics;
  - low risk of discovery;
  - legal and regulatory framework;
  - weak enforcement;
  - non-deterrent penalties.

Also, the OECD listed, from a *demand-side* perspective a set of detailed factors that should be considered as potential counterfeit drivers (OECD 2007a):

- product characteristics;
  - low prices;
  - acceptable perceived quality;
  - ability to conceal (fake) status;

- consumer characteristics;
  - no health concerns;
  - no safety concerns;
  - personal budget constraint;
  - low regard for intellectual property rights;
- institutional characteristics;
  - low risk of discovery and prosecution;
  - weak or no penalties;
  - availability and ease of acquisition;
  - socio-economic factors.

Counterfeit-History

It has been emphasized repeatedly in the literature that the strongest indicator of the risk of product counterfeiting was past evidence of product counterfeiting (FDA 2004; Hopkins et al. 2003; O'Kane and Gopalkrishnan 2005; OECD 2007b; Phillips 2005). For example, the OECD's Economic Impact of Counterfeiting and Piracy Report specifically identifies the first of three steps in evaluating counterfeit magnitude as "identification of those goods that have been detected in international trade as counterfeit or pirated" (OECD 2007c).

That being said, the opposite is *not* true—a *lack* of evidence of past counterfeiting is *not* indicative of a low risk of future counterfeiting. The attacks can be random, a very well protected product may get counterfeited and a completely unprotected product may never get counterfeited (deKieffer 2006; FDA 2003). This knowledge is important since the past counterfeiting or diversion is indicative of the potential for more counterfeiting and diversion.

Counterfeit-Ability

"Counterfeit-Ability" is a measure of the challenge of duping consumers (or retailers) into thinking they are buying the genuine product (deceptive counterfeits), or to just make the counterfeit product sale (non-deceptive counterfeits). This factor includes consideration of the product itself, the packaging, and even the authentication systems, such as warranty registration (Collins 2004; GBLAAC 2003; Kontnik 2003; OECD 2007b; Sodipo 1997; World Customs Organization 2007b). This factor is important since the ability to actually sell a counterfeit product is a major driver for counterfeiters.

Counterfeit-Attractiveness

"Counterfeit-Attractiveness" is a measure of the ability to profit, taking into consideration the costs of production and of getting products to, and into, a market (GBLAAC 2003; Kontnik 2003; Sodipo 1997; World Customs Organization 2007a). A very large market size may not always be a positive if it contains very educated consumers (who would recognize minor variations between the genuine and fake product) or if there are very stringent deterrents in place (Lukasik 2008). A very small market size may not always be a negative if the set-up and execution costs and hurdles are low. This factor is important since the ability to make a financial profit is a major driver for counterfeiters.

Counterfeit-Hurdles

"Counterfeit-Hurdles" are measures that includes deterrent activities—whether designed or inherent in a system—that create challenges to any of the other counterfeit factors (Collins 2004; GBLAAC 2003; Kontnik 2003; OECD 2007b; Sodipo 1997; World Customs Organization 2007b). The easiest hurdles for a brand owner to implement are packaging components, since they only take one management decision, versus integrated supply chain or enforcement solutions (McNeely 2007, 2008). A hurdle could be a routine lab inspection of incoming raw materials, which would increase the chance of a counterfeit additive being rooted out. A characteristic of a supply chain that could reduce the impact of counterfeiting is establishing exclusive regional licensing agreements, where the local company pursues local enforcement of the brand (Kennedy 2005; Kontnik 2006; Studler 2008). This factor is important since fewer barriers or hurdles equate to a higher Counterfeit-Ability or Counterfeit-Attractiveness, but Counterfeit-Hurdles are unique and should be considered in their own factor.

For continuity of low always being good in the model, if a product has many or complex hurdles, then the risk ranking would be "low" for the contribution of this factor to the overall risk of product counterfeiting.

Market Profile

"Market Profile" is a measure of the infinite aspects of product development, production, supply chain specifics, retailing, consumer demographics, and geographic marketplace dynamics that contribute to the Counterfeit Product Risk Model. This factor is important since it considers characteristics that impact the counterfeit risk that are not included in the other factors.

Counterfeit Risk Sub-factors

The literature emphasized the value of breaking overall risk into components, such as likelihood and severity. The components of likelihood and severity have been broken down into qualitative characteristics such as high, medium, and low (Todd 2008). Further, it was recommended that the overall components, such as likelihood, be broken down into unique factors and sub-factors.

In many cases, key attributes play a much more significant role than other attributes, and the specific model sub-factors should be chosen by a specific company's expert panel (Aven 2003; Cruz 2002; Todd 2008). The literature and reports that were reviewed discussed some of the market factors in detail, for example, in the OECD report that covered specific industries, a common set of sub-factors was not found that would cover all products (OECD 2007d). The identification and application of sub-factors that are unique to each product or company are explored more in the "Model Development" section below.

Risk Factor Selection Analysis Summary

In this risk factor analysis section, five factors that comprise the counterfeit risks, Counterfeit-History, Counterfeit-Attractiveness, Counterfeit-Ability, Counterfeit-Hurdles, and Market Profile, were selected based on the information covered. The research supported using, but not explicitly universally defining, sub-factors.

Model Development

This section covers the methods used to develop the Counterfeit Product Risk Model, and an industry survey that was presented to industry professionals to review current industry actions and trends. The propositions and the model concepts were tested in an Expert Panel Survey and a practical example of the use of the model is included in the Case Study (see Chapter 15, this volume).

Model Overview

The model focuses on the probability portion of the general probability versus severity matrix—the term likelihood is used for harmonization with the International Standards Organization's "ISO 31000:2009 Risk Management." The likelihood is based on a series of five factors that, in turn, are based on a series of three (or more) sub-factors. The five factors are proposed to be used with all counterfeit products, whereas the sub-factors are to be developed for the specific products being reviewed.

Figure 7.1 is an overview of the Counterfeit Product Risk Model. The overview shows how the sub-factor derivation leads to the ranking of the sub-factors, which feeds up

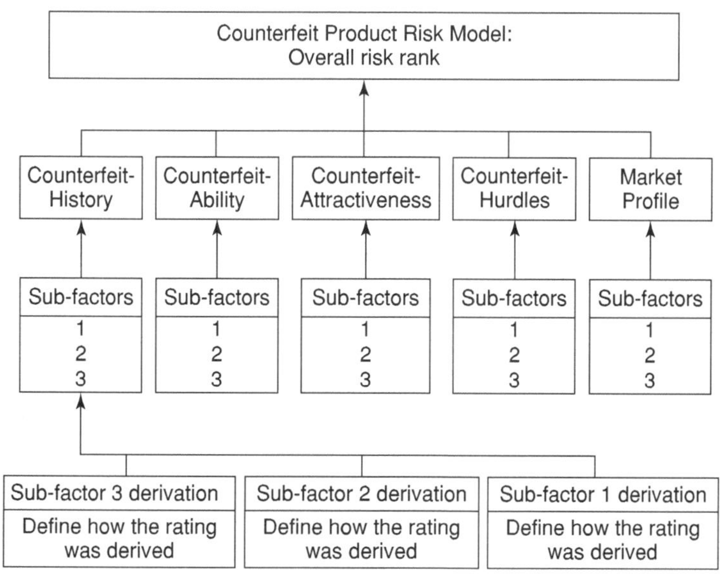

Figure 7.1 Counterfeit Product Risk Model overview

to the ranking of the factors, which finally feeds into the overall ranking of likelihood. The Counterfeit Product Risk Model uses a cascading approach to derive the sub-factors, which cascade up to the factors, which cascade up to the overall risk rank. Vice versa, by cascading down from the overall risk rank to ever increasing orders of detail can facilitate understanding between the parties developing or using the information. This concept is similar to the model cascading approach discussed by Howick et al. (2008) in "Building Confidence in Models for Multiple Audiences: The Modeling Cascade."

Summary

Product counterfeiting is a growing criminal threat. Traditional tactics applied from within one discipline or agency do not broadly address this unique, expanding, and evolving nature of the crime and the criminals. This chapter presented the unique aspects of product counterfeiting that require a non-traditional approach to determining the most efficient and effective countermeasures. Through the 2010 Joint Strategic Plan by the US Intellectual Property Enforcement Coordinator Office, this interdisciplinary and inter-agency approach is being championed. The Joint Strategic Plan will reinforce and coordinate interagency efforts and enthusiasm for applying non-traditional risk analysis and methods to combat the travesty that is product counterfeiting. We cannot incarcerate our way to safety but we may be able to prevent many incidents through further development of this risk model coupled with effective risk management strategies.

References

Acheson, David. 2007. "Food Protection, Food Safety and Food Defense." Paper presented at the Association of Food and Drug Officials (AFDO) Annual Conference. Retrieved November 3, 2010 from http://www.afdo.org/conferences/presentations2007.cfm.

Alcock, Lucy, Paul Chen, Hui Min Ch'ng, Sarah Hodson et al. 2003. "Counterfeiting: Tricks and Trends." *Journal of Brand Management 11*(2):133.

Arruñada, Benito and Xosé H. Vázquez. 2006. "When Your Contract Manufacturer Becomes Your Competitor, Best Practice." *Harvard Business Review 84*(19):135–144.

Aven, Terje. 2003. *Foundations of Risk Analysis.* Hoboken, NJ: John Wiley & Sons, Inc.

Bialik, Carl. 2007, February 19. "Efforts to Quantify Sales of Pirated Goods Lead to Fuzzy Figures." *Wall Street Journal On-line*, p. 1.

Budima, Gjeneza. 2006. "Can Corruption and Economic Crime Be Controlled in Developing Economies, and if so, Is the Cost Worth it?" *Journal of Financial Crime 13*(4):408.

CIB (Counterfeiting Intelligence Bureau). 1997. *Countering Counterfeiting. A Guide to Protecting & Enforcing Intellectual Property Rights.* Counterfeiting Intelligence Bureau (CIB), International Chamber of Commerce (ICC).

——. 2007. *Overview of Counterfeiting.* Retrieved April 11, 2007, from http://www.icc-ccs.org/cib/overview.php.

Closs, David J. and Edmund F. McGarrell. 2004. *Enhancing Security Throughout the Supply Chain [Full Report].* East Lansing, MI: Michigan State University.

Collins, David. 2004. *Brand Protection.* Leatherhead: PIRA International.

Cruz, Marcelo G. 2002. *Modeling, Measuring and Hedging Operational Risk.* New York: John Wiley & Sons, Inc.

deKieffer, Donald. 2006. "Potential Liability for Counterfeit Medications." *Journal of Pharmacy Practice* 19(4):215–220.

European Commission. 2005. "Customs: Commission and Member States Step Up Action against Counterfeiting and Piracy." Press release [electronic version]. European Commission (EC), Reference: IP/05/147, Brussels. Retrieved February 17, 2007 from http://europa.eu/rapid/pressReleasesAction.do?reference=IP/05/147&format=HTML&aged=0&language=en&guiLanguage=en.

——. 2007. "Counterfeit and Piracy, Combating Counterfeit & Piracy, A Serious Problem" [electronic version]. European Commission—Taxation and Customs Union (EC/TCU). Retrieved November 3, 2010 from http://ec.europa.eu/taxation_customs/customs/customs_controls/counterfeit_piracy/combating/index_en.htm.

European Union. 2003. "Council Regulation (EC) No 1383/2003." *Official Journal of the European Union 46*(L 196):7–11.

FDA (Food and Drug Administration). 2002. "A Safe Food Supply" [author's hardcopy]. *Performance Plan and Summary U. S. Food and Drug Administration.* Silver Spring, MD: FDA.

——. 2003. "Counterfeit Drug Task Force Interim Report" [author's hardcopy]. Silver Spring, MD: FDA.

——. 2004. *Combating Counterfeit Drugs* [author's hardcopy]. Silver Spring, MD: FDA.

——. 2009. "Economically Motivated Adulteration; Public Meeting; Request for Comment" [electronic version]. FDA-2009-N-0166. Department of Health and Human Services, Food and Drug Administration, 74, 15497–15499. Retrieved April 6, 2009 from the Federal Register Online via GPO Access [wais.access.gpo.gov, DOCID:fr06ap09-76].

Fourtou, Jean-René. 2006, November 26. "Opening Speech". Paper presented at the BASCAP (Business Action to Stop Counterfeiting and Piracy) meeting, Paris.

GBLAAC. 2003. "Product Counterfeiting Overview." Global Counterfeiting, Background Document, Global Business Leaders Alliance Against Counterfeiting (GBLAAC).

Hopkins, David M., Lewis T. Kontnik, and Mark T. Turnage. 2003. *Counterfeiting Exposed: How to Protect Your Brand and Market Share.* Hoboken, NJ: John Wiley & Sons.

Howick, Susan, Colin Eden, Fran Ackermann, and Terry Williams. 2008. "Building confidence in models for multiple audiences: The modelling cascade." *European Journal of Operational Research 186*(3):1068–1083.

INTERPOL. 2007. "Intellectual Property Crimes: Trademark Counterfeiting & Copyright Piracy." Interpol Intellectual Property Crimes Action Group (IIPCAG).

Ivanova, Kate. 2007. "Corruption, Illegal Trade and Compliance with the Montreal Protocol." *Environmental and Resource Economics 38*(4):475–496.

Kelly, Robert. 2005. *Illicit Trafficking: A Reference Handbook.* Santa Barbara, CA: ABC-CLIO.

Kennedy, Jim. 2005. "Defining a Successful Global Brand Protection Strategy." PIRA, Innovations in Security Technology Conference.

Kontnik, Lewis. 2003. "Pharmaceutical Counterfeiting: Preventing the Perfect Crime." White paper for FFF Enterprises.

Kontnik, L. T. 2006. "Manufacturer and Industry Responses to the Counterfeiting Challenge." *Journal of Pharmacy Practice 19*(3):140–145.

Los Angeles County Development Corporation. 2007. *A False Bargain: The Los Angeles County Economic Consequences of Counterfeit Products [Final Draft].* Los Angeles, CA.

Lukasik, Dave. 2008. "Motorola, Motorola Inc. Brand Protection." Paper presented at the Product Authentication and Brand Protection (PABS08). Hardcopy.

McNeely, Susan. 2007. "Selecting Brand-Security Technology Suppliers." *Packaging Digest 44*(3):39.
——. 2008. "Protecting Health Care Products—Lessons Learned with Eli Lilly." Paper presented at the Product Authentication and Brand Security Conference (PABS08). Hardcopy.
Morris, Julian and Philip Stevens. 2006. *Counterfeit Medicines in Less Developed Countries: Problems and Solutions.* London: International Policy Network.
Naim, Moises. 2005. *Illicit: How Smugglers, Traffickers, and Copycats are Hijacking the Global Economy.* New York: Doubleday.
New York City Comptroller's Office. 2004. *Bootleg Billions: The Impact of the Counterfeit Goods Trade on New York City.* NYC Comptroller's Office, November 2004.
O'Kane, Gerry and Shanker Gopalkrishnan. 2005. *The Future of Anti-Counterfeiting, Brand Protection and Security Packaging III.* Leatherhead: PIRA International Ltd.
OECD (Organization for Economic Co-Operation and Development). 2007a. *The Economic Impact of Counterfeiting and Piracy, Executive Summary* [Draft September 2007].
——. 2007b. *The Economic Impact of Counterfeiting and Piracy, Executive Summary* [Draft June 2007].
——. 2007c. *The Economic Impact of Counterfeiting and Piracy, Part I: Overall Assessment [Draft].*
——. 2007d. *The Economic Impact of Counterfeiting and Piracy, Part III: Industry Sectors* [Draft September 2007].
Paradise, Paul R. 1999. *Trademark Counterfeiting, Product Piracy, and the Billion Dollar Threat to the U.S. Economy.* Westport, CT: Quorum Books.
Phillips, Tim. 2005. *Knockoff: The Deadly Trade in Counterfeit Goods: The True Story of the World's Fastest Growing Crime Wave.* London: Kogan Page.
PriceWaterhouseCooper. 2007. *Economic Crime: People, Culture and Controls.*
Schwarz, Brian and Vanessa Wong. 2006. "Counterfeit CURES" [electronic version]. *INSIGHT, Center for Medicine in the Public Interest (CMPI).* Retrieved November 3, 2010 from http://www.amcham-shanghai.org/NR/rdonlyres/9D628132-7DEA-432E-B18B-C0E27AC73E37/2362/09.pdf.
Sodipo, Bankole. 1997. *Piracy and Counterfeiting: GATT, TRIPS and Developing Countries.* London: Kluwer Law International.
Studler, Nicholas. 2008. "Brand and IP Protection at the Coca-Cola Company." Paper presented at the Product Authentication and Brand Strategy (PABS08). Hardcopy.
Todd, Ewen. 2008. "Discussion of Counterfeit Risk Model." Personal communication.
Union des Fabricants (on INTERPOL.int). 2004. *Counterfeiting & Organized Crime Report.* Paris: Union des Fabricants.
United Nations International Narcotics Control Board. 2006. *International Narcotics Control Board Report.* International Narcotics Control Board, United Nations (UN/ NCB).
United States Department of Justice. 2006. "Prosecuting Intellectual Property Crimes" [electronic version]. CCIPS, Criminal Division, 2007. Retrieved September 2006 from USDOJ.gov, http://www.usdoj.gov/criminal/cybercrime/ipmanual/01ipma.pdf.
United States General Accounting Office. 1995, November. GAO report to Congressional Committees, "Fruit Juice Adulteration: Detection Is Difficult, and Enhanced Efforts Would Be Costly." GAO/RCED-96–18.
——. 2008. "Risk Management - Strengthening the Use of Risk Management Principles in Homeland Security." GAO-08-904T.
US Customs and Border Protection Bureau. 2008. "National Workload Statistics" [electronic

version]. Retrieved November 3, 2010 from http://www.cbp.gov/xp/cgov/about/accomplish/previous_year/national_workload_stats.xml.

van Schendel, Willem and Itty Abraham. 2005. *Illicit Flows and Criminal Things: States, Borders, and the Other Side of Globalization.* Bloomington, IN: Indiana University Press.

WHO (World Health Organization). 2007. "Counterfeit Medicines." Retrieved November 3, 2010 from http://www.who.int/medicines/services/counterfeit/en/.

Withington, Neil. 2005. "Second Global Congress on Combating Counterfeiting and Piracy: Getting it Right in the Fight against Fakes." Vol. Speech by Director, Legal & Security and General Counsel British American Tobacco.

World Customs Organization. 2006. "The First Review: Customs and Counterfeiting 2004." Press release [electronic version]. World Customs Organization (WCO). Retrieved August 18, 2008 from http://www.wcoomd.org/press/?v=1&lid=1&cid=1&id=125.

——. 2007a. "Counterfeiting and Piracy—Crime of the 21st Century?" *WCO NEWS*, 54. World Customs Organization (WCO).

——. 2007b. *Review of the Fight against Counterfeiting in 2006.* Brussels: World Customs Organization.

Yar, Majid. 2005. "A Deadly Faith in Fakes: Trademark Theft and the Global Trade in Counterfeit Automotive Components" [electronic version]. *Journal of Criminology* from www.internetjournalofcriminology.com.

Questions

1. When we consider clustering of events, does this mean that we assume that the risk of similar types of event occurring in the future are the same or are there factors that raise or lower these risks?
2. Is making a risk assessment the same thing as making a prediction about a future event? How do these differ?

PART II

PRINCIPLES OF RISK MANAGEMENT

Risk management builds on and extends risk assessment through an explicit focus on the applications of risk analysis, risk management, risk communication, and on decisions, values, and policy surrounding risk assessment and management (International Risk Governance Council 2005). Most well developed in the private sector, where it is also referred to as enterprise risk governance (Marsh and Noonan 2005; Tew and Elkins 2006), many corporations have moved risk analysis from specialized financial risk units, to an enterprise-wide focus that seeks to "ensure that the risk management mission is instituted, facilitates enterprise-wide risk management standards and practices, and provides communication channels to decision-makers" (Marsh and Noonan 2005: 4). In addition to responding to probabilistic modeling of risk hazard, vulnerability, and consequence, governance seeks to assess risk tolerance and risk acceptance, and to implement management practices to optimize prevention, response, recovery, and mitigation.

In the context of the US Department of Homeland Security and law enforcement agencies, a risk management approach to the threat of terrorism requires building information tools and systems, analytic capabilities, information dissemination, preparedness systems, and the leadership and management processes for effective implementation across federal, state, and local levels of government and in coordination with the private sector. Additionally, risk balance includes policy decisions related to investment strategies and tolerance and risk acceptance assessment. Thus, for example, citizens in the United Kingdom have grown accepting of the widespread use of closed-circuit television (CCTV) technology. In considering the expansion of CCTV in the US, risk management emphasizes the need to assess the public's tolerance for the civil liberties infringements of CCTV, as well as a public policy question of the relative benefits given the costs of widespread dissemination of CCTV.

The movement within the private sector toward risk management reflects a recognition that the risks facing enterprises in the risk society go well beyond financial risk to include strategic, operational, and hazard risks. Thus, there emerges a need for the adoption of strategies, policies, metrics, and processes throughout the entire organization. Essentially, risk becomes "everyone's responsibility" (Tew and Elkins 2006). Although risk tolerance and risk acceptance is important in the private sector, it is particularly critical in the public sector. In the private sector, the bottom line of minimizing risks to profitability and to maximizing new opportunities ultimately defines risk tolerance and acceptance. In the public sector, the goal of maximizing general welfare is likely to include competing interests in the context of uncertainty (Stern and Fineberg, 1996; Slovic, 2003). Thus, for example, should a vaccine of limited availability be prioritized for health care workers, the young rather than the old? Are tools for assessing tolerance and acceptability included in the risk management models being developed by governmental agencies in the US or other countries? Who is involved in such decision making?

The concerns about risk governance lead to a set of general questions that need to be addressed:

1. Are standardized risk management methodologies and mechanisms observed across security agencies?
2. How are threat, vulnerability, and consequence data defined?
3. To what extent are risk communication strategies and processes part of the risk management framework?
4. To what extent is risk tolerance and risk acceptability included in the risk management framework?
5. Which risk modeling approaches are being utilized (e.g., probabilistic risk assessment, stochastic models, event-based models) with what perceived benefits, costs, and challenges?
6. Are procedures in place for assessing and validating models? Are internal and/or external research processes to assess risk practices in place or planned?

8. SETTING LAW ENFORCEMENT PRIORITIES

Risk management addresses the ways in which agencies plan for and react to threats and incidents, in particular, in the judgments that they make about resource allocations. This process has important effects as well, depending on the level of impact of incidents where different levels of jurisdiction may influence the ways in which resources are managed, the ways in which information about threat is used, and the authority that is applied in responding to and coordinating response. Management also addresses a variety of issues related to the legitimate exercise of authority and oversight of the agencies delegated responsibility for risk assessment.

The chapter by McGarrell, Freilich, and Chermak examines the challenges facing law enforcement in managing the risk of terrorism. They look at this problem through the lens of intelligence-led policing, and draw on the lessons learned in programs developed to respond to crime problems, which emphasize the need to coordinate information within the law enforcement agency in a way that connects threats to response. It is in this way that agencies have found that they can achieve real success in preparing for and reacting to the challenges posed by terror groups.

INTELLIGENCE-LED POLICING AS A FRAMEWORK FOR RESPONDING TO TERRORISM[1]

Edmund F. McGarrell, Joshua D. Freilich, and Steven Chermak

Leading police associations in the United States and the United Kingdom have advocated that law enforcement adopt an intelligence-led policing (ILP) model. Much like the situation with community policing, there does not appear to be a commonly accepted definition of ILP or of the practical implications for police agencies' mission, structure, and processes. This chapter presents a model of ILP that builds on community policing, problem solving, and continuous improvement business models that have been adopted by police departments. Examples of these practices are reviewed as a method of illustrating the promise of an ILP approach. A broad conceptualization of ILP is presented under the belief that ILP will be most likely integrated into law enforcement and will have the greatest impact if it is adopted from an "all crimes" perspective. The chapter concludes with illustrations of the utility of ILP for addressing threats of domestic and international terrorism.

The terrorist attacks of September 11, 2001 (9/11) affected society in dramatic ways. Although scholars caution that the impact of such "focusing events" may be overstated and their overall impact is best assessed only after significant time has elapsed (Birkland 2004; Sabatier and Jenkins-Smith 1993), it appears that these attacks dramatically influenced society. For example, the 9/11 attacks led to the creation of the Department of Homeland Security as well as numerous attempts to reform the Federal Bureau of Investigation (FBI). There have also been increased demands on law enforcement to build global partnerships (that lead to the sharing of intelligence and data about terrorism and other crime issues) and to work more closely with other agencies at home (i.e., public health agencies, private agencies, fire services, etc.). Many states have developed or are in the process of developing fusion centers to increase the exchange of information and data across government sectors to "improve the ability to fight crime and terrorism" (see, e.g., the 9/11 commission report, Fusion Center Guidelines, 2005: 3).

Pressure to change is not new to law enforcement; indeed, it could be argued that no other public bureaucracy has changed as dramatically as law enforcement has since the early 1960s. Law enforcement organizations are constantly responding to external demands, implementing and evaluating new initiatives, and adopting strategies to ensure public safety. Recent years have witnessed calls for law enforcement to move toward an ILP model of policing. Of particular importance are the positions of highly influential commissions in the United Kingdom and in the United States. In the United

Kingdom, the National Criminal Intelligence Service developed the national intelligence model (NIM), which advocates for ILP. The NIM has been endorsed by the Association of Chief Police Officers, the Home Office, and Her Majesty's Inspectorate of Constabulary (Hale, Heaton, and Uglow 2004; Tilley 2003). Similarly, the National Criminal Intelligence Sharing Plan, a product of the Global Intelligence Working Group, calls for all US law enforcement agencies to develop an ILP model and has been endorsed by the US Department of Justice (DOJ), the International Association of Chiefs of Police, and every key professional law enforcement association in the United States (Bureau of Justice Assistance 2005a, 2005b; Carter 2004). Clearly there is momentum to move toward an ILP framework. Yet like community policing two decades ago, there is a lack of clarity about what is meant by ILP; its mission, goals, and objectives; how such a model should be implemented; and importantly, how this model could be used strategically to combat terrorism.

This chapter seeks to add to our understanding of ILP and its application to terrorism. We briefly consider the origins of ILP and provide a few examples of police practices that highlight the promise of the ILP model. We argue that the best practices identified in responding to gang violence, drug trafficking, and organized crime could be successfully used in the struggle against domestic and international terrorism.

Traditionally, ILP has been viewed as a specialized police function targeted primarily at terrorism and homeland security. We argue, in contrast, for a broad conceptualization of ILP that embraces community policing, problem-solving policing, a continuous improvement managerial philosophy, and an "all crimes" focus. Paradoxically, we argue, the greatest potential for responding to terrorism using this model lies in this very broad conceptualization.

Origins of ILP

Although the National Advisory Commission on Criminal Justice Standards and Goals (1973, 1976) recommended that all law enforcement entities in the United States develop an intelligence capacity, and that agencies with more than 75 personnel should have a full-time intelligence capacity, the concept of ILP is a more recent development. The term ILP can be traced to the United Kingdom and the Kent and Northumbria constabularies in particular (Bureau of Justice Assistance 2005a; Hale et al. 2004). The Kent Constabulary, for example, formalized a model of problem solving to focus on burglary and motor vehicle theft. Moving beyond reactive responses to specific incidents, officers engaged in systematic analysis of these offenses and found patterns of small numbers of chronic offenders responsible for large numbers of incidents as well as patterns of repeat victims and target locations. Utilizing this intelligence, a variety of strategic interventions were implemented, and the area experienced a significant decline in crime.

Similarly, the Northumbria Constabulary reported consistent annual drops in crime following its implementation of an ILP model focused on identifying chronic high-rate offenders (Anderson 1997; Hale et al. 2004; but see Ratcliffe 2002: 61–62).[2] As evident in the UK experience, the adoption of ILP did not arise solely from the concern with terrorism but rather builds on major developments in law enforcement that emerged in the past two decades of the twentieth century. These include community policing, problem-solving policing, and continuous improvement business models. In the next section we briefly discuss the evolution of law enforcement and examine how these models serve as an important foundation of ILP. We then look at examples of several "promising practices" of these models that appear to hold promise for the application of ILP to terrorism.

Community Policing

The community policing movement that has influenced law enforcement practice globally—though with varying levels of meaning, commitment, and implementation—sought to reform the professional model of policing, which had relied heavily on technology and motorized patrols, by reestablishing connections between the police and community members. Carter (2004) articulated the platform provided by community-oriented policing (COP) for ILP along a series of key dimensions. First, both COP and ILP are dependent on two-way communications with the public. Indeed, a key community policing strategy in the era of homeland security is to provide the public with information about activities and signs to be aware of and to explain what to do with information related to potential threats. Second, both COP and ILP are dependent on analysis of information and the transformation of information to actionable intelligence. This is true whether the analysis involves a patrol officer analyzing a series of break-ins or intelligence analysts in a fusion center conducting link analysis to assess terrorist funding. Third, in the absence of ethical decision making and a commitment to individual rights, both COP and ILP are likely to repeat mistakes that have resulted in significant setbacks for law enforcement.

Problem Solving

As Nick Tilley (2003) and Marilyn Peterson (Bureau of Justice Assistance 2005a) noted, ILP also builds on developments associated with problem-oriented policing (POP).[3] Based on the pioneering work of Herman Goldstein (1990), the POP model seeks to use analysis of recurring problems the police must manage to proactively craft interventions to address the problem. POP also draws on concepts generated by routine activities theory (Felson 1986, 1987), opportunity theory, and situational crime prevention (Clarke 1997). Crime problems are analyzed according to the crime triangle with the idea that interventions be designed to minimize the likely confluence of motivated offenders, suitable victims, or targets in the absence of effective guardianship.

Implementation of POP has often followed the problem-solving model that consists of scanning, analysis, response, and assessment (SARA), very similar to the intelligence process advocated by the National Criminal Intelligence Sharing Plan (Bureau of Justice Assistance 2005a, 2005b; Carter 2004). Greater infusion of the ILP philosophy within policing could be an important step in strengthening the analytic component of POP. One development that has emerged through the problem-solving movement has been a number of police–research partnerships to promote both the analysis and assessment components of POP. An additional foundation for ILP is the attempt to apply cutting-edge business practices to public-sector policing. Within the United States, the most well-known application of this business model is the New York Police Department's Compstat program developed during the Giuliani and Bratton administration (Silverman 1999).[4]

During the early 1990s, the New York Police Department gained the attention of police leaders and scholars through the implementation of a crime analysis and managerial accountability crime prevention program known as Compstat ("compare statistics"; Silverman 1999). Regular meetings of the police command staff, area commanders, special units, and prosecutors were convened to review current crime trends, to deploy officers and develop responses to effectively manage crime problems, and to hold commanders accountable for the level and trend in crime in their precincts. Other key elements of the success of Compstat include the reliance of information from line-level personnel, flexibility to adapt to the nature of the crime problem, and constant feedback that assessed the successes and failures of any strategic responses. The dramatic and unexpected decline in crime in New York City that occurred during this period generated significant interest and the widespread national adoption of Compstat (Weisburd et al. 2003).

Project Safe Neighborhoods (PSN)[5]

Two major initiatives to respond to gun crime illustrate the potential of problem-solving and ILP for terrorism. The PSN model calls for multiagency task forces to employ a strategic problem-solving model whereby local gun crime patterns are analyzed, strategies are crafted to respond to these patterns, and ongoing assessment and refinement occurs. PSN built on the crime-analysis and resource-deployment approach embodied in Compstat as well as the problem-solving model that was part of Boston's Ceasefire program and later part of the DOJ's Strategic Approaches to Community Safety Initiative (McGarrell 2005; US DOJ 2004).

Boston Ceasefire

Boston's Ceasefire program, also referred to as the Boston Gun Project, was a strategic problem-solving initiative intended to reduce the high level of youth gun violence in the city. Ceasefire was initiated by a multiagency working group involving the US Attorney's Office, local prosecutors, the Boston Police Department, probation officers, youth service workers, and a team of researchers from Harvard's Kennedy School of Government. The problem analysis revealed that youth violence was driven by a relatively small number of chronic offenders involved in networks of known offenders. The strategy that emerged was a deterrence-based model whereby the threat of federal prosecution was directly communicated to these groups of known offenders. Following crackdowns on several of the most violent groups and ongoing communication through meetings with probationers and parolees connected to these offending networks, youth violence declined dramatically (Braga et al. 2001; Kennedy, Braga, & Piehl 2001).

The Boston project was characterized by several distinctive features. First, a small working group was convened from multiple agencies and linked to a research team that conducted systematic analysis of the firearms crime problem. Second, the deterrence threat was coupled with attempts to link potential offenders to legitimate services offered by youth service workers, traditional service providers (e.g., jobs, education, drug treatment), and nontraditional providers, including the faith community. Third, several distinctive strategies emerged to communicate the deterrence message to potential offenders. These included offender notification meetings and police-probation teams conducting visits to high-risk offenders (Operation Nightlife). Fourth, the Bureau of Alcohol, Tobacco, and Firearms and the Boston Police Department developed supply-side strategies to disrupt illegal gun markets (Braga et al. 2002; Kennedy, Piehl, & Braga 1996). Finally, like in Project Exile,[6] the US Attorney's Office played a key leadership role by convening local, state, and federal resources and by bringing the threat of federal prosecution to the issue of illegal gun possession and use.

Strategic Approaches to Community Safety Initiative (SACSI)

Building on the Boston project, the DOJ developed SACSI in the late 1990s (Coleman et al. 1999; Dalton 2003; Roehl et al. 2004). SACSI demonstrated the utility of using strategic problem solving to tailor a federal initiative to local contexts that varied considerably across the ten SACSI sites. It also provided the opportunity for further testing of problem-solving approaches and strategies initially developed in Boston's Ceasefire. Thus, for example, a number of SACSI jurisdictions found value in using systematic crime incident reviews of homicide incidents and gun assaults to dig deeper into crime data to uncover patterns of offenders, victims, locations, and network connections that could then suggest intervention strategies. Many of the SACSI sites implemented offender notification meetings to communicate the deterrence message and offer the opportunity for linkage to legitimate services. The offender notification meetings were coupled with Richmond-style billboards, bus posters, and public service advertisements warning against illegal gun possession and use.

Using ILP to Respond to Terrorism

Although the ILP examples addressed above focused only on traditional street crime and gun crime in particular, we believe there is great potential in using the ILP model to respond to terrorism. The arrests following the London subway and the Madrid train bombings, for instance, suggest that intelligence gathered from multiple sources, including surveillance cameras, community informants, and prison officials, was instrumental in postevent investigations. Recent arrests of suspected terrorists in the United States (Florida), Canada, and Britain suggest that intelligence may be able to identify terrorist groups and allow for intervention prior to an attack. This is consistent with reports from countries, such as Israel, with long experience in combating terrorism. Unfortunately, there is little empirical research that assesses the effectiveness of counterterrorism strategies generally and specific strategies involving law enforcement organizations. Statistical analysis is rarely used in terrorism research (Merari 1991), and Silke's (2001) review indicates that only 3 percent of articles in terrorism journals used inferential analysis, compared to 86 percent in forensic psychology and 60 percent in criminology. Lum, Kennedy, and Sherley's (2006) recent Campbell systematic review of terrorism articles published between 1971 and 2003 similarly found that only 3 percent were empirical (p. 8) and only "seven ... contained moderately rigorous evaluations of counterterrorism programs" (p. 4). Lum et al. note that little has been written about the effectiveness of law enforcement strategies in responding to terrorist events. Next, we provide examples of how the best practices identified above may be useful in responding to domestic and international terrorism.

Terrorism and Far-right Extremists

Many law enforcement officials, academics, and watch groups claim that the domestic far-right poses a significant threat to national security. The movement has been linked to many crimes that range from acts of domestic terrorism and violent hate crimes to gun charges. Some of these crimes (e.g., the 1995 Oklahoma City bombing) meet most definitions of terrorism. But other crimes (e.g., nonviolent tax refusal cases) do not. It is important to note that it has been empirically demonstrated that within the United States, domestic terrorism attacks generally outnumber international ones seven to one, and among domestic groups, far-right-wing groups are especially dangerous (LaFree et al. 2006). There is also evidence that the number of attacks committed by the far-right has increased with time, and some scholars conclude that far-right extremists, especially groups motivated by religious ideology, are strong candidates to commit future acts using weapons of mass destruction (Gurr and Cole 2002; Tucker 2000). It is not surprising that local and state law enforcement agencies have concluded that the domestic far-right poses a major threat to public safety (Carlson 1995; FBI 1999; Riley et al. 2005).

Despite clear evidence of a threat, there is little empirical study of the far-right, and most of the research that has been conducted does not rely on data. Instead, this body of research makes sweeping generalizations and conclusions with no empirical support, and the most frequently cited works have been produced by journalists. The empirical research that has been published, however, shows the value of systematically understanding the activities of these groups and the potential, then, of law enforcement agencies' use of the analytic elements of the ILP to respond to terrorism.

One of the most important domestic terrorism data collection efforts is the work by Brent Smith and his colleagues for the American Terrorism Study (ATS; Smith 1994; Smith and Damphousse 1996, 1998; Smith and Morgan 1994; Smith and Orvis 1993; Smith et al. 2002). The data collection effort, conducted in cooperation with the FBI's Terrorist Research and Analytical Center, includes persons indicted federally as a result of an investigation under the FBI's Counterterrorism Program. These researchers were provided lists (1980 to 1989; 1990 to 1996; 1997 to 2002) of persons indicted, and then they traveled to federal courthouses to collect data from trial transcripts and docket information. The data have been used by the principal investigators to answer important questions, including regarding the prosecution and punishment of international and domestic terrorists (Smith et al. 2002), the effects of political motive on sentencing outcomes (Smith and Damphousse 1996), the testing of competing theoretical models of sentencing outcomes (Smith and Damphousse 1998), empirical validation of the growth of leaderless resistance tactics (Damphousse and Smith 2004), the profiling of American terrorists (Smith and Morgan 1994), and prosecutorial strategies in terrorism cases (Smith and Orvis 1993).

These researchers have also made their database available for additional analysis to the scholarly community, and their efforts are starting to bear fruit. For example, Mark Hamm (2005) recently completed a project that examined international and domestic terrorist involvement in a wide variety of crimes. An important element of this study was using the ATS database to compare international jihad groups and domestic right-wing groups. Although he found some differences comparing the groups (international jihad groups were more likely to commit aircraft and motor vehicle crimes, use explosive materials, and commit firearms violations, and right-wing groups were more likely to commit robbery, mail fraud, and some types of firearm violations, such as with machine guns), he concluded with the following:

> Best practices of investigators should be directed to crimes that facilitate the operation of complex organizations. A clear pattern of precursor activity across diverse terrorist organizations is discovered. The study shows that all terrorist organizations require money, material, transportation, identity documents, communication systems, and safe havens to accomplish their aims. Crimes to finance these operations should be the top priority for investigators. (Hamm 2005: vi)

Recently, Smith, Damphousse, and Roberts (2006) published a study examining preincident indicators of terrorist incidents. This analysis was based on 45 cases from the ATS and 20 additional cases selected by subject-matter experts. The study provides support for the conclusion that law enforcement agencies can prevent a terrorist attack if able to access the right types of data and intelligence in timely manner. There are many valuable insights provided in the final report, but two are particularly relevant to the ILP model. First, planning a terrorist incident does not occur spontaneously—there is a period of time between when the planning process begins to when the incident is completed. On average, approximately 2 to 3 months elapse between planning and commission. Second, terrorism is a very local event. They find that nearly one-half of all terrorists lived within 30 miles of the target and that their preparatory behaviors all occurred within this geographic range. It is likely that the individuals and/or groups that are most likely to commit an incident are "known, chronic offenders," and thus it is logical to conclude that local law enforcement has vital information about the groups that are most likely to commit terrorist acts in its jurisdiction. The challenge for law enforcement is in the systematic analysis and application of problem-solving principles to identify such threats before an incident occurs.

It is important to note that the ATS database focuses only on individuals indicted federally for domestic security and terrorism crimes and excludes other incidents, even if they are ideologically motivated. Thus, nonviolent incidents, such as those involving far-rightists who refuse to pay taxes because of their ideological belief system or crimes falling under state jurisdiction, are excluded from the ATS. The extant research, however, indicates that there would be great value in examining this broader range of cases. Freilich, Chermak, and Gamarra (2006) reviewed more than 250 studies on far-right-wing extremism and political crimes in the fields of criminology, political science, sociology, and terrorism and concluded that far-right extremists have been investigated by both federal and state authorities and that they are involved in a variety of criminal acts, including (1) tax refusal; (2) gun charges; (3) land use violations; (4) violations of motor vehicle regulations requiring license plates and driving licenses and other similar regulations; (5) false liens and financial schemes, (6) abortion-related crimes; (7) hate crimes against non-whites, gays, foreigners, and Jewish, Islamic, and other non-Christian faiths; (8) antipornography attacks; (9) antigovernment strikes; (10) antiglobal attacks; (11) attacks against leftists and other ideological enemies; (12) attacks against opponents within the movement (i.e., crimes committed because of intramovement strife); (13) possession and planned use of chemical and biological weapons; (14) preparatory crimes, such as bank robberies, counterfeiting, credit card theft, weapons procurement and the manufacture of illegal firearms, drug dealing, cybercrime, and identity theft; (15) other crimes that further the movement's survival or ideology; and (16) nonideological "routine" crimes.

Some intriguing anecdotal evidence suggests that ideologically motivated tax refusal, for example, may act as a "trigger" or "gateway" to subsequent, more serious criminal behavior. This thesis is consistent with part of the hate-crimes literature that argues that bias-crime perpetrators escalate from less to more violent hate crimes (Shively 2005). Gordon Kahl (who murdered three law enforcement officials before being killed himself in 1983), Robert Matthews (the leader of the notorious terrorist group The Order, responsible for five murders, in the 1980s), Richard Snell (who murdered an African American law enforcement officer and a pawn broker in the 1990s), and Frank Nelson (the leader of the Minnesota Patriots Council and one of first to be convicted under the Biological Weapons Anti-Terrorism Act in 1995) were all involved in antitax actions before their later, more serious crimes (Corcoran 1990; Flynn and Gerhardt [1989]1995; Hamm 2005; Tucker and Pate 2000).

Thus, a problem-solving evaluation of the full range of crimes committed by the far-right would help inform law enforcement practice by systematically documenting the types of crime committed by far-rightists, how frequently these crimes occur, whether these ideologically motivated offenders specialize (i.e., land use violators usually do not commit other crimes) or are generalists, whether a small number of perpetrators are chronic ideological offenders responsible for large numbers of incidents, what groups commit which crimes, and regional variations in incidents. This might include variations in what different groups have committed each type of offense as well as similarities and differences between groups. Such comparisons would give law enforcement additional knowledge to better assess threats and prioritize ideologies and groups. It would also provide additional information about the spatial and temporal patterns of these perpetrators and events to supplement the preincident indicator study (Smith et al. 2006).

ILP and International Terrorism

In the following section, we consider the lessons provided by a specific investigation by the Turkish National Police (TNP) of a series of bombings in Istanbul.[7] Components of the case have reaffirmed TNP's commitment to intelligence gathering and analysis as a key step in prevention and suggest that terrorist operations are vulnerable to an ILP approach.

In November 2003, a series of coordinated suicide bombings were carried out by al-Qaeda in Istanbul. The targets represented Israel and the West, including two synagogues, an HSBC bank, and the British consulate. The attacks resulted in 68 deaths and more than 700 injured. The investigation and arrests that ensued revealed that the network involved in the bombings had trained in Afghanistan. Of particular interest was the interpersonal web that grew from the four suicide bombers as well as the range of materials confiscated in the investigation. Specifically, nearly 300 people were identified who had some knowledge of the planned attack. Of these, 48 were viewed as hard-core committed terrorists, leaving approximately 250 community members who were not ideologically committed to al-Qaeda's goals and who had some information that potentially could have been used in preventive action. Similarly, the materials confiscated included items such as 800 pounds of hexamethylene tetramine, 150 pounds of pentaerythritol-M, and 150 pounds of sodium carbonate, all ingredients used for the explosives. Additional items included sticks of dynamite, AK-47 rifles and shotguns, 22 hand radios, forged passports and drivers' licenses, and 89 large detergent barrels. The point was that given all these materials, as well as the number of people in the community with some knowledge or suspicions, the opportunity existed to acquire intelligence that could be used in a predictive intelligence fashion aimed at prevention.

These findings were reinforced through interviews conducted by TNP officers with detained terrorists. The results suggested that the majority of terrorist members became involved with terrorist organizations through family and friendship networks and were not ideologically committed. Many expressed the desire to leave the terrorist network but feared retribution. Only approximately one-third of those interviewed were assessed to be ideologically committed to extremist views. Consistent with the findings from the investigation, the interviews suggested to the TNP that establishing connections to the community could result in informants and intelligence that could form the basis for predictive intelligence.

Challenges and Key Components

Perhaps the greatest challenge to implementing ILP is conceptual. The model of ILP presented herein is based on a broad conceptualization that views ILP as an overarching framework that builds on community policing, problem solving, and partnerships and takes an "all crimes" approach. Alternative conceptualizations may include viewing ILP as focused solely on homeland security and terrorism or following the UK model of ILP as involving the identification and targeting of high-rate, chronic offenders. Applying the model only to terrorism would be unfortunate because of terrorists' involvement in such a wide variety of routine and preparatory crimes. Scholars studying ILP in the United Kingdom have noted the movement toward conceptual integration of ILP, community policing, problem solving, and partnerships but have also noted some of the inherent tensions of the styles of policing (Hale et al. 2004; Tilley 2003). For example, community policing calls for decentralized, geographic-based structure, but ILP is likely to include a centralized intelligence function.[8] The danger, given the challenges of organizational change, is that integration of ILP, COP, POP, and partnerships results in a watered-down model whereby little meaningful change actually occurs (Hale et al. 2004). Despite these challenges and obstacles, there do appear to be identifiable components of what needs to occur for an integrated model of ILP to be implemented in a meaningful fashion.[9] Zimmermann's (2006) research, a series of case studies of high-performing PSN task forces (PSN case study cites) as well as organizational research (Morash & Ford, 2002), suggests several critical components for the implementation of ILP. Foremost is leadership. As noted above, organizational change is difficult and is unlikely absent commitment by the organizational leaders at the highest levels as well as those in key positions throughout the organization. ILP must be seen as a top priority and as something that is not going away. Such leadership is also essential to any hope of restructuring the organization in a way that allows for the integration of intelligence throughout the organization (Carter, 2004).

A second key ingredient is a multiunit, multiagency structure.[10] Within the law enforcement agency, ILP cannot be seen as the sole province of an intelligence or crime analysis unit. Rather, it must be integrated throughout the organization. This is critical for both input into the intelligence process—as information held by street-level officers; gang, drug, vice, and major crime investigators; community policing officers; and others will form the basis of much intelligence analysis—and serving as key consumers of intelligence products (Riley et al. 2005). Beyond the single agency, a multiagency structure is essential, particularly in the United States, with its decentralized, fragmented law enforcement structure but also through the connections to other criminal justice agencies such as prosecutors, corrections officials (particularly gang and threat assessment units), probation and parole.[11] For example, it was discussed earlier how tax code violations might be a trigger to more serious and violent crimes and that evidence exists that far-right extremists commit a large number of preparatory crimes, such as credit card fraud, gun violations, and identity theft (Corcoran 1990; Flynn and Gerhardt [1989]1995; Hamm 2005; Tucker and Pate 2000). The ILP model would then suggest, if such a pattern were uncovered, that Internal Revenue Service and local tax officials (or Environmental Protection Agency and local zoning officers, if land use violations are found to precede more serious offenses on the individual or regional level) should play an integral role on domestic terrorism task forces.

Similar to the multiunit, multiagency structure are the ties to the community. As the Istanbul investigations revealed, members of the community are likely to have information that can be critical to the prevention of terrorist attacks. A key component of the community is the business sector. Business officials, and private-sector security officials in particular, may also be key sources of information about emerging threats. In many respects, at least within the United States, the focus on homeland security represents an unfunded mandate for local law enforcement. The development of an intelligence capacity, whether a full-blown unit or an intelligence liaison officer in small agencies, becomes one additional function for resource-limited agencies (see Riley et al. 2005). This is one reason for an "all crimes" approach to ILP as suggested by the National Criminal Intelligence Sharing Plan (Bureau of Justice Assistance, 2005b). The investment in ILP is likely to create economies of scale when applied to all types of crime and terrorist threats. Additionally, the "all crimes" approach brings the benefit of increasing the likelihood of identifying links between criminal activity and terrorist threats (e.g., drug trafficking as a fund-raising activity).[12]

Beyond an "all crimes" approach, ILP is most likely to be effective when implemented in a way that focuses resources on particular crime types, criminal organizations, and terrorist threats. This is a cornerstone of Goldstein's (1990) problem-solving model, which argues for breaking crimes into specific problem categories that can be analyzed and subjected to tailored interventions. It also appears to be a key to the success of directed patrol operations aimed at a specific crime (e.g., gun crime) in hot spot locations (Cohen and Ludwig 2003; McGarrell et al. 2001; Sherman and Rogan 1995). Similarly, the promise of the deterrence-based approach of Boston Ceasefire, SACSI, and PSN appears to relate to the focused deterrence model that seeks direct communication with those at high risk for being involved in gun violence (Braga et al. 2001; Kennedy et al. 2001; McGarrell et al. 2006; Papachristos, Meares, and Fagan 2005).

Within the intelligence community, different definitions of different forms of intelligence are utilized. For example, within the United Kingdom, agents working in Her Majesty's Customs and Excise (HMCE) distinguish between operational, tactical, and strategic intelligence, whereas the UK National Crime Squad distinguishes between tactical and strategic intelligence (Sheptycki 2004). Carter (2004: 83) and Cope (2004: 188) also make the distinction between tactical or operational and strategic intelligence. In a slight variation, Peterson (Bureau of Justice Assistance, 2005a) refers to crime analysis, investigative (evidential) analysis, and strategic analysis. The key point is the need for both broad levels of analysis of current and emerging threats (strategic intelligence) and more case-specific intelligence on individuals and groups believed to be actively involved in criminal activity (tactical intelligence). Thus, for example, strategic understanding of the role of gangs, drugs, and hot spot patterns in gun violence is critical to developing strategic plans for reducing gun violence. Timely knowledge of specific gangs, drug markets, and hot spots is the type of tactical intelligence produced in Compstat and Chicago's Violence Initiative Strategy Evaluation and Deployment Operations Center meetings that can result in strategic interventions intended to reduce the likelihood of crime. Similarly, TNP's discovery that many in the community had suspicions about a group of suicide bombers, as well as its discovery that a substantial portion of those involved in terrorist networks were not ideologically committed to the terrorist cause, was highly useful strategic intelligence. To use this in a predictive strategy required tactical intelligence on who posed a threat and who within these groups could be pulled away from the network and/or recruited as informants. The 9/11 commission criticized the intelligence community for an overreliance on technology (National Commission on Terrorist Attacks Upon the United States 2004). Clearly, technology is not a sufficient condition for effective ILP. On the other hand, it is likely a necessary condition.[13] As Zimmermann's (2006) research indicated, an adequate technological infrastructure was essential for the integration of analysis and problem solving in PSN.

As noted at the outset, ILP may be at a stage in 2006 similar to COP 15 to 20 years ago.[14] The concept is being endorsed by all the key law enforcement professional organizations and there are reports of promising practices such those reviewed herein, but it remains a fairly nebulous concept and most agencies are just toying with implementation. For the field of policing, an essential step is evaluation of ILP so that its foundation is truly evidence based (Ratcliffe 2002).

For specific agencies, the most important step is likely to be to actually experiment with ILP. Police officials should select the most important crime problem or threat facing the community (e.g., gangs, meth labs, gun crime, drug markets, domestic terrorist groups, international terrorist financing, human trafficking). A multiunit, multiagency task force could be established that meets regularly and is held accountable.

The task force would work through the intelligence process (see Bureau of Justice Assistance 2005a, 2005b; Carter 2004; Center for Problem-Oriented Policing 2006; Ratcliffe 2002) and employ the Compstat philosophy of continuous assessment, timely deployment of resources, and accountability. It is likely through practice, hopefully informed by research partnerships,[15] that ILP will move to the point of evidence-based best practice that improves the effectiveness and efficiency of law enforcement and the safety of communities and nations.

Notes

1. Edmund F. McGarrell, Joshua D. Freilich, and Steven Chermak. "Intelligence-led Policing as a Framework for Responding to Terrorism" from *Journal of Contemporary Criminal Justice*. Copyright © 2007 by Sage Publications. Reprinted with the permission of Sage Publications, via Copyright Clearance Center.
2. Researchers at both sites (see Hale et al. 2004; Ratcliffe 2002) linked the emergence of ILP at Kent and Northumbria to two influential commission reports within the United Kingdom. The first, issued by the Audit Commission and titled "Helping with Enquiries—Tackling Crime Effectively," called for using analysis as a tool for more effective and efficient policing (Audit Commission 1993). The second, issued by Her Majesty's Inspectorate of Constabulary (HMIC 1997), endorsed ILP and called for information collection and intelligence analysis as key components of modern policing. Hale et al. (2004) note the importance of a subsequent HMIC report in 1998 linking ILP, problem solving, and partnerships as keys to crime reduction.
3. Tilley (2003) provides a very thoughtful discussion of the similarities and points of departure between ILP and problem-oriented policing.
4. Although writing for the broad intelligence community, Kamarck (2005) has developed a thoughtful statement on the application of business principles to the collection and management of information.
5. These sections are adapted from McGarrell (2005).
6. Project Exile is also an important foundation of PSN. Established in the Northern District of Virginia to address problems of homicide in Richmond, the program involved a commitment to significantly increased federal prosecution of gun crime offenders coupled with a media campaign to communicate a deterrent message to potential offenders. A recent evaluation concludes that Project Exile significantly reduced homicide (Rosenfeld, Forango, and Baumer 2005; in contrast, see Raphael and Ludwig 2003).
7. The author is thankful to the leadership of the TNP which have provided briefings from the intelligence and antiterrorism divisions. Particular thanks to Dr. Ahmet Yayla and Majors Samih

Teymur and Bilal Sevinc, who are pursuing PhD study at North Texas State University and Michigan State University and who provided the data presented herein.

8. Riley et al. (2005) also note that decentralized multiagency structures can create problems in counterterrorism intelligence sharing because of security clearance issues.
9. Meaningful fashion is obviously a loosely defined term but would suggest changes in vision and values, organizational structure, and street-level operations.
10. Within the United States, given restrictions on local, state, and tribal law enforcement's involvement in national intelligence activity, the involvement of local, state, and tribal law enforcement agencies in national intelligence is typically through involvement in joint terrorism task forces operated by the FBI (Riley et al. 2005). This inherently involves multiagency collaboration and information sharing.
11. Multiagency collaboration is also a key component of the NIM in the United Kingdom (Sheptycki 2004).
12. Riley et al. (2005) report that most of the US law enforcement agencies they studied included both criminal intelligence and counterterrorism intelligence within the same unit.
13. Hale et al.'s (2004: 297) review of annual audits of the police in England and Wales provides examples of symbolic implementation of ILP, where senior managers mention ILP but where the information technology was not adequate to support ILP.
14. In the same review of audits (see note 12), Hale et al. (2004) report on findings of organizational resistance and the challenge of organizational change in moving to a community-based and problem-oriented style of policing, similar to that found within the United States (Greene and Mastrofski 1988).
15. Many of the problem-solving best practices reviewed in this article were supported through action research partnerships between university researchers and law enforcement agencies working in a collaborative fashion.

References

Anderson, R. 1997. "Intelligence-led policing: A British perspective." Pp. 5–8 in *Intelligence Led Policing: International Perspectives on Policing in the 21st Century*, edited by A. Smith. Lawrenceville, NJ: International Association of Law Enforcement Intelligence Analysts.

Audit Commission. 1993. *Helping with Enquiries: Tackling Crime Effectively.* London: HMSO.

Birkland, T. A. 2004. "'The World Changed Today': Agenda-setting and Policy Change in the Wake of the September 11 Terrorist Attacks." *Review of Policy Research* 21:179–200.

Braga, A., P. J. Cook, D. M. Kennedy, and M. H. Moore. 2002. "The Illegal Supply of Firearms." Pp. 319–352 in *Crime and Justice: An Annual Review*, edited by M. Tonry. Chicago: University of Chicago Press.

Braga, A., D. M. Kennedy, E. J. Waring, and A. M. Piehl. 2001. "Problem Oriented Policing, Deterrence, and Youth Violence: An Evaluation of Boston's Operation Ceasefire." *Journal of Research in Crime and Delinquency* 38(3):195–225.

Bureau of Justice Assistance. 2005a. Intelligence-led Policing: The New Intelligence Architecture. Washington, DC: US Department of Justice, Bureau of Justice Assistance.

——. 2005b. *National Criminal Intelligence Sharing Plan.* Washington, DC: US Department of Justice, Bureau of Justice Assistance.

Carlson, J. R. 1995. "The Future Terrorists in America." *American Journal of Police* 14(3):71–91.

Carter, D. L. 2004. *Law Enforcement Intelligence: A Guide for State, Local, and Tribal Law Enforcement Agencies.* Washington, DC: US Department of Justice, Office of Community Oriented Policing Services.

Center for Problem-Oriented Policing. 2006. *Problem-oriented Policing Guides.* Washington, DC: US Department of Justice. Available from http://www.popcenter.org.

Clarke, R. V. G. 1997. *Situational Crime Prevention: Successful Case Studies.* New York: Harrow and Heston.

Cohen, J. and J. Ludwig. 2003. "Policing Gun Crimes." Pp. 217–239 in *Evaluating Gun Policy: Effects on Crime and Violence*, edited by J. Ludwig and P. Cook. Washington, DC: Brookings Institution.

Coleman, V., W. C. Holton Jr., K. Olson, S. C. Robinson, and J. Stewart. 1999, October. "Using Knowledge and Teamwork to Reduce Crime." *National Institute of Justice Journal* 16–23.

Cope, N. 2004. "Intelligence Led Policing or Policing Led Intelligence?" *British Journal of Criminology* 44:188–203.

Corcoran, J. 1990. *Bitter Harvest, Gordon Kahl, and the Posse Comitatus: Murder in the Heartland.* New York: Penguin.

Dalton, E. 2003. *Lessons in Preventing Homicide.* East Lansing: Michigan State University. Retrieved, January 9, 2005, from http://www.cj.msu.edu/~outreach/psn/erins_report_jan_2004.pdf.

Damphousse, K. R. and B. Smith. 2004. "Terrorism and Empirical Testing: Using Indictment Data to Assess Changes in Terrorist Conduct." *Sociology of Crime, Law, and Deviance* 5:75–90.

FBI. 1999. Project Megiddo. Available from http://www.cesnur.org/testi/FBI_004.htm.

Felson, M. 1986. "Routine Activities, Social Controls, Rational Decisions, and Criminal Outcomes." Pp. 119–128 in *The Reasoning Criminal*, edited by D. Cornish and R. Clarke. New York: Springer-Verlag.

——. 1987. "Routine Activities and Crime Prevention in the Developing Metropolis." *Criminology* 25:911–931.

Flynn, K. and G. Gerhardt. (1995). *The Silent Brotherhood: The Chilling Inside Story of America's Violent Anti-government Militia Movement.* New York: Signet/Penguin. (Original work published 1989.)

Freilich, J., S. Chermak, and A. Gamarra. 2006. *Empirical Research on Far Right Extremism.* Baltimore, MD: Academy of Criminal Justice Sciences.

Fusion Center Guidelines. 2005. *Fusion Center Guidelines: Developing and Sharing Information in a New Era.* Washington, DC: Bureau of Justice Assistance.

Goldstein, H. 1990. *Problem Oriented Policing.* Philadelphia: Temple University Press.

Greene, J. and S. Mastrofski. 1988. *Community Policing: Rhetoric or Reality?* New York: Praeger.

Gurr, N. and B. Cole. 2002. *The New Face of Terrorism: Threats from Weapons of Mass Destruction.* New York: I. B. Tauris.

Hale, C., R. Heaton, and S. Uglow. 2004. "Uniform Styles? Aspects of Police Centralization in England and Wales." *Policing and Society* 14(4):291–312.

Hamm, M. 2005. "Crimes Committed by Terrorist Groups: Theory, Research, and Prevention." Final report to the National Institute of Justice, US Department of Justice, Washington, DC.

HMIC. 1997. *Policing with Intelligence.* London: HMIC.

International Risk Governance Council. 2005. *Risk Governance: Towards an Integrative Approach.* Geneva: International Risk Governance Council.

Kamarck, E. C. 2005. *Transforming the Intelligence Community: Improving the Collection and Management of Information* (Transformation of Organizations series). Washington, DC: IBM Center for the Business of Government.

Kennedy, D. M., A. A. Braga, and A. M. Piehl. 2001. "Developing and Implementing Operation Ceasefire." Pp. 5–54 in *Reducing Gun Violence: The Boston Gun Project's Operation Ceasefire.* Washington, DC: National Institute of Justice,.

Kennedy, D. M., A. M. Piehl, and A. A. Braga. 1996. "Youth Violence in Boston: Gun Markets, Serious Youth Offenders, and a Use-reduction Strategy." *Law and Contemporary Problems,* 59:147–196.

LaFree, G., L. Dugan, H. Fogg, and J. Scott. 2006. "Building a global Terrorism Database." Final report to the National Institute of Justice, US Department of Justice, Washington, DC.

Lum, C., L. W. Kennedy, and A. J. Sherley. 2006. "Are Counter-terrorism Strategies Effective? The Results of the Campbell Systematic Review on Counter-terrorism Evaluation Research." *Journal of Experimental Criminology* 2(4):489–516.

McGarrell, E. F. 2005. *Strategic Problem Solving and Project Safe Neighborhoods.* East Lansing: Michigan State University. Available from http://www.cj.msu.edu/%7Eoutreach/psn/Strategic_Problem_Solving_PSN_wp1%5B1%5D.pdf.

McGarrell, E. F., S. Chermak, A. Weiss, and J. Wilson. 2001. "Reducing Firearms Violence through Directed Police Patrol." *Criminology and Public Policy* 1(1):119–148.

McGarrell, E. F., S. Chermak, J. Wilson, and N. Corsaro. 2006. "Reducing Homicide through a 'Lever-pulling' Strategy." *Justice Quarterly* 23(2):214–231.

Marsh, M. H. and T. E. Noonan. 2005. *Risk Management Approaches to Protection: Final Report and Recommendations by the Council.* Washington, DC: National Infrastructure Advisory Council.

Merari, A. 1991. "Academic Research and Government Policy on Terrorism." *Terrorism and Political Violence* 3(1):88–102.

Morash, M. and J. K. Ford, eds. 2002. *The Transformation to Community Policing.* Thousand Oaks, CA: Sage.

National Advisory Commission on Criminal Justice Standards and Goals. 1973. *Police.* Washington, DC: US Government Printing Office.

——. 1976. *Report of the Task Force on Organized Crime.* Washington, DC: US Government Printing Office.

National Commission on Terrorist Attacks Upon the United States. 2004. *The 9/11 Commission Report.* Available from http://www.9-11commission.gov/.

Papachristos, A., T. Meares, and J. Fagan. 2005. "Attention Felons: Evaluating Project Safe Neighborhoods in Chicago." Available from Social Science Research Network, http://ssrn.com/abstract=860685.

Raphael, S. and J. Ludwig. 2003. "Prison Sentence Enhancements: The Case of Project Exile." Pp. 251–286 in *Evaluating Gun Policy: Effects on Crime and Violence*, edited by J. Ludwig and P. Cook. Washington, DC: Brookings Institution.

Ratcliffe, J. H. 2002. "Intelligence-led Policing and the Problems of Turning Rhetoric into Practice." *Policing and Society* 12(1):53–66.

Riley, K. J., G. F. Treverton, J. M. Wilson, and L. M. Davis. 2005. *State and Local Intelligence in the War on Terrorism.* Santa Monica, CA: Rand.

Roehl, J., D. Rosenbaum, S. Costello, J. Coldren, A. Schuck, L. Kunard et al. 2004. *Strategic Approaches to Community Safety Initiative (SACSI) in 10 U.S. Cities: The Building Blocks for Project Safe Neighborhoods.* Chicago: University of Illinois at Chicago.

Rosenfeld, R., R. Forango, and E. Baumer. 2005. "Did Ceasefire, Compstat, and Exile Reduce Homicide?" *Criminology and Public Policy* 4(3):419–450.

Sabatier, P. and H. C. Jenkins-Smith. 1993. *Policy Change and Learning: An Advocacy Coalition Approach.* Boulder, CO: Westview.

Sheptycki, J. 2004. "Organizational Pathologies in Police Intelligence Systems." *European Journal of Criminology* 1(3):307–332.

Sherman, L. W. and D. P. Rogan. 1995. "The Effects of Gun Seizures on Gun Violence: 'Hot spots' Patrol in Kansas City." *Justice Quarterly* 12:673–693.

Shively, M. 2005. "Study of Literature and Legislation on Hate Crime in America." Final report to the National Institute of Justice, US Department of Justice, Washington, DC.

Silke, A. 2001. "The Devil You Know: Continuing Problems with Research on Terrorism." *Terrorism and Political Violence* 13(4):1–14.

Silverman, E. 1999. *NYPD Battles Crime: Innovative Strategies in Policing.* Boston: Northeastern University Press.

Slovic, P. 2003. "Going Beyond the Red Book: The Sociopolitics of Risk." *Human and Ecological Risk Assessment* 9:1181–1190.

Smith, B. L. 1994. *Terrorism in America: Pipe Bombs and Pipe Dreams.* New York: State University of New York Press.

Smith, B. L. and K. R. Damphousse. 1996. "Punishing Political Offenders: The Effect of Political Motive on Federal Sentencing Decisions." *Criminology* 34(3):289–321.

Smith, B. L. and K. R. Damphousse. 1998. "Terrorism, Politics, and Punishment: A Test of Structural-contextual Theory and the Liberation Hypothesis." *Criminology* 36:67–92.

Smith, B. L., K. R. Damphousse, F. Jackson, and A. Sellers. 2002. "The Prosecution and Punishment of International Terrorists in Federal Courts: 1980–1998." *Criminology and Public Policy* 1:311–338.

Smith, B. L., K. R. Damphousse, and P. Roberts. 2006. *Pre-incident Indicators of Terrorist Incidents: The Identification of Behavioral, Geographic, and Temporal Patterns of Preparatory Conduct.* Washington, DC: US Department of Justice, National Institute of Justice.

Smith, B. L. and K. D. Morgan. 1994. "Terrorists Right and Left: Empirical Issues in Profiling American Terrorists." *Studies in Conflict and Terrorism* 17:39–57.

Smith, B. L. and G. Orvis. 1993. "America's Response to Terrorism: An Empirical Analysis of Federal Intervention Strategies during the 1980s." *Justice Quarterly* 10:661–681.

Stern, P. and H. Fineberg. 1996. *Understanding Risk: Informing Decisions in a Democratic Society.* Washington, DC: The National Academies.

Tew, J. and D. Elkins. 2006. "Managing Enterprise Risks in Global Automotive Manufacturing Operations." Presentation to Government–University–Industry Research Roundtable Meeting. Washington, DC: National Academy Press.

Tilley, N. 2003. *Problem-oriented Policing, Intelligence-led Policing, and the National Intelligence Model* (Crime Science Short Report series). London: Jill Dando Institute of Crime Science.

Tucker, J. B., ed. 2000. *Toxic Terror.* Cambridge, MA: MIT Press.

Tucker, J. B. and J. Pate. 2000. "The Minnesota Patriots Council." Pp. 159–183 in *Toxic Terror,* edited by J. Tucker. Cambridge, MA: MIT Press.

US DOJ. 2004. *Project Safe Neighborhoods: America's Network against Gun Violence.* Washington, DC: US Department of Justice, Bureau of Justice Assistance.

Weisburd, D., S.D. Mastrofski, A. M. McNally, R. Greenspan, and J. Willis. 2003. "Reforming to Preserve: Compstat and Strategic Problem Solving in American Policing." *Criminology and Public Policy* 2(3):421–456.

Zimmermann, C. 2006. "Federal Incentives to Address Gun Violence: A Model of Success and Failure." Unpublished doctoral dissertation, Michigan State University, East Lansing.

Questions

1. How does ILP improve on conventional, reactive policing?
2. How do the challenges posed by terrorism create new efforts by police to deal with these threats in ways that are different from the control of crime?

9.
POLICE ORGANIZATIONS AND THE ADOPTION OF RISK MANAGEMENT STRATEGIES

Experience in the private sector indicates that organizations with effective risk management approaches demonstrate strong leadership and integrate risk management throughout the culture and structure of the organization (Marsh and Noonan 2005). Within agencies such as the Department of Homeland Security and law enforcement, the experience with innovations such as community policing and problem-oriented policing suggests a tendency for organizational resistance and the isolation of risk management responsibilities into "innovation ghettos" (Toch and Grant 1991). Yet, the isolation of risk management to particular units or particular levels of government will inhibit the information gathering, intelligence analysis and sharing, and preparedness processes necessary to address the threats posed by terrorism. Additional organizational challenges, identified by the National Research Council's Committee on Risk Characterization, emerge from the tendencies for organizations to let down their guard against low probability risks and to operate sub-optimally in crisis situations (Stern and Fineberg 1996).

The Hancerli, Kennedy, and Kirkpinar chapter examines how risk is used in the organizational decision making of an integrated police agency, the Turkish National Police (TNP). In investigating the ways in which different offices within the TNP conceptualize and act on risk, the authors speculate that a more rigorous approach to the planning of risk management strategies would improve current practice while still conforming to what is currently being done procedurally. An important aspect of this approach relates to the attempt to get the police to move to a more proactive stance rather than simply reacting to circumstances that arise.

A MODEL OF RISK ASSESSMENT, MANAGEMENT, AND DECISION MAKING STRATEGIES USED BY THE TURKISH NATIONAL POLICE

Suleyman Hancerli, Leslie W. Kennedy, and A. Rahmi Kirkpinar

Introduction: The Origins of Risk Assessment in Law Enforcement

Increasingly, law enforcement agencies are looking to develop ways in which they can improve on decision making both from the standpoint of the leadership of these organization and from the perspective of line officers. This change is being encouraged by the emphasis that is being placed on intelligence-led policing, supporting the integration of wide ranging information sources into the strategic planning of the organization that can then be used for tactical purposes. There is, in addition, a realization that information or "intelligence" that is collected but not properly managed is not of much use to decision making. This is particularly important in the deliberations around resource management and prioritization of programs. In looking at ways to assist in their decisions, security agencies, including the police, have begun to discuss how they might incorporate risk assessment and risk management into their operations (Van Brunschot and Kennedy 2008). While this approach has been widely used throughout the private sector, labeled enterprise risk management, in managing corporate assets and insuring security for employees, it has been slow to catch on in the public sector. While it is not uncommon for law enforcement or security agencies to talk about doing threat analysis, the broader issues around risk assessment appear to be underdeveloped. It appears, as well, that while this approach has become much more the operational modus operandi of national security agencies (see the Department of Homeland Security and, in particular, the reports from the US Coast Guard's embracing of risk assessment), it is less likely to be identified as an integral part of the management strategies used in local and regional law enforcement.

In this chapter, we intend to review the different aspects of risk assessment, drawing on the literature that has examined the component parts of this approach and offer an integrated model that would help us tie this literature together. We will then describe a research project that incorporated interviews with key informants from the Turkish National Police (TNP). In these interviews we sought to establish the extent to which risk assessment is already practiced in this agency, across different functional departments: we examine where it is being used (if at all), how it is applied, and the means by which the leadership in this organization has adapted to this form of assessment in terms of organizational change.

We learned in our preliminary conversations with police leadership in charge of operational units of TNP prior to conducting the research that they believe that they are currently involved in risk assessment in their daily activities. What is missing, they say, are the labels that give these practices legitimacy within the organization. Given this belief, this makes the research that follows that much more compelling as we seek to establish how the current practice conforms to or differs from what the research literature suggests form the central aspects of risk assessment and risk management.

Intelligence-led Policing and Risk Assessment

The most important reason given for the development of intelligence-led policing and the introduction of risk analysis as a part of this is to move police away from reactive to proactive policing strategies. Ratcliffe explains the integration of the intelligence model in policing as "the application of criminal intelligence analysis as an objective decision making tool in order to facilitate crime reduction and prevention through effective policing strategies and external partnership projects drawn from an evidential base" (2003: 3). Ratcliffe emphasizes that three components contribute to this form of policing: the interpretation of intelligence; the influence that this has on decision makers; and the ultimate impact that this has on the criminal environment. These all contribute to crime prevention. Intelligence is a constant process of data collection, analysis, distribution, and assessment. The collection of intelligence can be done by specialized groups, crime analysts, or the police members themselves.

This shift in focus from reactive to proactive policing is given as a major reason for the widespread support offered to the implementation in the UK of an intelligence-led policing program that has as its core a National Intelligence Model. This model purports to standardize police reporting, analysis, and intelligence-based actions in ways that have direct impacts on day-to-day police operations. As John and McGuire (2004) state, police leadership has begun to use this approach believing that they will attain more effective outcomes particularly if they are convinced of its ability to address risk. In fact, as John and Maguire observe, the National Intelligence Model is designed to allow police leadership to move beyond a focus on particular crimes to a more strategic, risk-based approach in identifying areas of concern and developing resource allocation strategies. The intelligence that is provided by analysts, police officers on the street, and targeted surveillance combines to form a risk assessment that sets future police action. This assessment can include much more than reports on crime but also the scrutiny and assessment of business operations, review of demographic changes, identification of crime networks, and so on (John and Maguire 2004).

Maguire (2000: 326) points out that this shift from passive surveillance to active risk management has its dangers, as the concerns about privacy and control of invasive practices are harder to manage with such a wide scope of police action. This danger is further enhanced as intelligence, unlike general surveillance, is specifically targeted toward particular individuals and groups. It is an aggressive form of information collection and decision making process, requiring huge police resources. These problems have created some implementation difficulties of this national model and there is some discussion that its application has not been widely successful. Nonetheless, the advocacy in support of the greater scope of control that is promised by this approach and its promise of greater accountability has provided strong impetus for its continued implementation across the police agencies in the UK. It is important for the prioritization of policing programs and the problems that they set out to address (McGuire and John 2006)

McGarrell, Freilich, and Chermak (2007) (see Chapter 8) have reviewed the importance of intelligence-led policing strategies in dealing with terrorism threats. They reiterate the importance of tying intelligence-led policing into an agency-wide structure where there is a commitment on the part of the leadership to use this information to set priorities within the organization and to manage this in light of community concerns. This agency-wide perspective is an important one that we will address in this chapter.

A Scheme for Developing a Risk Model

This study is informed by a series of research questions that deal with the elements of Assessment, Connections, Tasks, Information, Organization, and Notification (ACTION), where the acronym helps make sense of each of the steps involved in risk process (Kennedy and Van Brunschot 2009). In the assessment stage, it is important for us to understand how we connect exposure and vulnerability to actual crime outcomes. This is where the risk analyst would be asked to make connections between the characteristics of the problem, the likely motivation of the actors, the likelihood of success, and the deterrent effect of police action. Theories of crime causation are helpful here and provide guidelines for how these elements will come together.

In the risk process, the tasks that each individual in the organization undertakes in addressing risk become important in understanding how the organization will respond to these threats. As has been clearly demonstrated in the application of the Compstat model in American policing, the presentation of the data analysis from different parts of the police jurisdiction are not simply provided as a way of understanding crime patterns but they are an important aspect of accountability. A key product of these sessions is the assignment of tasks to address the problems that are identified in the data. In a review cycle that becomes part of the operation of the organization, this information analysis, tasking, accountability, and assessment form an essential tool in managing the resources and targeting problem areas.

Following from the analysis of risk, it is assumed that there would be plans made for intervention, prevention, and mitigation but this cannot be done without an evaluation strategy that provides information, not on the risk this time, but on the effect that the programs have on their intended targets. This information can feed back through the intelligence cycle. It can also be used to help redesign aspects of the organization in creating more effective delivery and programmatic outcomes.

Finally, the risk process relies on notifying others about the success of the programs that it promotes. It is also important that the organization is aware of and communicates to others about future threats and the limitations that may have emerged in the application of prevention strategies. This is a key element in encouraging the development of innovation within the police agency, using risk assessment as a vehicle for promoting proactive programming and the integration of information and intelligence into police operations.

Methodology

With the impetus coming from intelligence-led policing and the proactive aspects of risk assessment, incorporating the identification of hazards, vulnerabilities, and exposure in the context of policing intervention, the scene is set for police leadership to make use of these tools in their strategic and tactical decision making strategies. The question is to what degree and to what extent the elements of the risk assessment with the ACTION model are actually practiced. In other words, we examined how the respondents use risk assessment and the ways in which the items outlined in the ACTION model connect to their daily policing practices. By doing this, we intended to review different aspects of the risk assessment and management approach affecting and constructing the decision making strategies in three departments in the TNP.

Organizational Structure of the TNP

The TNP was founded on June 24, 1920. On July 4, 1934, Act on the Duties and Competencies of the Police was adopted, specifying basic authorities and duties of the TNP. This act has been amended several times, but it is still affecting the way police fulfill its tasks today. On June 4, 1937 the TNP Act was introduced. With this act, the TNP gained a centralized character and has kept it until today. This new organization act shaped national police in many aspects, ranging from ranks to promotion and from annual leaves to salary as well as set out the principles for its functioning.

In 1937, the Police Academy, and in 1938, the Police College were founded as the initial training institutions of the TNP. Over the course of the time, the TNP has developed in line with new technologies and concept of policing. Today, the TNP is the national agency in Turkey with more than 200,000 staff. It is headquartered in Ankara. At the top of the organizational pyramid, there is the Minister of the Internal Affairs with the greatest authority. The General Director of the TNP works under the Ministry of Internal Affairs. The General Director of the TNP is the ultimate authority over the police in Turkey. The TNP consists of the Central Organization and Provincial Police Departments in 81 different provinces throughout the Turkey. All annual budgets of central organization and provincial police departments are given by central government unlike in US police department.

Research Design

In May 2008, two of the authors conducted a training session on risk analysis and management at the Turkish International Academy Against Drugs and Organized Crime (TADOC) training center of the TNP in order to have some insight into the risk analysis and management approaches of the police officers working in anti-smuggling and organized crime units in eight different provinces in Turkey. We engaged the participants in this program that laid out a framework for risk assessments and then walked them through the process: probing with them their major challenges in their operations; discussing what they see as major threats to social order in their jurisdictions; reviewing how they have managed these threats to date (both successfully and unsuccessfully); reviewing what information and intelligence they use and how they use it; and examining how they make allocation decisions (deploying officers, buying more equipments, teaming with other agencies, or adding technological solutions) based on this information.

Then, we examined where it was being used (if at all), how it was applied, and the means by which the leadership in this organization has adapted to this form of assessment in terms of organizational change. After that, we talked about the changes in governance structures that would accommodate better risk assessment, along the lines of intelligence-based policing, and more informed resource allocation decisions. Using this approach, we engaged everyone in an ongoing problem solving exercise to highlight problems and suggest resolutions.

We found that the group was able to understand each component of the risk assessment process but admitted that they rarely were involved in this type of detailed exercise in their work. They also admitted to having a limited understanding of the steps that they needed to follow in doing this type of analysis. This group was drawn from the leadership within the organized crime and drug abatement department of the TNP. The questions that these findings raised for us concerned the awareness within the whole organization of the applications of risk assessment; the actual extent to which this might be currently practiced both within the organized crime unit but also in other parts of the agency; and the readiness of the organization to adapt these strategies in programmatic terms that would lead to changes in the ways in which tasks are assigned, intelligence is used, and accountability is determined.

To address these questions, we set out to collect data from the key decision makers in the three major departments of the TNP, including but not limited to (1) the Department of Anti Smuggling and Organized Crime; (2) the Department of Anti Terrorism; and (3) the Department of Public Order, with the intention of establishing the extent to which risk analysis and management is performed in each and the similarities and differences that appear in this implementation.

We conducted this research project by using our focus interviews with the key informants from these certain departments of the TNP. In the interview process, the 12 Turkish police officers participated in the face-to-face and/or phone interviews conducted by the two of the authors of this study under the auspices of the TADOC. The interview questionnaire consisted of 25 open- and close-ended questions related to risk analysis, management, and decision making strategies used by the Turkish police officers. The interview questions were formulated through getting some insights into conducting the training session on the risk analysis, management, and decision making strategies for the police officers at TADOC in May 2008.

Research Questions

There are six research questions in this study and each of the research questions addresses one letter from the word of ACTION. In this context, each of the research questions has been answered by evaluating the relevant part of the interview statements. For the purpose of measuring the six research interests in the study, we prepared the 25 interview questions, which have been classified into the six categories: (1) Assessment; (2) Connection; (3) Tasks; (4) Information; (5) Organization; and (6) Notification (ACTION). This acronym helped us make sense of each of the six steps that the TNP officers may follow in the risk assessment process. Since we have evaluated the answers of the 12 interviewees by classifying them into the six categories in accordance with the elements of the ACTION model, we have also answered the research questions as well. The research questions are:

1. To what extent the officers apply risk analysis approach to their everyday activities and/or operations? (*Assessment* of Risk)
2. What analytical strategies are available to interpret data and construct meaningful products for use in risk assessment? (Making *Connections*)
3. What frameworks exist to guide the framing of strategic questions and to inform decision making? (Setting *Tasks*)
4. How do analysts and the officers collect information for threat analysis and distribute it to decision makers? (Collecting *Information*)
5. In the aim of refining the organization, what do the officers do to incorporate the results from the risk analysis into their operational programs? (Refining the *Organization*)
6. What notification programs are used to disseminate successes and failures of implementation strategies? (*Notifying* Others)

Findings and Analysis

During the interviews, we specifically tried to examine how the officers use risk assessment and the ways in which the items outlined in the ACTION model connect to their daily practices in their works. We are interested in the extent to which the elements of the ACTION model are found in those three departments and the degree to which they use this method to guide their policing activities and operations as well. To this end, we examined below each element of the ACTION risk model by looking at the interview statements.

Assessing *Uncertainty, Vulnerability, Exposure, and Threat*

The interviewees clearly claimed that there are some standard ways and main roads in any case to go after some specific risk assessment strategies. To do this, the officers use some methods that are agreed on. And all the methods can only be discussed and granted as well through the collective thinking at meetings.

With this in mind, some of the interviewees from the two departments, (1) the Anti-Smuggling and Organized Crime Department and (2) Public Order Department, stated that their departments organize yearly evaluation meeting with participation of the 81 provinces' anti-smuggling and organized crime/public order unit chiefs in order to evaluate all threats/hazards in the field, to focus on the factorial reasons of successes and failures, the solution proposals, and to determine action plans for the further years. For example, one participant from the Department of Public Order responded:

> We believe that coordination plays an important role to assess risks and to increase the awareness towards risks. Our department provides operational coordination between the provincial police departments. We have periodic meetings with those departments. During the meetings we gather information about criminal groups' structures, and try to track down their movement from one place to another. In light of the collected information we try to evaluate which city will have higher risk in terms of illegal activities of identified criminal groups. We share this information with related province departments to inform them to take necessary precautions. We do similar cooperation international level as well.

Some interviewees from the two departments, (1) the Anti-Smuggling and Organized Crime Department and (2) Anti-Terrorism Department, also stated that they make their evaluations in their monthly and/or weekly meetings of their departments by looking at events occurred in the past and/or events that might be occurring in the future. In such meetings, they usually talk about their strategies, action (future) plans, what they have done before, what they are supposed to do for the future, what was wrong and what was right for the past incidents. It is obvious that the officers believe in that such meetings help the officers standardize their day-to-day policing activities that are impacted by some components, such as risk-based intelligence-led policing approach, data analysis with the crime pattern, effective decision making process, and legitimacy of the organizational change.

Some participants also stated that while making evaluations with uncertainty, vulnerability, exposure, and threat and conducting the risk assessment, the officers usually evaluate several factors, such as crime statistics, trends on crime, existence hot spots, using their past policing experiences, media and public voices, gathering information about criminal groups and their activities, sharing intelligence with the other agencies to take necessary cautions, and making coordination between the agencies from different provinces.

Making Connections

Using the assessment tools, we are able to search for connections between the risks and the ways in which the risks are addressed. This also helps us search a way of identifying the types of data that are needed and the underlying theoretical causal explanations related to crime emergence. In this manner, we here searched for the collateral aspects of the risk problem and the ways in which the risks are addressed. While evaluating the interview statements, we have realized that the TNP has a very distinctive tool, which is named POLNET (Intranet Police Network in Turkey), to make communication effectively workable between the central departments at the headquarters and the police departments in the 81 provinces. Almost all interviewees mentioned how they have been using the POLNET for activities from communicating risk messages to distributing circular order in real-time base. The unique feature of the POLNET provides policy makers of the TNP with real-time data nationwide. Each unit uses these data to evaluate increased risks or crime in any given part of the country then prepare responses accordingly. Turkish police officers use the POLNET not only as a data collection tool but also as an effective and uninterrupted communication tool.

The participants have also revealed that while conducting risk analyses and making connections between risks and their responses to those risks, besides POLNET, they have been utilizing some other skills, tools, and technologies, such as computer programs, software systems, internet, public open sources, digital databases of crime/criminals, every kind of communicative device, online and/or hard copy publications of criminal organizations, interviews, interrogations, experiences gained through either field activities or in-service trainings, experts from either inside agencies or outside agencies, and statistics reading skills of the experts.

Setting Tasks *to Respond to and Prevent Risks*

It is here that we search for the risk problems and different forms of responds to these risks. Drawing from the assessment of risks, we focused on the elements of prevention, preparedness, and response strategies of the three departments. Interestingly, more or less all of the participants from the three departments stated that, generally speaking, both action plans in aiming to prevent crime and action plans in aiming to respond crime are included in their risk assessment process.

The participants from the Anti-Terrorism Department revealed that both strategies, preventive and responsive, eliminate the actual risks, but the officers are careful that eradicating the encountered risks may sometimes cause some other risks. So, encountered risk levels should be evaluated by the officers carefully and then either preventive or responsive strategies must be used to deal with the evaluated risks. Concerning this, one participant from this department stated:

> Our risk assessment is the action plan. For instance, we have worked on university students. We have seen that some terrorist groups recruit their members mostly from among the university students between the ages of 18–24. After that, we have identified the university students as high risk group to be targeted by this terrorist group to recruit. Therefore, we developed our action plan to follow for the other activities.

The participants stated that each unit in this department uses its own risk assessment process while either responding to crimes or preventing crimes. In fact, each unit can make the better evaluation with the vulnerabilities in the area in which they work. For instance, one participant stated:

> The division of human resources does its own risk assessment while making the personnel planning; the logistic division does its own risk assessment while preparing the budget; the training division does its own risk assessment while planning training programs; and the operational units do their own risk assessments while planning their operations as well.

So, each unit knows what the actual risks and possible threats for itself better than the others. This does not mean that they do not get any help from inside and/or outside experts on crime analyses. They definitely do get the help from others.

We also discovered that the officers from the Public Order Department use their risk assessment strategies, preventive or responsive, depending on the encountered risk levels. But they mostly preferred to name their responses as preventive strategies by disregarding responsive strategies. One respondent revealed that

> responsibilities for risk assessment are definitely shared. Each division does their own risk assessment based upon the field of the activities. ... If these risk assessment requires different divisions' opinions, then the Research and Development division coordinates efforts and collects information from the other divisions.

Collecting **Information** *about Events*

We note that the officers from the three departments are already aware of the importance of information and intelligence sharing in order to do accurate risk assessments in policing activities. The participants see the information and intelligence sharing as an essential part of the department's overall successes. One of the participants from the Anti-Terrorism Department stated that

> every piece of information is a part of the puzzle in a game. One might be holding a part of the puzzle that might be the missing part for another one. So, if one holds a piece of information that is the only part of the fact and never shares with the others, that person might be missing the rest of the fact anyway. And also that piece of the information might be useless since it is the only part of the whole picture. It is the way that sharing information and knowledge with the other agencies has a key role.

As stated above, the participants stated that they share and collaborate in risk assessment information and knowledge, but there are some criteria in sharing intelligence and information with the others. They decide to share intelligence and information with the agencies depending on the area of activity. The participants also revealed that unfortunately they do not have any standard written manual that can be used in conducting the risk assessment procedure, but they have some specific research studies, guidelines, and manuals in order to standardize policing activities. Relevantly, one of the participants stated:

> We have some inter-departmental guidelines to standardize our services among branches. These guidelines describe the responsibilities and duties of each anti-terrorism division and office. We have also manuals for the policing operations. These manuals also describe processes step by step during an operation. The department follows the criteria stated in those manuals to allocate its monetary resource and personnel resource appropriately.

Similarly, the four participants from the Public Order Department stated that information sharing and collaboration depends on the risk assessment subjects. They, in general, share their risk assessment information and intelligence within two different categories. For the operational purposes, they share risk assessment information with the Organized Crime, Intelligence, Foreigners, Border and Immigration, and Anti-Terrorism Departments. For the budgeted and personnel allocations, they share their risk assessment information with Finance, Supply and Maintenance, and Personnel Departments as well. Having said that, one of the participants revealed that

> intelligence and information sharing is playing an essential role to overcome risks and threats. However, sometimes agencies might be skeptical to share information and intelligence. We used to be more skeptical to share information and intelligence with the other governmental agencies even within the TNP departments. But, fortunately, nowadays we have become much more open organization in terms of intelligence sharing that makes us more effective and accountable.

Refining the Organization

It is here that drawing from the assessment of risks, we focus on the elements of "refining the organization," which are developing the monitoring capacity within the organization, setting out guidelines for reporting relations, strategic and tactical decision making, developing accountability schemes in organizational restructuring, and so on.

According to the participants from the Anti-Smuggling and Organized Crime Department, the activities of the officers are supervised and inspected by their superiors and authorities regarding their department's risk assessment approach. Three of the participants stated that since they do not have any special status, they are subject to all administrative and judicial inspections like the other parts of the TNP Directorate. So, their risk modeling is operating in accordance with the rules and guidance of the institutional standards as it is supposed to be in Turkey.

The management in this department supports the officers using their own methods in refining their organization. They believe that although they are not authorized to make the changes directly, they are definitely supported in planning and managing strategies for their prioritized resource allocations all the time. One of the participants stated that "after setting up the priorities and responsibilities, sometimes the situations and trends may change, so strategic action plans are made accordingly, and let the authorities take an alternative action in the case of a change." The strategic action plan is prepared by the higher level commission and this enables the officers to realize their responsibilities, duties, and priorities. The participants also stated that they sometimes work in close cooperation with the experts from civilian institutions and never feel that they have any problem with this. Furthermore, their annual report on the drug situation of Turkey is published for the general public as well.

As for the participants from the Anti-Terrorism Department, they revealed that there is not any determined risk model that they use constantly. They try to work in accordance with the existing level of risks and threats. Relevantly, one of the participants stated that

> the officers get together for each critical event to evaluate them properly (either before the event occurs or after it occurs). The officers take into consider the risk level for each event all the time to be able to prevent the cases. Let's assume that we have heard that one terrorist organization will attack to a place guarded by the police by using a flying plane model. The officers get together with the experts on flying plane models and airports in order to evaluate the situation.

In addition, one participant disclosed that

> the Anti-Terrorism Department does not use outside expertise during risk assessment process very often. They only use outside expertise if there is vital or unavoidable situations. They use outside expertise to produce action plans mostly after they have addressed the risk.

Like the other two departments stated above, we found that all of the participants from the Public Order Department stated that they do not have any particular authorities that oversee their risk assessment process in their policing activities, but there is a hierarchical structure; thus, there is always higher authority to oversee the operations since any failure affects the success of the department. For instance, when the officers identify an event with a high risk, it increases the attention of the high-level executive officers working in the department anyway.

In addition, one of the participants also stated that there are two important audits, (1) Police Internal Investigators and (2) the Ministry of Interiors' Internal Investigators' Review Boards in which the departments in the TNP generally go through two separate audits. Another participant also stated that they have a "Board of Inspectors Presidency" under the direct command of the Director General of the TNP. The supervisors from this board review the projects, employees, and training programs in order to evaluate the effectiveness and appropriateness. They fairly examine whether the projects, programs, and trainings have produced the expected outcomes or not. The same participant continuously stated that they allocate their resources based on the risk levels in the provinces all over the country. So, the risk assessment approach helps them shape their resource allocations. Last but not least, while having risk assessments they take advantage of some experts from the other agencies and institutions.

Notifying *Others*

Drawing from the assessment of risks, it is here that we focus on the elements of notifying others in creating risk awareness, risk communication, and developing current organizational status. For the purpose of promoting risk awareness in the Anti-Terrorism Department, the participants stated that they communicate risk messages to their employees through several ways, which include some meetings, workshops, trainings, seminars, official reports, fact sheets, telephone, e-mail, official letters, fax, websites, making direct contacts, hard copy and/or on-line publications, preparing informative flyers and printing brochures, and POLNET. The POLNET helps the officers make communication workable between the central departments at the headquarters and the provincial departments in the 81 provinces. Each provincial and central branch has access to the POLNET, so they can receive messages, and follow recent reports and developments. It also allows policy makers of the TNP to have real-time data nationwide. The officers use these data to evaluate the actual and possible risks any given part of the country then prepare responses accordingly. The officers use the POLNET not only as a data collection tool but also as an effective communication tool. The officers also use an interactive map of Turkey, on which each city is marked. The officers enter different data to this software and this program is operated on the POLNET system. When high-risk possibilities get closer or increase in a particular city or district, the system warns each proper authority and/or branch automatically.

The participants revealed that they promote risk awareness in their department by evaluating the information collected from open-sources, mass media, websites of some terrorist organizations, and the intelligence gathered from the public agencies with which they have cooperation. They also promote risk awareness in society by printing some brochures on the issues that they want to share with the public and distributing them in the right places. They use some publications, press releases, open-source publications and announcements made by official agencies, internet, mail systems, and telephone services for the same purpose. They also create some official websites that are accessible to the general public. The aim of this is to inform the general public about risk messages.

For the purpose of promoting risk awareness in the Anti-Smuggling and Organized Crime Department, the officers communicate risk messages to their employees in almost the same ways as the Anti-Terrorism Department. The officers also use some annual reports that the department prepares every year.

For the purpose of promoting risk awareness in the Public Order Department, the officers also communicate risk messages to their employees in the same ways as the other two departments. Unlike the other two departments, the participants from this department stated:

> We work on a "Safe School Project" as successful policy implementation based on our risk assessment. We have started this project because we have evaluated according to our statistics and news on the media that criminal activities around the schools will be bigger problem in the near future. Therefore, we have targeted school neighborhoods, warned city police departments; they deploy additional police patrol during crowded time. We also establish strong relations with school managements and student families to increase risk awareness, and the result is significant success.

Discussion and Conclusion

The officers mostly make evaluations with uncertainty, vulnerability, exposure, and threat through the collective thinking ways accomplished at official meetings. Through such meetings, they ensure coordination between the central departments at the headquarters and the provincial departments in the 81 provinces all over the country with the purpose of sharing information, knowledge, experience, data, and statistics on crime. Risk evaluations are based on the information and intelligence gathered by both inside and outside public agencies. After analyzing the gathered data, the officers inform their superiors, authorities, other relevant agencies, and departments about the attained results. It seems that the TNP has very intensive information sharing structure within the organization.

The officers use both risk assessment strategies, preventive and responsive, depending on the encountered risk levels that they have to deal with. They revealed that each unit is responsible for making the evaluations with the risk assessment depending on the field activity they have to handle. But, the departments also have some crime analysis and crime prevention offices, in which the risk assessment effort is coordinated among the units and/or departments. In making evaluations with the risk assessment, there is no particular authority that oversees the risk assessment processes in their policing activities, but there is a hierarchical structure in the TNP. This means that there is always a higher authority to oversee the activities. There are also some official boards to review the projects, employees, and training in promoting the effectiveness and correctness of the activities. While doing all these activities, officers stated that they work well with the outside experts involved.

The officers communicate risk messages in several ways, such as meetings, workshops, trainings, seminars, official reports, fact sheets, telephone calls, internet, official letters, fax, websites, making direct contacts, hard copy and/or on-line publications, informative flyers and printing brochures, POLNET, and some other skills and tools. They also promote risk awareness in their departments by evaluating the information collected from open-sources, local dailies, mass media, and websites of some terrorist organizations. In addition, the officers use some annual reports prepared by the police departments, periodicals, publications, open-source pamphlets, announcements, internet, post system, brochures, telephone services, and press releases in promoting risk awareness in the society. They also create some websites that are accessible to the general public.

Lastly, as stated earlier, of course intelligence is a constant process of data collection, analysis, distribution, and assessment. The intelligence collection and analysis can be done by specialized groups, crime analysts, and the police members only. When the gathered intelligence is properly managed by the police, it can be described as an effective decision making process of the risk assessment in which the police actually incorporate risk assessment and management into their daily work. So, in the decision making process of the risk assessment, the real task for the officers to accomplish is to understand how the whole police organization responds to the actual risks and possible threats.

The interviewees believe that they are currently involved in the decision making process of the risk assessment in their daily work. This definitely helps the officers to move away from reactive to proactive policing strategies, which is a part of developing the risk-based intelligence-led policing. But, we believe, through analyzing the interview statements, that the officers are missing the labels that give these practices legitimacy within the organization. In other words, the officers were rarely involved in preparing a written and standardized risk analysis model that they constantly use in their daily work. This finding raised for us concern about the awareness within the whole organization of the applications of the risk assessment. The reason behind this is that we interviewed with the leadership of the three departments, but not the field officers.

References

John T. and M. Maguire. 2004. *The National Intelligence Model: Key Lessons from Early Research.* London: Home Office.

Kennedy, L. W. and E. Van Brunschot. 2009. *The Risk in Crime.* New York: Rowman and Littlefield.

McGarrell, E. F., J. D. Freilich, and S. Chermak. 2007. "Intelligence-led Policing as a Framework for Responding to Terrorism." *Journal of Contemporary Criminal Justice* 23:142–158.

Maguire, M. 2000. "Policing By Risks and Targets: Some Dimensions and Implications of Intelligence-Led Crime Control." *Policing and Society* 9:315–336.

Maguire, M. and T. John. 2006, March. "Intelligence Led Policing, Managerialism and Community Engagement: Competing Priorities and the Role of the National Intelligence Model in the UK." *Policing & Society* 16(1):67 85.

Marsh, M. H. and T. E. Noonan. 2005. Risk Management Approaches to Protection: Final Report and Recommendations. Washington, DC: National Infrastructure Advisory Council.

Ratcliffe, J. 2003. *Intelligence Led Policing. Report 248.* Canberra: Australian Institute of Criminology.

Stern, P. and H. Fineberg. 1996. *Understanding Risk: Informing Decisions in a Democratic Society.* Washington, DC: National Academy Press.

Toch, H. and J. Grant. 1991. *Police as Problem-solvers.* New York: Plenum.

Van Brunschot, E. and L.W. Kennedy. 2008. *Risk Balance and Security.* Thousand Oaks, CA: Sage.

Questions

1. Given the issues raised concerning the use of information to influence control strategies of police in counter-terrorism, how do police make use of good intelligence while protecting the privacy rights of citizens?
2. When police become more focused on forecasting future events based on risk-based approaches, how does this affect (positively or negatively) their ability to respond to ongoing demands for service from the community?

10.
SHARING INFORMATION TO MANAGE AND REDUCE RISK

The need to share information within an organization is highlighted in the previous chapter. Managing risk also requires collecting information from outside and sharing it with other agencies. In the Carter chapter, he explores how this process can be facilitated through the use of interagency reporting techniques, in particular, suspicious activity reports in monitoring illegal activities.

REDUCING RISK THROUGH SUSPICIOUS ACTIVITY REPORTING

David L. Carter

Since the terrorists' attacks of 9/11, a wide range of new initiatives has been developed by the Department of Justice (DOJ) and Department of Homeland Security (DHS) to reduce the risk of future attacks against the US. One of the most comprehensive initiatives has been the wide integration of new law enforcement intelligence standards to enhance the sharing of substantive threat information among America's law enforcement agencies.[1] The fundamental premise of intelligence initiatives is to identify threats and prevent them from reaching fruition.[2] However, to truly stop a threat, good information is needed to "feed" the intelligence process. While law enforcement agencies have a diverse array of information resources, perhaps the consistently most valuable information comes from observations made by community members. However, community members often do not recognize some behaviors as criminally suspicious, may not feel the need to report such behavior, or may mistake lawful behavior as "suspicious" because it does not conform to their values or perceptions. As a result, there must be guidance as to what types of information need to be collected and how that information may be shared. These are the foundations on which new initiatives of suspicious activity reporting are based.

Suspicious Activity Reporting

Law enforcement intelligence has long used information, both solicited and unsolicited, to learn about criminal threats. Traditionally referred to as "tips and leads," this information was most commonly collected by officers from informants and citizens, but often not managed well, typically with few policies and procedures on the maintenance, sharing, and review of this information. Similarly, the information was frequently not stored or organized in a manner that provided ready access to officers to search for similar types of suspicious activity in order to determine if there was a crime trend or a repeat offender in a community. There typically was no specific training on the types of behaviors that should be included in suspicious activity reports and limited effort in distinguishing reports of behaviors that truly posed threats versus suspicious activity reports that had no criminal nexus or were little more than racial profiling.

As the philosophy and processes of intelligence have changed, so have the processes related to tips and leads. It has been recognized that there is value in a proactive stimulus to generate more of this raw information. Furthermore, a more structured process was needed to capture, evaluate, store, and share this information. The process is evolving with coordinated development by the Global Intelligence Working Group on behalf of the DOJ, the Program Manager's Office for the Information Sharing Environment (PM-ISE), the Information and Analysis Directorate (I&A) of the DHS, the new Program Manager's Office for Suspicious Activity Reporting in DOJ, and the new Program Manager's Office for Fusion Centers in DHS. These new initiatives collectively seek to manage documentation of suspicious behaviors on a consistent, structured basis through the use of suspicious activity reports (SARs):

> The SAR process ... focuses on what law enforcement agencies have been doing for years—gathering information regarding behaviors and incidents associated with crime—and establishing a process whereby information can be shared to detect and prevent criminal activity, including that associated with domestic and international terrorism.[3]

What Is Suspicious Activity Reporting?

Suspicious activity reporting is a formalized process to document and share observed behaviors that are indicative of criminal activity. Information—including "tips and leads"—may come from law enforcement personnel, private sector partners, or citizens. This information is placed in written form and processed through the law enforcement agency, including verifying facts and confirming there was suspicious behavior that potentially has a criminal nexus, in order to have the SAR integrated into the analytic process of the intelligence function.

There are three types of SAR. The first is the financial SAR. This was mandated by the Bank Secrecy Act wherein financial institutions must report certain types of transactions to the Treasury Department. The suspicious activity most commonly associated with financial SARs is money laundering or trafficking in unlawful commodities.[4]

The second is the "all crimes" SAR. This is simply the documentation and reporting of suspicious activity related to any crime. The activity may be observed by a law enforcement officer or reported to an officer. This is similar to what many law enforcement agencies have used traditionally known by various names such as a field intelligence report, field interview report, or miscellaneous investigation report.

The third type is the information sharing environment-suspicious activity report (ISE-SAR). This is the documentation of suspicious behavior specifically related to terrorism or crimes that support or facilitate terrorist planning and acts. The reason for having a distinct SAR for terrorism is based on statutory provisions emanating from the Intelligence Reform and Terrorism Prevention Act of 2004.

At a law enforcement agency, the form and processes for line officers are typically the same for both the "all crimes" SARs and ISE-SARs. However, when the SAR is processed through the intelligence unit or fusion center, the ISE-SAR is processed differently and shared much more widely. The reason for the different processing is based on the responsibility of the information sharing environment (ISE) to deal only with terrorism and crimes supporting terrorism. The standards and processes for dealing with ISE-SARs will be discussed later.

At this point, there is value in understanding what is needed for a SAR process in a law enforcement agency to ensure substantive accuracy and a lawful foundation. There are several organizational components that need to be in place in order to have an effective suspicious activity reporting process. These include:

1. An agency-wide process for documenting suspicious activity. A policy guiding the format, use of case numbers, and processing of the SAR must be a formal part of the agency's records system.
2. An agency-wide process of sharing SARs with other components of the agency and shared externally with other agencies—such as a fusion center—as appropriate. A report of suspicious activity is of little value if it is just recorded, but not shared.
3. Training of personnel on the indicators of the crimes of concern. Indicators are the signs, symbols, materiel, and behaviors that are known to be linked to crimes. For example, there are certain symbols known to be linked expressly to right-wing extremist groups just like the possession of certain precursor chemicals are known to be associated with "cooking" methamphetamine. Virtually any type of crime will have some type of indicators. As a result, personnel must be trained to recognize these indicators to support suspicious activity reporting. Importantly, training should include the fact that activities should be viewed as suspicious in light of the "totality of the circumstances," not just on the observation of an indicator alone.
4. An SAR review and purge process. Not all activities that initially appear suspicious are criminal. Consequently, the SAR policy should have a process to review the SAR, and ensure there is a reasonable criminal nexus. If not, the SAR should be purged. Importantly, the agency's privacy policy should be a key component in the review and purge process.

SARs are a critical source of information for the intelligence cycle. Oftentimes the reports identify behavior related to previously unknown threats. In other cases the SAR provides information about an inquiry that helps more clearly understand a known threat or new dimensions of a known threat. The SAR also aids in defining new intelligence requirements. While the SAR is certainly not the only source of new information for the intelligence cycle, it fulfills a critical role, particularly when supported by officers who are comprehensively trained on terrorism and crime indicators and who consistently report that suspicious activity.

Suspicious Activity Reporting Processes

An important part of the suspicious activity reporting process is the underlying assumption that many people observe suspicious activity in their daily lives but either do not recognize it or think the suspicious behavior is not of sufficient consequence to report. As will be seen, a goal of recent initiatives is to generate increased reporting of suspicious activity from the public with the belief that as these reports increase, the probability of gaining critical threat information will also increase. The keys to success include:

- Recognizing suspicious behaviors that have a criminal nexus.
- Developing substantive information about the characteristics and possible intent of the suspicious behavior.
- Ensuring that all behavior is reported to the intelligence unit or a fusion center (instead of simply being recorded in a law enforcement agency's records management system).
- Reviewing and analyzing the information and, if creditable, sharing it or storing it in a searchable database.

Embracing this vision, the Los Angeles Police Department implemented a new policy for SARs related to both international terrorism and domestic criminal extremism. The policy defines a SAR as:

> [A] report used to document any reported or observed activity or any criminal act or attempted criminal act, which an officer believes may reveal a nexus to foreign or domestic terrorism. The information reported in a SAR may be the result of observations or investigations by police officers, or may be reported to them by private parties.[5]

The directive further provides specific illustrations of behavior that may be deemed suspicious, as well as guidance for reporting the behavior. This new vision of suspicious activity reporting has two integrated dimensions. The first dimension is development of a model for refining suspicious activity processes. The second dimension is broadening the scope of persons who may report suspicious activity.

With the first dimension, "suspicious activity" can encompass a broad array of behavior. Moreover, what a person defines as suspicious will vary, depending on a person's life experiences, values, and other social factors. Direction must be given so that suspicious activity is defined (and viewed) in the context of the types of information that has a criminal nexus and that is needed for the intelligence process. To accomplish this, a three-part model can be used: Observe, Document, and Report.[6]

Observe

People must be observant, but observations need to be targeted. Awareness training is needed to ensure that the behavior that has been observed has a potential criminal nexus. Often, training includes providing information about "indicators"—behaviors that, based on experience, are reasonably linked to potentially criminal activities. There are two critical components of indicators: first, the focus must be on behavior—not the attributes of an individual. Moreover, there must be a reasonable likelihood that the behavior could be related to a crime; that the behavior is not just "odd" or "unusual." "Attributes of an individual" would include perceived[7] race, ethnicity, gender, religion, and so forth. That is, a person should not be deemed "suspicious" simply because he or she possesses these attributes—suspicion must be based on behavior. If suspicious behaviors are observed and documented as they relate to the preparation or commission of a crime, then the attributes can be properly used as part of the description of the suspicious person. Of course, these guidelines are general and must be applied in light of circumstances and other information the observer has.

The second critical component is for the observer to understand the types of behaviors and symbols (e.g., gang markings, tattoos, slogans of criminal extremists, etc.) that are particularly indicative of criminal activity. This means that observers can benefit from some type of training or information, such as intelligence reports, about suspicious activities.

For law enforcement officers, training programs are available, both in class and online, that can provide important contemporary information about threat indicators. One such program that has been highly successful is the State and Local Anti-Terrorism Training (SLATT) program funded by the US Bureau of Justice Assistance (BJA). SLATT training provides detailed information to officers on indicators of both international and domestic terrorism/criminal extremism from the perspective of how patrol officers, for example, may encounter and observe such behaviors during their regular work shift.[8]

Communities Against Terrorism (CAT), also funded by the BJA, is a program directed toward the non-law enforcement community to provide suspicious activity indicators. The CAT program has templates of flyers containing potential criminal indicators that law enforcement agencies can distribute to specific industries.[9]

Document

When suspicious activity is observed, the value of the SAR is increased if the observer provides explicit information about the suspicious persons and the specific nature of the suspicious activity. While law enforcement officers are trained to collect such information, the public, which reports a significant amount of suspicious activity, typically does not have this knowledge. Public education programs need to inform community members of the kinds of information that is most valuable: detailed descriptions of the suspicious persons and their vehicles, location of the activity, the actions that appeared suspicious, and any objects the person might have that added to the suspicious activity are all types of information that must be documented. Once again, the CAT program provides the details that are specific to different kinds of businesses.

Report

For law enforcement officers, new processes are being developed to establish common standards for capturing and reporting SARs that will enhance SAR information sharing. The process is being driven by the ISE with a particular focus on terrorism information.[10]

Reporting suspicious activities has been more problematic for community members. In many cases, information has not been reported because the community simply did not know how or to whom to report this information. Law enforcement agencies have been improving their capacity to assist citizen reporting; for example, the Connecticut Department of Public Safety, among other agencies, has a comprehensive online SAR that citizens can submit to report suspicious activities. SAR websites for the public may not be the best alternative for all jurisdictions and there is a concern in major cities and urban areas that public submissions of SARs through a website will generate too many unfounded reports. Conversely, in rural and moderately populated areas the websites may be a valuable tool. The best avenue is to determine if public submission of SARs through a webpage is a mechanism that will best serve the suspicious activity reporting needs of an intelligence unit or fusion center in light of the jurisdiction's characteristics.

As another mechanism for reporting suspicious activity, the BJA CAT program designed the suspicious activity brochures so that each law enforcement agency could place its reporting and contact information at the bottom of the brochure.

The goal of these programs, of course, is to encourage people to be aware of behavior that is likely to be criminal, capture critical facts about the behavior, and submit the information to a law enforcement agency. This is at the heart of the "See Something, Say Something" program of the New York (City) Police Department (NYPD). In many cases, the information can provide important missing puzzle pieces to the intelligence analyst. In other cases, the information may inform the law enforcement agency about potential criminal activity about which the agency was previously unaware.

Some agencies have formalized the process further. For example, the NYPD created Operation Nexus that "actively encourages business owners, operators, and their employees to apply their particular business and industry knowledge and experience against each customer transaction or encounter to discern anything unusual or suspicious and to report such instances to authorities."[11]

As initiatives are developed to encourage people to report suspicious behavior, law enforcement is more likely to identify and mitigate threats. A significant element of this process is to ensure that suspicions are based on behavior and that the observer captures as much accurate information as possible to report the suspicious activity. While these goals are important, they are nonetheless the beginning of the process. Once this information is in the hands of the intelligence function, it must be integrated to learn what else is known about the suspicious activity and to define future information needs in order to clearly understand if a threat exists and, if so, how it can be stopped.

Private Sector and Community Engagement

Lessons learned from community policing have demonstrated the value of engaging the community to help protect their neighborhoods from predatory crime. The same philosophy can be applied to the broader transjurisdictional problem of protecting communities against terrorism.

To maximize the quality and quantity of information provided by the community, law enforcement must provide a framework of knowledge. The more that law enforcement can educate the community, the more robust the feedback from the community.[12] Engaging the community can significantly increase the information-collection parameters for the intelligence process. Several key factors make these partnerships work effectively:

1. The law enforcement agency must establish trusted relationships with people through local law enforcement officers. Typically, a "meet and greet" alone will not suffice. Information that may be sensitive or confidential may need to be exchanged, or the member of the public passing on the information may be fearful. As a lesson learned from community policing, the bond of trust established through an ongoing dialog becomes a critical element.
2. Once trust is established, an effective means of two-way communication must be established with each individual. In some cases, this may be technological, such as e-mail or telephone. In other cases, individuals may feel comfortable sharing the information only on a one-to-one basis with their trusted law enforcement partner.
3. Just as with officers, training—or public awareness—about the signs and symbols of terrorism and organized crime is important for enabling the partner to identify substantive information that should be reported. For example, a community partner may see a symbol, tattoo, or foreign word that is symbolic of a terrorist or criminal entity, but if he or she has not been trained to recognize these symbols, the information will not be reported. Similarly, community and private partners should receive information about behaviors that are unusual and which suggest criminality in order to enhance their specific awareness.[13]
4. Partner training should also include, as noted previously, instruction on the type of information that should be documented for reporting to law enforcement. While documentation of behaviors, evidence, vehicles, and personal descriptions are second nature to law enforcement officers, it is not a customary practice for non-law enforcement personnel. Explaining what kinds of information should be reported—along with date, time, and location—is an important element in partner training.
5. The agency should develop a policy and process to protect the privacy and safety of community partners. This includes developing a privacy policy to protect business partners' proprietary information that may need to be disclosed during the information sharing process. Partners need to be informed of these policies in order to reinforce the trusted relationship.

A number of good partnerships have emerged, notably in larger law enforcement agencies. The NYPD's Shield[14] is a comprehensive information sharing and public information program. The Delaware Information Analysis Center[15] has established formal relationships with non-law enforcement government organizations and targeted private sector entities for information sharing over the secure Automated Trusted Information Exchange network. The Nassau County (New York) Police Department created the Security/Police Information Network as a comprehensive public–private information sharing network that is stratified by sectors and divided between vetted and non-vetted information sharing partners using e-mail and periodic meetings. There are certainly other examples, but the point is clear: partnerships are critical, there are models to provide guidance, and partnerships must be established throughout America's communities, not just in major metropolitan areas to increase the reporting of suspicious behavior to law enforcement.

Partnering with Community Members

Just as with any other crime or community problem, it is important to enlist the support of community members. In the intelligence arena, two factors make community partnerships more challenging. First, the agency is concerned about threats, and second, many community members are uncomfortable with the intelligence role in law enforcement. On the issue of threats, previous discussions have pointed out the need to inform citizens of the types of behavior they need to report.

With respect to the intelligence role, some people have expressed concern that forming community partnerships for intelligence is akin to turning citizens into informants on their neighbors. Many individuals assume that intelligence activities routinely violate citizens' privacy and civil rights. The challenge is to educate skeptics about the process and the importance of citizen involvement in suspicious activity reporting simply because citizens are often in a place to observe suspicious behaviors more often than law enforcement officers. The notion of citizen participation goes back to one of the most fundamental principles of modern law enforcement articulated by Sir Robert Peel in developing the London Metropolitan Police in 1829: "The police are the public and the public are the police."

The SARs and Risk Relationship

The relationship between SARs and the reduction of risk is conceptually quite simple. As more people report suspicious activity and that information is integrated with other known threat-related information, the greater the probability that sufficient intelligence can be generated to identify criminal offenders to apprehend them and eliminate the threat. While the concept is simple, operationalizing it is more difficult for several reasons. Citizens must be taught what to look for and encouraged to report it; the law enforcement agency must have a process for assessing, integrating, and sharing the information; there must be an analytic function to give meaning to the information; and there must be controls built into all phases of the process to protect the privacy and civil liberties of persons who are reported as performing suspicious behavior. Developing these processes is one challenge—another is to have community members and law enforcement personnel consistently use the processes. Dogmatism, which is inherent in any new endeavor, typically becomes the greatest challenge to overcome.

Establishing Controls on SARs/Information

Information contained in SARs must be managed carefully for a fundamental reason: the basis or motivation for reporting the information and the inherent accuracy of the information is often unknown. As such, law enforcement must determine whether lawful activity was misperceived as being illegitimate and whether the information was reported accurately. Despite these potentially limiting factors, when suspicious activity is reported to law enforcement, the agency has a public safety responsibility to determine if there is a reasonable suspicion that a crime is being planned or has been committed. If so, the information should be retained and serve as the basis for a further criminal inquiry and operational planning to protect the public.

Because of this sensitivity, the Global Justice Information Sharing Initiative's Privacy Committee identified six critical factors, or points in the SAR process, wherein information management controls should be implemented to ensure the security and privacy of personal identifying information (PII) while at the same time permitting the agency to further its inquiry into the behavior. These controls are based on the same principles of information management used to protect information in a criminal intelligence records system. With some modification from the Privacy Committee's report, the six controls include:

1. Receipt/Collection. At the time of receipt or collection, suspicious activity information should be assessed and reviewed for sensitivity and confidentiality using corroborative information, if possible. Every effort should be made to validate or refute the information in order to resolve as quickly as possible whether there is a criminal nexus or whether the SAR is unfounded. Collecting information that is purely expressive, an activity protected by the First Amendment, should be prohibited.
2. Storage. Storage of suspicious activity information should be handled similarly to data that rise to the level of reasonable suspicion. Those requirements should include an audit and inspection process, supporting documentation, and logical separation or labeling of the unconfirmed SAR from other information.
3. Access. Because of uncertainty about the information's credibility and/or accuracy, access to suspicious activity information should be handled similarly to access to information that rises to the level of reasonable suspicion. Access should be allowed only where there is a right to know and a need to know the information in the performance of a law enforcement, homeland security, or public safety activity.
4. Dissemination. Suspicious activity information, if systematically collected and stored for interagency distribution, should be disseminated primarily in response to an inquiry, and only for law enforcement, homeland security, and public safety purposes. Uncorroborated suspicious activity with PII should not be regularly disseminated in bulletins or other similar products, but it may be included in secure information databases and disseminated to relevant law enforcement, homeland security, and public safety agencies that have the right to know and need to know the information for public safety purposes.
5. Retention. The retention period for suspicious activity information should be established by policy and be of sufficient time to determine the veracity of the information in light of the agency's expertise and resources. Suspicious activity records should have a "disposition" label, such as "undetermined/unresolved" or "cleared/unfounded," to clearly notify the user of the information's status. Agencies should also consider the need for maintaining some type of suspicious activity data for purposes of statistical reporting and performance measurement when setting retention and purge procedures.

6. Security. Physical and electronic security measures should be similar to those used for information that is rising to the level of reasonable suspicion.[16]

Permeating each factor in this process should be constant attention to privacy, accuracy, and corroborating the information in the SAR, as reinforced by a report to Congress by the PM-ISE:

> Protecting the information privacy and legal rights of Americans is a top priority: At the local level, SARs will be incorporated into existing processes and systems used to manage other crime-related information and criminal intelligence so as to leverage existing policies and protocols utilized to protect the information privacy, civil liberties, and other legal rights of the general public. Multiple levels of review and vetting will be established to ensure that information is legally gathered and managed, and reports containing personally identifiable information that are unfounded, or that cannot be reasonably associated with criminal activity, will not be shared beyond the originating entity.[17]

Complicating these processes among many law enforcement officers is that all information may be useful sometime in the unknown future. This belief produces a great reluctance to destroy any information or reports. It must be remembered that SARs are largely unsolicited tips and leads from sources that have a wide range of credibility. If the follow-up shows that the allegation in the SAR is unfounded, there is no reason to retain any PII.

Policy Recommendations for Developing and Managing a Suspicious Activity Reporting System

As has been evident in this discussion, law enforcement agencies are increasingly developing a more comprehensive system to stimulate the reporting of suspicious activity as well as to have a more structured process to manage SAR records. Although the PM-ISE is developing national standards for SARs that report terrorism information, there are no current national standards for policies related to criminal activity that is not related to terrorism. As a result, the Major Cities Chiefs Association and the Global Justice Information Sharing Initiative approached this task with support from the DOJ and the DHS. A number of substantive recommendations were made from this project and best practices were identified. The following is a summary of some of the project's findings and recommendations related to the management of SARs:

- Agencies should educate and gain the support of policymakers about issues and processes related to suspicious activity reporting.
- All privacy and civil liberties policies and training should be reviewed to verify that they are consistent with current law and national standards.
- Information management of the SAR process should be integrated with existing processes and systems that manage criminal investigative information and criminal intelligence, thereby leveraging existing policies and protocols to protect privacy and civil rights.

- The SAR policy should be communicated to the public because transparency is the key to acceptance.
- Gathering, processing, reporting, analyzing, and sharing suspicious activity information is critical to preventing crimes, including those associated with domestic and international terrorism. The SAR process should have an "all crimes" orientation.
- The SAR process should use criminal information management processes, applying SARs to the appropriate criminal records management system or criminal intelligence records system.
- Suspicious activity submissions should not bypass the local law enforcement agency or agency of original jurisdiction and the standard 911 reporting systems.
- When an agency receives information that affects another jurisdiction, it is the responsibility of the receiving agency to immediately notify the affected agency and discuss coordination, deconfliction, investigation, and vetting procedures with the affected agency. Once vetted, further dissemination of the information would be the responsibility of the affected agency.
- A process should be established to ensure that SARs are made available to fusion centers and local Joint Terrorism Task Forces (JTTF) in a timely manner.
- An ongoing emphasis should be placed on defining and communicating trends in terrorism activity, geographically specific threat reporting, dangers to critical infrastructure, and general situational awareness.
- There is a need for a common national methodology, including common data codes, for sharing suspicious activity data in order to discern patterns across the country.
- Training is a key component of the SAR process. All relevant agency personnel must be trained to recognize behavior and incidents indicative of criminal activity associated with international and domestic terrorism.
- Incorporating outreach to the public, law enforcement, and the private sector in the collection process is important to the success of the program.
- Develop a common national methodology to horizontally and vertically share SAR data in a timely manner that is consistent with privacy and civil liberty guarantees.
- Develop a standardized training program to provide consistent nationwide SAR training.[18]

The intent of these recommendations is to maximize the utility of suspicious activity information while at the same time protecting civil liberties. While these recommendations are intended to bring about standardization of the SAR process, the report notes that every jurisdiction will have to develop policies and procedures that take into account the unique circumstances and relationships within that community.[19] Regardless of the policies are implemented, they represent a new responsibility that has implications throughout the organization.

Overcoming Dogmatism

The human resistance to change vexes virtually every new endeavor facing an organization—suspicious activity reporting is no different. Uncertainty of what is expected, how to complete the SAR form, the expectation that this will be "more work," cynicism about the effectiveness of the SAR process, and the mere burden of "having to learn something new" are all obstacles that have to be overcome within the law enforcement organization. Community members will also have resistance to overcome: suspiciousness of the intent of the process, concerns about civil rights and privacy, fear of reprisal from those persons about whom suspicious activity has been reported, distrust of the police, and the feeling that "citizens are being asked to do the police department's work" are all also obstacles. Despite these hurdles, they can be overcome by implementing transition initiatives and, most importantly, showing value to participating in the process.

For the community, value must show that the SAR process, with community participation, will help make the community safer from crime, and reduce disorder. Building on the idea that citizens have an obligation to contribute to the safety of their community, providing public education programs, enlisting support of community leaders, proactively increasing the dialog between the police and community, and having a transparent process for handling SARs, are all important pathways to gain public participation.

For officers, value may be demonstrated by showing how the SAR process will make their jobs easier, help prevent crime, help solve crime, and increase citizen satisfaction of the police. The pathway in the law enforcement agency is to first show commitment by leaders, develop forms and processes to make suspicious activity reporting "easy," provide a detailed and substantive training program showing the value of SARs and how to perform the tasks of the process, and ensure supervision to both reinforce policy and provide guidance to ensure that policy is not only being followed but is also being executed properly.

Information Sharing Environment-Suspicious Activity Reports (ISE-SARs)[20]

As noted previously, initiatives related to the ISE are formalizing suspicious activity reporting in an effort to ensure comprehensive information sharing specifically related to terrorism information as a mandate from the Intelligence Reform and Terrorism Prevention Act of 2004. While many detailed components of this process are being developed and are beyond the scope of the current discussion, there is value in having some familiarity with the ISE-SAR initiatives. Remember, the ISE-SAR applies only to terrorism information.

The planned use of the ISE-SAR focuses on

> the DOJ, DHS, DoD [Department of Defense], and the FBI [Federal Bureau of Investigation], working in partnership with state and local officials, [to] institute a standardized approach to gathering, documenting, processing, analyzing, and sharing terrorism-related suspicious activities reports. Front-line law enforcement personnel will be trained to

> recognize behaviors and incidents indicative of criminal activity associated with domestic and international terrorism. Once documented, SARs will be evaluated by trained personnel to determine if they have a terrorism nexus. If a terrorism nexus is established, the SAR will be made available to the local JTTF, regional, and/or statewide fusion centers, and DHS.[21]

To further this process, the ISE created the Common Terrorism Information Sharing Standards (CTISS) program. The CTISS allows for business process-driven, performance-based common standards for preparing terrorism information for maximum distribution and access. Two categories of common standards are formally identified under CTISS:

1. Functional Standards: Set forth rules, conditions, guidelines, and characteristics of data and mission products supporting ISE business process areas.
2. Technical Standards: Document specific technical methodologies and practices to design and implement information sharing capability into ISE systems.[22]

The standards are designed to increase effective sharing of the ISE-SAR with the intent of enhancing the discovery and analysis of potential terrorism-related patterns or trends on a regional and national basis beyond what would be recognized within a single organization, jurisdiction, state, or territory. In applying this process, the PM-ISE issued the first version of the ISE-SAR Functional Standard. The ISE-SAR Functional Standard supports sharing information among federal, state, local, and tribal partners that relates to suspicious activity or incident information that has a potential terrorism nexus, in a manner that protects the privacy and legal rights of Americans.

Nationwide Suspicious Activity Reporting Initiative

Following development of the ISE-SAR, the next step was to determine the best process for collecting and integrating SAR information from state, local, and tribal law enforcement (SLTLE) agencies. The BJA, FBI, and the I&A Directorate of DHS, working through the office of the PM-ISE, is approaching this challenge through the Nationwide Suspicious Activity Reporting Initiative (NSI) to establish a set of shared databases that allow all SLTLE agencies to network on a national level. This initiative, which builds on a project developed by the Major Cities Chiefs Association,[23] seeks to develop a process that ensures that all suspicious activity information is compatible; that standard processes are being developed in the way that information is collected, formatted, and exchanged; and that suspicious activity reporting is done in a manner consistent with constitutional protections. The NSI is a process for making sure there is a common mechanism to get SARs through fusion centers that will, in turn, facilitate the exchange of regional and national analyses among ISE partners.[24]

Activities in the NSI have created a framework to support the reporting of suspicious activity from the point of initial observation to the point where the information is available in the ISE. A key element is that the NSI incorporates the SAR process of SLTLE agencies into a nationwide effort.

All agencies, regardless of size or jurisdiction, have a role in the nationwide SAR process. Those that do not participate represent a weakness in our system of identifying pre-incident indicators of criminality. Each law enforcement agency's internal SAR process, once developed, can be incorporated into the nationwide SAR initiative. Clearly, without the support and participation by SLTLE agencies, the amount of valuable terrorism-related SARs in the ISE would be limited.

Part of this initiative is the development of a searchable library that would allow law enforcement agencies to access the information on summary SAR reports through a web-based portal to which all law enforcement officers have easy access. The current model provides that officers using Law Enforcement Online, RISS.net, or the Homeland Security Information Network-Intelligence can access the National Criminal Intelligence Resource Center (NCIRC) as a portal to enter and search for ISE-SARs entered by law enforcement agencies all over the US. The NCIRC is networked to secure servers throughout the US that serve as the "shared space" to store, search for, and retrieve SARs. The shared space concept is a networked data and information repository for making standardized terrorism-related information available through common terrorism information sharing standards, applications, and services. Involvement in the shared space will provide agencies with the ability to access local, regional, state, and national SAR data.

While there is widespread recognition of the importance of identifying and reporting suspicious activities, the initiative is not without its critics. The American Civil Liberties Union stated:

> This overbroad reporting authority gives law enforcement officers justification to harass practically anyone they choose, to collect personal information, and to pass such information along to the intelligence community. Suspicious activity report policing opens the door to racial profiling and other improper police behavior, and exposes law-abiding people to government prying into their private affairs without just cause.[25]

Because of these concerns, training is essential. It must be emphasized that the focus is on collecting information about behaviors that represent pre-operational indicators of criminal activity. Unusual or odd behavior would not be included unless there is some articulable relationship between that behavior and a potential crime. Beyond training, it is essential to have ongoing supervision and a substantive assessment mechanism to document the effectiveness of the process.

Summary

Simply stated, risk can be reduced when law enforcement agencies have comprehensive, validated information about threats. The SAR process has been designed to collect, process, and share threat information with America's law enforcement agencies. While a significant amount of work has been devoted to ensure an efficient, effective, and constitutional system, challenges remain. Law enforcement agencies must be diligent in their implementation and marketing of suspicious activity reporting if true risk reduction is to be achieved.

Notes

1. A wide range of government documents and reports on these news intelligence initiatives can be found at http://it.ojp.gov/default.aspx?area=home&page=1224#cat_Intell.
2. Carter, David L. (2009).
3. Suspicious Activity Report Support and Implementation Project (2008: 1).
4. The financial SAR deals specifically with financial transactions where financial institutions must report large cash transactions as suspicious activity. For more information about the financial SAR see http://www.fincen.gov/reg_sar.html.
5. Los Angeles Police Department (March 5, 2008: 1–2).
6. This model is based on a community training model for counter-terrorism developed by the Regional Community Policing Training Institute at Wichita State University, Dr. Andra Katz-Bannister, Director.
7. The word "perceived" is used here because often an observer sees a characteristic that is mistaken for an attribute. For example, an untrained observer may see a person that the observer concludes is from the Middle East when, in fact, the person is of a different ethnicity.
8. See http://www.slatt.org. The website not only provides training schedules but also online training and a wide range of resource materials. For full access to the website, one must register with employment verification that the user has a bona fide need for access to the site. Because the program is funded by BJA, there is no charge for the training.
9. CAT brochures may be downloaded by registered users of www.slatt.org at https://www.slatt.org/secured/cat.aspx.
10. See http://www.ise.gov/docs/NSI_PCRCL_Analysis_July2010.pdf for more information on the ISE initiatives and standards related to SARs.
11. See https://secure.nypdshield.org/Pages1/operationnexus.aspx. The NYPD Shield website requires users to register to have access to this site. In most instances, an individual will be given access upon a request to NYPD on the site.
12. Ideally, the law enforcement agency would be able to provide feedback to the citizen about information that was reported. Many times this is not feasible in the intelligence environment; however, it serves as positive reinforcement to the citizen when feedback is provided.
13. A good example of behaviors is found in the Michigan State Police video, *Seven Signs of Terrorism*, which is available for public distribution. See http://www.michigan.gov/msp/0,1607,7-123-1564-155763--,00.html.
14. See http://www.nypdshield.org.
15. See http://dsp.delaware.gov/Intelligence.shtml.
16. These factors are based on the following document with some modification: Global Intelligence Working Group (2007: 7–8).
17. McNamara, Thomas E. (2008: 29).
18. Suspicious Activity Report Support and Implementation Project (2008: 2–5).
19. Ibid., p. 6.
20. See http://www.ise.gov/docs/ISE_AR-2010_Final_2010-07-29.pdf.
21. McNamara (2008: 29).
22. See http://www.ise.gov/docs/NSI_PCRCL_Analysis_July2010.pdf.
23. See *Findings and Recommendations of the Suspicious Activity Report (SAR) Support and Implementation Project*, http://it.ojp.gov/default.aspx?area=home&page=1224, under the Intelligence/Fusion Centers heading.
24. Remarks by DHS Secretary Michael Chertoff to the International Association of Chiefs of Police, San Diego, California, November 11, 2008, http://www.dhs.gov/xnews/speeches/sp_1227117095921.shtm.
25. German, Mike and Jay Stanley (2008: 2).

References

Carter, David L. 2009. *Law Enforcement Intelligence: A Guide for State, Local and Tribal Law Enforcement Agencies.* 2nd ed. Washington, DC: Office of Community Oriented Policing Services.

German, Mike and Jay Stanley. 2008. *Fusion Center Update.* Washington, DC: American Civil Liberties Union.

Global Intelligence Working Group. 2007. *Privacy Committee Report: Tips and Leads Issues Paper.* Washington, DC: Global Justice Information Sharing Initiative, US DOJ.

Los Angeles Police Department. March 5, 2008. *Reporting Incidents Potentially Related to Foreign or Domestic Terrorism.* Special Order Number 11.

McNamara, Thomas E. 2008. *Annual Report to the Congress on the Information Sharing Environment.* Washington, DC: PM-ISE.

Suspicious Activity Report Support and Implementation Project. 2008. *Final Report.* Washington, DC: Major Cities Chiefs Association; US DOJ and DHS.

Questions

1. How does the SAR process complement the investigative work done already by law enforcement in their investigations of organized crime?
2. Does the SAR process, in your opinion, impinge on the privacy of citizens' financial transactions?

11. COORDINATING COUNTER-TERRORISM RISK ASSESSMENT AND MANAGEMENT

Looking at coordinated efforts among law enforcement in assessing and managing terrorism threats, there is a strong case for the coordination of effort across police agencies at all levels, from small units to large metropolitan forces. Also, there needs to be a strategic alliance that develops around the response from local, state, and national law enforcement. This approach reinforces the view that risk of such events as terrorism reaches down right into communities and the need to develop a common scheme of response is underlined in this approach. Most importantly, Carter discusses the role that intelligence plays in defining the risks at all jurisdictional levels and outlines the most effective ways of responding to these risks. This effort includes a strategy that would educate officers at all levels to both understand what the intelligence means and how the terrorist risk can be manifested. With these tools in hand, the desire for a more coordinated reaction to the dangers posed can be realized.

LAW ENFORCEMENT INTELLIGENCE

Suspicious Activity Reporting and Preventing Self-radicalized Terrorism[1]

Jeremy Carter

> Since the profile of a would-be terrorist is becoming less and less obvious ... In that kind of fog, small behaviors necessarily loom large.
>
> (Ripley 2007)

Introduction

Adaptation is perhaps the best word to describe the approach law enforcement has taken to respond to emerging threats within our communities in the wake of the terrorist attacks of September 11, 2001 (9/11). However this adaptation is twofold. The threat of terrorism is adapting as well. Beyond the evolution of terrorist methods and targets, law enforcement is faced with an emerging threat rooted in terrorist motives and self-radicalization. The heightened emergence of self-radicalized terrorists has resulted in an environment of increased risk within our communities. Law enforcement must rely on best practices and lessons learned from intelligence and proactive information sharing while continuing to think critically of the risks these emerging threats pose. These best practices, lessons learned, and critical thinking will be discussed, specifically as they relate to suspicious activity reporting, and illustrated using case examples from the plotted attack of Fort Dix, New Jersey and the attack at Fort Hood, Texas military bases.

Crime and Terrorism Risk: Determination and Prevention

The intent of this chapter is to focus on specific case examples to demonstrate the application and utility of law enforcement intelligence practices precludes an in-depth discussion of risk in the criminal/terrorism context. However, a few critical components when determining risk associated with crime and terrorism are necessary to point out. The mission of law enforcement intelligence is to prevent or mitigate risks. This mission requires, or assumes, certain knowledge to be available to law enforcement—such as information on the actors involved along with their motives, methods, and targets. Without this information we are asking law enforcement for impossible outcomes.

Paramount to this discussion is that the prevention of crime and terrorism is a process based upon a set of operational assumptions. This process requires a commitment to a philosophy of practice that is counterintuitive of traditional law enforcement operations—those that are proactive rather than reactive in nature. This is not to say what happens after crime or terrorism occurs is not of significant importance, the vital role of first responders and emergency management goes without question. Ratcliffe (in press) sums up the operational assumptions that law enforcement must recognize when determining threats and identifying ways to mitigate or prevent threats from reaching fruition. Simply put, in order to prevent crime or terrorism it is operationally assumed that law enforcement has committed to proactive action. Moreover, in order for law enforcement to be proactive it must rely on a certain degree of predictability of criminal and terrorist actions. Lastly, for actions to be predictable, a pattern of these actions must be identifiable (Ratcliffe in press). These assumptions are illustrated in Figure 11.1.

Coupled with these operational assumptions is a scale of opportunity and time that has implications for risk identification and thus prevention. Opportunity and time play different roles in street crimes than in complex (or enterprise) crimes and pertains differently to terrorism as well. Street crimes, such as robbery and sexual assault, rely upon little (if any) planning and time to carry out the criminal act. Offenders of street crimes rely more heavily on an opportunistic target and the time required to prepare for the criminal act is minimal while having multiple targets for the criminal act. Complex criminality, such as organized crime and white-collar crime, require more planning before the criminal act occurs. For example, identifying channels through which money can be laundered and the processes of laundering the money through these channels is time-consuming. Moreover, the targets available for complex criminality are less than those of street crimes—it is difficult to identify organizations to embezzle money from or racketeer money. Terrorism, both acts by lone-wolf terrorists[2] and networks alike, require significant planning prior to an attack. Examples abound, such as the sophistication and logistics of the 9/11 attacks on the World Trade Center and the detailed planning of the July 7, 2005 bombings in London, England. The availability of suitable targets is most difficult for terrorism based upon accessibility, security, and

Figure 11.1 Operational assumptions for risk prevention
Source: Taken directly from Ratcliffe (in press) with permission.

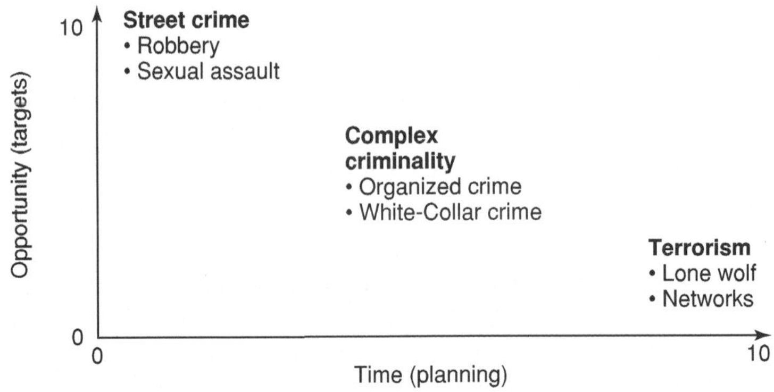

Figure 11.2 Planning and opportunity for crime and terrorism

physical and emotional impact. Figure 11.2 illustrates the opportunity and time differences of crime and terrorism.

The major implication for crime and terrorism risk prevention is twofold: (1) the more planning that an act requires, the more information will potentially be available to law enforcement; and (2) the more planning an act requires, the more likely law enforcement will be able to intervene.

Law Enforcement Intelligence

The law enforcement process for preventing or mitigating criminal and terrorist risks is intelligence-led policing (ILP). Once again, space considerations preclude a comprehensive discussion of ILP for further context; however ILP can be defined as:

> The collection and analysis of information related to crime and conditions that contribute to crime, resulting in an actionable intelligence product intended to aid law enforcement in developing tactical responses to threats and/or strategic planning related to emerging or changing threats. (Carter and Carter 2009: 317)

More specifically, ILP relies upon an active intelligence cycle. In its most basic form, the intelligence cycle is comprised of six steps: (1) planning and direction; (2) collection; (3) processing and collation; (4) analysis; (5) dissemination; and (6) re-evaluation. These six steps outline the process by which raw information enters the ILP capability and how ILP is able to inform decision makers and thus influence actions to guide proactive efforts. Interwoven within these steps are a variety of informal and formal partnerships for two-way communication between law enforcement and other entities, policies guiding ethical police behavior and the protection of civil liberties, channels for raw information to be pushed toward law enforcement, and methods for critical thinking and analysis of raw information. Figure 11.3 is a more complex diagram of the intelligence cycle. A pivotal component of information flow is the management of suspicious activity reports (SARs).[3]

Simply put, SARs are a form of tips and leads provided by members of the community—such as business employees, community members, emergency personnel, etc.—

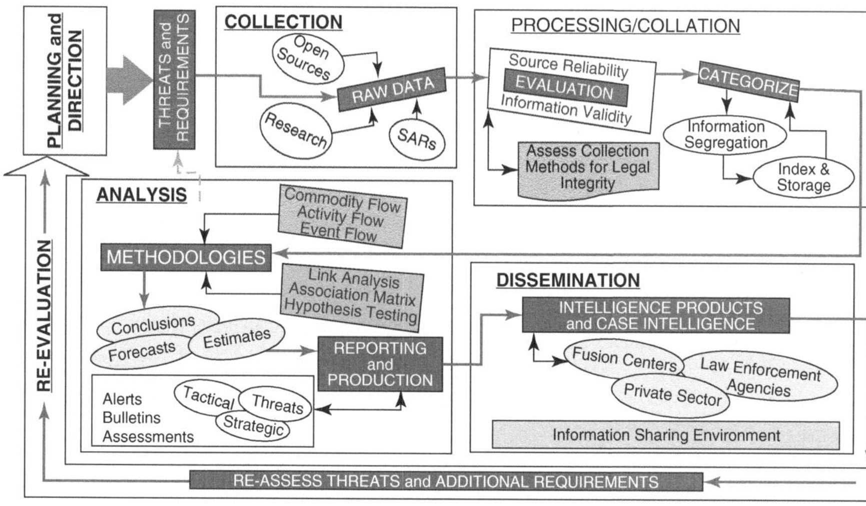

Figure 11.3 Detailed law enforcement cycle
Source: Taken directly from Carter (2009) with permission.

based on an individual's behavior they believe to be suspicious of criminal or terrorist activity. The application of ILP will be illustrated using specific case examples of self-radicalized terrorism incidents.

Self-radicalization of Terrorists

The presence of a single terrorist, or lone-wolf, was felt by the US prior to 9/11. Timothy McVeigh's attack on the Alfred P. Murrah Federal Building in Oklahoma City, OK and Theodore "Ted" Kaczynski's, the "Unabomber," attacks on airlines, universities, and other targets across the country from the late 1970s to the mid-1990s serve as prominent examples. Post-9/11, the presence of lone-wolf terrorism has come in the form of Umar Farouk AbdulMutallab's attempt to detonate a bomb on Northwest Airlines flight 253 from Amsterdam, Netherlands to Detroit, MI on Christmas Day 2009; a shooting attack on March 4, 2010 at the Pentagon in Washington, DC by John Patrick Bedell where he wounded two security guards; and Faisal Shahzad's May 1, 2010 attempt to kill hundreds in New York City's Times Square by leaving a large bomb and a variety of shrapnel in the back of a sports utility vehicle parked in the middle of Times Square.

Radicalization is a process wherein a non-threatening individual transforms to become a violent "true believer." The steps in the radicalization process can vary. Many experts (Silber and Bhatt 2007; Gartenstein-Ross and Grossman 2009) believe it involves four primary stages:

1. Pre-radicalization: an individual's lifestyle prior to radicalization (relationships, environment, jobs, social life, etc.).
2. Self-identification: an individual is influenced by internal and external factors attributing to the exploration of extreme philosophies, ideologies, and values.

3. Indoctrination: an individual intensifies their beliefs and adopts extreme philosophies, ideologies, and values with no exceptions.
4. Jihadization: an individual accepts their duty to participate in jihad as a holy warrior.

Others believe it involves a key motivator to influence beliefs and actions, referred to as mentorship (Whitelaw 2009; Sageman 2004). Regardless of the process by which an individual transforms from a non-risk to a violent risk, opportunities exist that lend themselves to the utility of law enforcement intelligence practices. The progression from stage to stage requires active exposure to extreme ideologies that are often identifiable in public places—such as libraries, places of worship, and universities. Such applications will be discussed in the context of the following Fort Dix, New Jersey and Fort Hood, Texas case examples.

Case Study: Fort Dix, New Jersey

On the evening of May 7, 2007, six men described as homegrown "Islamic militants" (FOX News 2007) with no apparent connection to international terrorist networks were arrested for conspiring to attack US armed forces personnel at Fort Dix, NJ. These men had planned to attack the base armed with a variety of firearms in an attempt to kill as many soldiers as possible. Of the six individuals involved in the Fort Dix terrorist plot, three were brothers. Dritan "Anthony" or "Tony" Duka, Shain Duka, and Eljvir "Elvis" Duka, illegal aliens from the former Yugoslavia, had been living illegally in the US for more than 23 years and were accepted as Americans by neighbors and friends who had no idea they would scheme to attack a military base (Anastasia 2007). The Duka brothers entered the United States near Brownsville, Texas, in 1984 when they crossed the border (FOX News 2009). The brothers had no criminal record, however reports indicate the three accumulated 19 traffic citations, but because they operated in "sanctuary cities," where law enforcement does not routinely report illegal immigrants to homeland security, none of the tickets raised red flags (FOX News 2009). The three brothers operated a roofing business together.

The brothers conspired with three other suspects in the plot—Mohamad Ibrahim Shnewer from Jordan; Serdar Tatar from Turkey; and, Agron Abdullahu from the former Yugoslavia. Shnewer was a legal US citizen and worked as a taxi cab driver, a construction laborer, and also at his family's supermarket. Tatar was in the US legally with a green card and worked as an assistant manager at a 7-Eleven store in Philadelphia and delivered pizzas for his family's pizzeria (FOX News 2008). Abdullahu was also in the US legally with a green card and worked as a baker in a supermarket (USA Today 2008). Five were charged with conspiracy to kill US military personnel while Abdullahu was charged with aiding and abetting illegal immigrants in obtaining weapons. The weapons were reportedly Abdullahu's personal weapons and included a SKS semi-automatic rifle, a Beretta Storm semi-automatic rifle, a Mossberg 12-gauge pump shotgun, and a Beretta 9-millimeter handgun.

During November 2006, the six suspects began planning the attack in more detail. Tatar's parents owned a pizzeria near Fort Dix where Tatar indicated he used to deliver on the base multiple times. Beyond the normal deliveries, Tatar mentioned that he noticed an increase in business when large concentrations of troops were stationed on the base at a given time. Due to the number of deliveries Tatar's family pizzeria made to Fort Dix, the pizzeria possessed a detailed map of Fort Dix labeled "Cantonment Area Fort Dix, NJ"—a map that Tatar took from the pizzeria and provided to the group for planning.

The plot to attack Fort Dix was prevented through a citizen tip. Specifically, in January 2006 the Duka brothers went to a local Circuit City store in Cherry Hill, NJ to get an 8-mm video converted into a DVD. They were greeted by store clerk Brian Morgenstern who later called the police after becoming troubled by what he saw on the video—ten long-bearded men of Middle Eastern descent shooting weapons at a firing range and calling for jihad (Inskeep 2008). The police arrived at Circuit City, watched the video with Morgenstern and determined the video was suspicious enough to call the County Counter Terrorism Coordinator who then contacted the local Federal Bureau of Investigation (FBI) field office. That same afternoon the Duka brothers returned to pick up their copied video without incident—the FBI received their copy of the video a week later. Despite this video, authorities maintain there was no direct evidence connecting the men to any international terror organizations such as al-Qaeda (Russakoff and Eggen 2007).

The tip from Morgenstern led to an investigation that included infiltration by a cooperating witness—Mahmoud Omar, a legal immigrant from Egypt. Omar's history included a 2001 conviction for bank fraud and a 2004 arrest for a fight with a neighbor. In 2006 the FBI recruited him to infiltrate this group. The group often watched terror training videos, clips featuring Osama bin Laden, a tape containing the last will and testament of some of the 9/11 hijackers, and tapes of armed attacks on US military personnel (MSNBC 2007). The men trained by playing paintball in the woods in New Jersey and taking target practice at a firing range in Pennsylvania's Pocono Mountains where they had rented a house (Mulvihill 2008).

In addition to plotting the attack on Fort Dix, the defendants spoke of assaulting a Navy installation in Philadelphia during the annual Army–Navy football game and conducted surveillance at other military installations in the region. The group also considered targeting Dover Air Base in Delaware, Fort Monmouth in New Jersey, the Coast Guard Building in Philadelphia as well as the Philadelphia Federal Building (The Washington Post 2007). The individuals involved in the terror plot were arrested when attempting to purchase additional firearms. The six were arrested trying to buy AK-47 assault weapons, M-16s, and other weapons from an FBI informant. It remains unclear when the attack was to take place.

Fort Dix Implications for Law Enforcement Intelligence and Risk Prevention

The Fort Dix case provides an excellent illustration of law enforcement intelligence practices in the prevention of terrorism risk. The initial awareness of the attack came from an alert citizen working at Circuit City (Morgenstern) who witnessed suspicious activity on the videotape. The existence of this al-Qaeda training video and other terrorism literature in the suspects' possession demonstrate the "self-identification" stage of the radicalization process wherein the individuals began exploring and learning this extreme philosophy and applied it in the form of training and planning for an attack on a US military base. This step in radicalization of the six Fort Dix suspects, which has been endorsed by terrorist experts (Emerson 2007), serves as a point of opportunity for law enforcement to learn of potential attacks through successful partnerships with non-law enforcement entities. This SAR from Circuit City's Morgenstern is one that came as a result of community partnerships in which community members were educated and made aware of such suspicious activity and asked to inform the police if such suspicious activity was observed. The importance of such partnerships for raw information collection to combat terrorism has been well documented (Jenkins 2010; NSI 2010; IACP 2004).

This information was then relayed to a local police department and then was turned over to the County Counter Terrorism Coordinator who forwarded it to the FBI. The local field office then initiated the intelligence-led investigation that ultimately connected the information and prevented the attack. Not only was the initial SAR from Morgenstern critical, but the local police department did not stall on the information but rather forwarded it appropriately. Local, state, and federal agencies then shared information and follow-up investigation, intelligence collection, and analysis. The case also illustrates what the FBI describes as "pre-emptive prosecution." This relied on an informant to infiltrate and provide information and then a pre-emptive arrest. Despite the success of this approach, it can result in a weaker case from a prosecution standpoint—an issue that has risen on several accounts as many reports argue the suspects had no intention of actually attacking Fort Dix (Piette 2009; FOX News 2009).

As aforementioned, terrorism relies on long-term planning and the right opportunity. As a result of the time it took for the six suspects to review maps of the Fort Dix military base, collect numerous weapons, train with these weapons, and do strategic training with paintball guns, law enforcement agencies had time to gather information and develop strategic and tactical intelligence to prevent the attack on Fort Dix. Moreover, the suspects had multiple targets in mind at the outset of the planning stages. Based upon a variety of factors—mainly accessibility and potential casualties—the six suspects chose Fort Dix as the target for the attack since it provided the most ideal opportunity for their desired outcome.

Case Study: Fort Hood, Texas

On November 5, 2009, Nidal Malik Hasan concealed his FN Herstal 5.7 × 28mm pistol along with 13 extra ammunition magazines and went to the Soldier Readiness Center at Fort Hood, Texas where hundreds of deploying soldiers were being given shots and eye examinations. Hasan jumped on a table and yelled "Allahu Akhbar!" (God is Great) in Arabic and began firing (CNN 2009b). The result was 13 people dead and 31 injured. Hasan's attack ended when two base police officers shot and seriously wounded him—initial reports indicated the suspect had been killed by the police officers. Born in the United States to Palestinian parents, Hasan graduated from Virginia Tech and joined the US Army against his family's wishes (Sherwell and Spillius 2009). The military paid for him to go to medical school and he became one of the few psychiatrists in the military when he was assigned to Fort Hood (The Washington Post 2010). Hasan had asked about resigning his commission and began to doubt his military commitment, a situation thought to have worsened after he received orders to deploy to Afghanistan.

Hasan had met Anwar al-Aulaqi, a radical Yemeni-American, while living in the Washington, DC area and later communicated with him via the Internet. Al-Aulaqi had left the United States for Yemen after being investigated in connection with the 9/11 attacks.[4] On July 15, 2009, al-Aulaqi had posted a message on his website urging the deaths of US army personnel (Hsu 2009). Three weeks later, Hasan purchased the weapon used on the November 5th attack at a Killeen, Texas gun store for $1,140 (Allen 2009). Hasan criticized the wars in Iraq and Afghanistan. He called himself a Muslim first and an American second. He spoke at a seminar at Walter Reed Military Hospital in Washington in June 2007, making a presentation entitled "Why the War on Terror is a War on Islam." During his presentation he said, "It's getting harder ... for Muslims ... to morally justify being in a military that seems constantly engaged against fellow Muslims" (Friedman 2009). He was reported to have "applauded the killing of a US soldier at an Arkansas recruiting center." Hasan is also believed to have had business cards which contained the abbreviation "SoA(SWT)," which means "Soldier of Allah," and "Subhanahu Wa Ta'all,"—or Glory to God (Gibbs 2009).

In December 2008 his emails to al-Aulaqi were discovered by a Joint Terrorist Task Force (JTTF) in San Diego. A report stated that the "content of those communications was consistent with research being conducted by Major Hasan in his position as a psychiatrist at ... Walter Reed ... and nothing else derogatory was found ... the JTTF concluded Hasan was not involved in terrorist activities or planning" (Marquise 2010: 22). This information was shared with members of the Washington JTTF. As a result of the consistencies between Hasan's research at Walter Reed and the content of the email exchanges with al-Aulaqi, the investigation was dismissed despite allegations that his emails to al-Aulaqi contained cryptic or coded communications explaining how to transfer money overseas so as to not attract the attention of law enforcement (Marquise 2010). Based on all that is known, many have concluded that the Fort Hood attack was a terrorist attack against the US military. Despite this, the FBI has said its investigation indicates that the alleged gunman acted alone and was not part of a broader terrorist plot (CNN 2009a).

Fort Hood Implications for Law Enforcement Intelligence and Risk Prevention

The attack on Fort Hood is a tragic and frustrating reminder to the law enforcement community as to the "nature of the beast" when attempting to combat terrorism in the United States. As harsh reality reminds us, despite the best efforts of dedicated law enforcement personnel, preventing every threat is simply beyond reach. As discussed previously, terrorism requires long-term planning and thus instances for law enforcement to identify suspicious behavior. However, terrorism also has limited opportunities. When compared to other complex criminality, such as organized crime, terrorism does not provide law enforcement the ability to pattern certain behavior and methods as a result of a "one chance to strike" approach taken by terrorists. Organized crime requires planning and time but once a niche or method is discovered that yields desired outcomes, the criminal group will continue to use these methods on targets they know will yield success. This allows law enforcement to identify patterns and increase the likelihood of predictability, and thus increasing the likelihood of prevention. Terrorism, on the other hand, requires long-term planning but with the intent to be put into action at only one time on the intended target specific to that planning. Figure 11.4 provides an illustrative comparison of the likelihood of preventing criminal and terrorist risks. Terrorist methods may be similar; however they are not consistent enough for law enforcement to identify specific patterns of behaviors to implement prevention methods. Law enforcement must then rely upon lessons learned from incidents that do reach fruition in an effort to identify precursors to such events and develop mechanisms to get this information into the intelligence cycle.

If there is a positive to be taken from the events at Fort Hood, it is lessons learned for the future application of intelligence practices—specifically the identification of an individual's behavior that may illicit further investigation as it challenges the boundaries of reasonable suspicion. Hasan's behavior certainly warranted further investigation. His views on the US military, his speech at Walter Reed hospital, and his business cards identifying him as a "Soldier of Allah" were enough to meet the threshold of reasonable suspicion for an investigation. Further compounding the risk he posed were his access to Fort Hood and access to weapons.

Just as hindsight is 20/20, these facts seem obvious and yet the attack came to fruition. However, the information simply was not there, or at least was not brought together in a fusion process, to guide law enforcement decision making. Without the mechanisms in place to identify Hasan's behavior as suspicious, there was no way for this raw information to be passed along to the necessary actors. Throughout the course of events leading to the attack, if the mechanisms had been in place to identify and

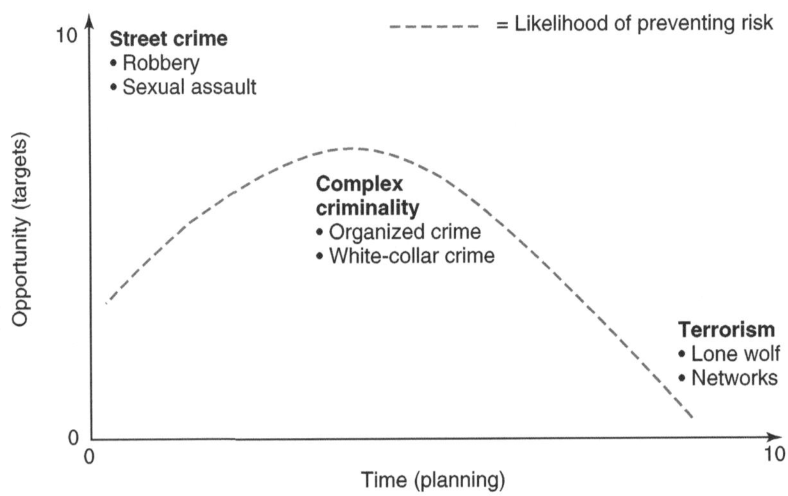

Figure 11.4 Likelihood of preventing criminal and terrorist risk

report his behavior, these suspicious activities would have been reported at different times and each within its own context. Moreover, as each suspicious behavior makes its way into the intelligence cycle, the "pieces of the puzzle" begin to paint a picture of what may occur. As intelligence analysts continue to input files on Hasan, his behaviors begin to shed light on a series of events leading up to what could possibly be a lethal attack. Thus, if the events had been reported through partnerships with personnel at the hospital, personnel on Fort Hood's military base, and an conscious awareness among citizens of what behaviors are suspicious law enforcement may have had the information necessary to act and prevent the threat.

Conclusions

The Fort Dix and Fort Hood cases examples provide real-world context to the application of law enforcement intelligence practices and, more specifically, suspicious activity reporting. Law enforcement's ability to prevent criminal and terrorist risks relies on the availability of raw information that is oftentimes outside the law enforcement purview. Preventing crimes and terrorism is not based so much on the highly sophisticated and technological tools available, but the simple, informal passing of information related to observed behavior that when taken in context of the totality of circumstances "simply doesn't seem right." State, local, and tribal law enforcement agencies must continue their efforts to establish relationships with their communities, businesses, and infrastructure entities to develop channels for raw information to get into the intelligence cycle. Without raw information, there cannot be analysis. Without analysis, there cannot be pattern identification. Without pattern identification, there cannot be predictability. Without predictability, there cannot be prevention.

Notes

1. Correspondence should be directed to: Jeremy G. Carter, School of Criminal Justice, Michigan State University, 560 Baker Hall, East Lansing, MI 48824-1118, P: (517) 355–6649, F: (517) 355–6646, E: carter64@msu.edu.
2. A lone-wolf terrorist is a single individual who is sympathetic to or affiliated with terrorist/extremist groups and who acts without direct support from a terrorist/extremist group.
3. For more information on SARs and the National SAR Initiative, visit: http://nsi.ncirc.gov/.
4. Al-Aulaqi has also been tied to Faisal Shahzad, the man who attempted to detonate a bomb in New York City's Times Square on May 1, 2010 (Huffington Post 2010; Shane and Mekhennet 2010).

References

Allen, N. 2009. "Fort Hood Massacre: Barack Obama Would Have to Sign Death Warrant." *Telegraph.* November 9. Retrieved from http://www.telegraph.co.uk/news/worldnews/northamerica/usa/barackobama/6531599/Fort-Hood-massacre-Barack-Obama-would-have-to-sign-death-warrant.html.

Anastasia, G. (2007). "Fort Dix 6: 'Good Boys' or Terrorists?" *The Seattle Times.* May 13. Retrieved from http://seattletimes.nwsource.com/html/nationworld/2003704914_dixprofiles13.html.

Carter, D. L. 2009. *Law Enforcement Intelligence: A Guide for State, Local and Tribal Law Enforcement Agencies.* 2nd ed. Washington, DC: Office of Community Oriented Policing Services.

Carter, D. L. and J. G. Carter. 2009. "Intelligence Led Policing: Conceptual Considerations for Public Policy." *Criminal Justice Policy Review* 20(3):310–325.

CNN. 2009a. "Investigators Look for Missed Signals in Fort Hood Probe." November 10. Retrieved from http://www.cnn.com/2009/CRIME/11/09/fort.hood.shootings/.

——. 2009b. "Fort Hood Soldier: I 'Started Doing What I Was Trained To Do'." November 9. Retrieved from http://www.cnn.com/2009/CRIME/11/09/fort.hood.foster/index.html.

Emerson, S. 2007. "Discussing the terrorist plot to attack Fort Dix." MSNBC. May 8. Retrieved from http://www.steveemerson.com/4284/discussing-the-terrorist-plot-to-attack-fort.

FOX News. 2007. "Store Clerk Helps Feds Bust 6 in Alleged 'Jihad' Plot to Kill U.S. Soldiers at Fort Dix." Associated Press. May 8. Retrieved from http://www.foxnews.com/story/0,2933,270601,00.html.

——. 2008. "RAW DATA: Charges in Fort Dix Terror Trial." Associated Press. December 22. Retrieved from http://www.foxnews.com/story/0,2933,470897,00.html.

——. 2009. "Fifth Man Convicted in Fort Dix Terror Plot Sentenced to 33 Years in Prison." Associated Press. April 29. Retrieved from http://www.foxnews.com/story/0,2933,518337,00.html.

Friedman, T. L. 2009. "American vs. The Narrative." The *New York Times.* November 28. Retrieved from http://www.nytimes.com/2009/11/29/opinion/29friedman.html.

Gartsenstein-Ross, D. and L. Grossman. 2009. *Homegrown Terrorists in the U.S. and U.K. An Empirical Examination of the Radicalization Process.* Washington, DC: Foundation for Defense of Democracies Press.

Gibbs, N. 2009. "The Fort Hood Killer: Terrified ... or Terrorist?" History News Network. November 11. Retrieved from http://hnn.us/roundup/entries/119883.html.

Huffington Post. 2010. "Faisal Shahzad Had Contact with Anwar al Awlaki, Key Figure in Fort Hood, Christmas Day Attacks." May 13. Retrieved from http://www.huffingtonpost.com/2010/05/06/faisal-shahzad-had-contac_n_566666.html.

Hsu, S. S. 2009. "Hasan's ties to radical cleric raise issues for law enforcement." *The Washington Post.* November 18. Retrieved from http://www.washingtonpost.com/wp-dyn/content/article/2009/11/17/AR2009111703830.html.

ICAP (International Association of Chiefs of Police). 2004. *National Policy Summit: Building Private Security/Public Policing Partnerships to Prevent and Respond to Terrorism and Public Disorder. Community Oriented Policing Services.* Retrieved from http://www.theiacp.org/Portals/0/pdfs/Publications/ACFAB5D.pdf.

Inskeep, S. 2008. "Opening Statements to Start in Fort Dix Plot." National Public Radio. October 20. Retrieved from http://www.npr.org/templates/story/story.php?storyId=95891389.

Jenkins, B. M. 2010. *Would-be Warriors: Incidents of Jihadist Terrorist Radicalization in the United States since September 11, 2001.* Santa Monica, CA: RAND Corporation.

Marquise, R. 2010. "Fort Hood Attack." *The Counterterrorist, Official Journal of the Homeland Security Professional* 3(1):20–28.

MSNBC. 2007. "6 Held on Terrorist Conspiracy Charges in N.J." May 8. Retrieved from http://www.msnbc.msn.com/id/18549005/.

Mulvihill, G. 2008. "Man Sentenced in Fort Dix Plot Case." USA Today. March 31. Retrieved from http://www.usatoday.com/news/nation/2008-03-31-619191134_x.htm.

NSI (Nationwide SAR Initiative). 2010. *Nationwide Suspicious Activity Reporting (SAR) Initiative.* Retrieved from http://www.nsi.ncirc.gov/.

Piette, J. 2009. "Meeting Exposes Fort Dix 5 Frame-up." Workers World. May 10. Retrieved from http://www.workers.org/2009/us/fort_dix_5_0514/.

Ratcliffe, J. H. In press. "Intelligence-led Policing: Anticipating Risk and Influencing Action." In *INTELLIGENCE 2010: Revising the Basic Elements*, edited by M. B. Peterson, B. Morehouse and R. Wright. Richmond, VA: International Association of Law Enforcement Intelligence Analysts.

Ripley, A. 2007. "The Fort Dix Conspiracy." *Time Magazine.* December 6. Retrieved from http://www.time.com/time/nation/article/0,8599,1691609-2,00.html.

Russakoff, D. and D. Eggen. 2007. "Six Charged in Plot to Attack Fort Dix." *The Washington Post.* May 9. Retrieved from http://www.washingtonpost.com/wp-dyn/content/article/2007/05/08/AR2007050800465.html.

Sageman, M. 2004. *Understanding Terror Networks.* Philadelphia: University of Pennsylvania Press.

Shane, S. and S. Mekhennet. 2010. "Imam's Path from Condemning Terror to Preaching Jihad." The *New York Times.* May 8. Retrieved from http://www.nytimes.com/2010/05/09/world/09awlaki.html.

Sherwell, P. and A. Spillius. 2009. "Fort Hood Shooting: Texas Army Killer Linked to September 11 Terrorist." *Telegraph.* November 7. Retrieved from http://www.telegraph.co.uk/news/worldnews/northamerica/usa/6521758/Fort-Hood-shooting-Texas-army-killer-linked-to-September-11-terrorists.html.

Silber, M. D. and A. Bhatt. 2007. *Radicalization in the West: The Homegrown Threat.* New York: New York Police Department, Intelligence Division.

USA Today. 2008. "Man Sentenced in Fort Dix Plot Case." Associated Press. March 31. Retrieved from http://www.usatoday.com/news/nation/2008-03-31-619191134_x.htm.

The Washington Post. 2007. "Six Charged in Plot to Attack Fort Dix: "Jihadists" Said to Have No Ties with Al-Qaeda." May 9. Retrieved from http://www.washingtonpost.com/wp-dyn/content/article/2007/05/08/AR2007050800465.html.

——. 2010. "Clues Left by Fort Hood Suspect Raise Haunting Question: Should Army Have Seen it Coming?" January 1. Retrieved from http://www.dallasnews.com/sharedcontent/dws/dn/latestnews/stories/010110dnmethasanclues.3f4b199.html.

Whitelaw, K. 2009. "U.S. Ponders How To Stop Homegrown Terrorism." National Public Radio. December 16. Retrieved from http://www.npr.org/templates/story/story.php?storyId=121510640.

Questions

1. Terrorism experts increasingly warn of so-called "home-grown domestic terrorists." To what extent do the Fort Hood and Fort Dix cases reflect this concern? What steps can be taken to try and identify these risks before they occur and prevent attacks?
2. From a risk governance standpoint, are there likely to be privacy issues that arise in assessing domestic terrorist groups in contrast to international terrorist groups? How might government officials try to minimize these concerns?

12.
EMERGENCY RESPONSE
Public–Private Sector Cooperation for Preparedness

In looking at risk, the compelling aspects of risk management come when agencies are actually confronted with emergencies. When in crisis mode, the ability to respond in a timely and organized fashion is partly a function of the steps that have been taken to prepare for and, at times, prevent the worst effects of the event. There have been many calls for a more resilient public emergency response program and much has been learned, as this has turned into actual programs, from the steps that private sector agencies have taken in preparing for the worst. Also, cooperation provides a greater likelihood that resources will be available for a rapid and fully funded response.

In the following chapter, Weber and Gade explore this private–public partnership in crisis management, showing how this collaboration can work effectively and reduce the long-term negative consequences of emergencies.

PUBLIC–PRIVATE PARTNERSHIPS FOR CRISIS MANAGEMENT

A Community Solution

Brit R. Weber and Susan J. Gade

Introduction

Critical incidents can have a profound impact upon communities across the nation. Traditionally, emergency preparedness was viewed as a primary function of law enforcement. Similarly, there was little collaboration regarding the nature of risks and threats between the public and private sectors. Private security and law enforcement often operated in separate silos and engaged in problem-solving in isolation. Furthermore, the attacks of September 11, 2001, and more recently, Hurricane Katrina, illustrate the inability of local government agencies to adequately respond to man-made and natural disasters. These events resulted in significant cultural, political, and social change in the daily lives of Americans. Furthermore, they demonstrate the need to engage in cooperative efforts to address safety and security. Therefore, understanding the dynamics between the sectors can lead to more thoughtful solutions for crisis management.

A critical incident is classified as any incident that endangers people or property, and occurs outside the scope of routine business activities (Jones 2000). This includes workplace violence, school shootings, acts of terrorism, and natural disasters or other significant events that may adversely affect a community. In contrast, crisis management refers to an organizational response to an incident in order to mitigate loss and effectively return to normal operations. Thus, joint crisis management involves the coordination of efforts from members of the public and private sectors for emergency planning and incident response. The terms emergency management and crisis management are used interchangeably throughout this chapter. Finally, a partnership consists of a cooperative relationship between individuals or groups who agree to share responsibility for a specific goal.

To assess the level of business preparedness, a survey of 461 companies was conducted by Urban Environmental Research in the greater Las Vegas metropolitan area. According to Sheila Conway, the group's managing partner, a one-week disruption would result in the closing of many local businesses (Miller 2008). Further, almost half of area businesses did not have a business continuity/disaster recovery plan (Miller 2008). Unfortunately, these findings are typical in other large metropolitan communities throughout the country. Jones (2000) reports that 93 percent of businesses that incur extensive data loss will close within five years, and 47 percent of companies will close after experiencing a serious fire or theft. As such, the preparedness capability of most businesses could be characterized as fragile.

Failing to plan for accidents, emergencies, and critical incidents may yield serious financial consequences for a community. In cities, counties, and regions, local governments face budget limitations with dwindling resources and the added pressure of continuing to provide essential services. The terrorist attacks on September 11 compounded this problem. Hurricanes Katrina and Rita exacerbated local, state, regional, and national resources and resulted in 1,300 deaths, $80 billon dollars in damage, and forced evacuations from five states (US Department of Homeland Security 2006). In addition, the coordination of resources to disaster areas is often fragmented, at best. After Hurricane Katrina, poor interagency communication and a lack of coordination resulted in the loss of resources targeted for disaster areas. Yet, many businesses and organizations have a vast amount of resources and are often willing to provide assistance to the public sector during times of crises, including human capital. From a security standpoint alone, according to the Hallcrest Report II, private security (proprietary and contract) outnumbers police three to one thus representing a significant potential resource (Cunningham, Strauchs, and Van Meter 1990).

One solution to enhancing our nation's safety and security is a public–private partnership for joint crisis management, which utilizes an all-hazards approach. Partnerships can benefit communities through enhancing their safety, social capital, and economic well-being. The value of diversified public–private collaboration for managing critical incidents can improve preparedness capabilities and response to critical incidents. Ultimately, protocols and agreements can be developed for emergency planning, response to, and recovery from the impact of man-made and natural disasters. Partnership models may vary as to purpose, construction, and extent. Unfortunately, there is little scholarly research available on public–private partnerships for emergency management that utilize an all-hazards approach. As such, we attempt to broaden the scope for the purpose of problem-solving on multiple levels. In this chapter, we explore the purpose of public–private partnerships, their implementation, and how they can increase the level of community preparedness and homeland security. In addition, we also discuss some of the challenges to establishing cooperative relationships and provide a model of success developed from an existing partnership program.

Historical Aspects of Public–Private Partnerships

Little documentation is available regarding the origins of public–private partnerships in the United States. However, various solicitations for community involvement date back to the early twentieth century, when the government requested public assistance to combat crime. In the 1930s, J. Edgar Hoover implemented the FBI's Most Wanted Program, which enlisted public support to help apprehend the nation's top fugitives. During the Great Depression, counterfeiting was also widespread, when the value of the US dollar had plummeted. In 1937, the US Secret service launched their "Know Your Money Campaign" to increase public awareness regarding fraudulent currency. This led to a substantial reduction in the number of counterfeit dollars in circulation and improved the public's perceptions of law enforcement.

Four decades ago, the concept of formalized partnerships between the public and private sectors began to emerge. During the early 1970s, the New York Partnership was formed to improve economic vitality. Interestingly, what spurred the development of this group were the financial problems that led to a stronger commitment by the New York Partnership to integrate their organization into the community (Gifford 1986). In 1972, the Private Security Advisory Council was established by the US Department of Justice to enhance cooperation between private security and law enforcement in effort to prevent and deter crime (IACP 2004).

Recognizing the volume of private security and their contributions to public safety, the Council influenced the development of subsequent training and education standards (Nemeth 2005).

During the 1980s and 1990s, several notable partnerships programs had been adopted both domestically and abroad. Founded by Chief Michael Shanahan in 1980, The Washington Law Enforcement Executive Forum (WLEEF) brought together members of law enforcement and local businesses to address crime problems (IACP 2004). Currently, the WLEEF is composed of 26 leaders from both the public and private sectors who maintain ties with local, state, and national government agencies, engage in legislative efforts, and promote business security (Nemeth 2005). In 1985, the Overseas Security Advisory Council (OSAC) was commissioned under President Reagan to protect US businesses operating on foreign soil. The primary objective of OSAC is to provide information regarding international threats and assist US businesses with developing safety and security measures in a global environment. The following year, the International Association of Chiefs of Police (IACP) formed the Private Sector Liaison Committee (PSLC) to promote cooperation and enhance relationships between public sector agencies and the business community. The PSLC currently has several subcommittees that collaborate with other groups to address emergency management and security issues. It is important to note that the American Society for Industrial Security International, a leading organization for security professionals, was also involved in several joint initiatives with the National Sheriffs' Association, the IACP, and other organizations.

In 1990, Hallcrest Report II—Private Security Trends (1970–2000) was published under a grant from the National Institute of Justice. The report indicated the need for public–private partnerships for the safety and security of communities. In 2000, the Hallcrest Division of Science Applications International Corporation published *Operation Cooperation, Guidelines for Partnerships between Law Enforcement & Private Security Organizations* (Connors et al. 2000), which provided an overview of public and private cooperative programs in the United States to enhance security and safety. Operation Cooperation was a "call for action" and outlined best practices based upon research and strategies from public and private sector practitioners.

The Federal Law Enforcement Training Center (FLETC) serves as a training facility for law enforcement agencies. With support from Mobil Corporation, FLETC implemented Operation Partnership in the 1990s, a cooperative training program for law enforcement and security executives. Various additional training programs are now offered by FLETC, such as cyber terrorism and critical infrastructure protection.

In 1998, the US Department of Justice awarded a federal grant to Michigan State University (MSU) to research the effectiveness of collaboration between government agencies and the business community for emergency preparedness. It was believed that larger businesses created dynamic security programs; therefore, law enforcement and private security could assist in emergency preparedness planning, response, and recovery. Under the direction of Radford Jones, focus groups with leaders from public agencies, businesses, and non-profit organizations were held to identify the challenges, possible solutions, and best practices of partnerships. The highlights of this information led to the publication of *The Critical Incident Protocol—A Public and Private Partnership* (Jones 2000), which contained best practices and lessons learned based upon input from both scholars and practitioners. Due to the interest in the publication, the federal government asked MSU to develop a program to facilitate partnerships at the community level. In 2002, MSU created Critical Incident Protocol (CIP)—Community Facilitation Program, which establishes public–private partnerships for joint crisis management. This program, which has served as a model in communities across the nation, will be discussed in greater depth in the following section.

While cooperation between agencies and organizations has existed for some time, formalized partnerships for emergency preparedness and homeland security is a more recent phenomenon. The advent of new technology, modern terrorism, and transnational crime has influenced the nature of their development. Thus, partnerships have evolved over the years to reflect the current dynamics and challenges faced by law enforcement and security professionals worldwide.

The CIP Program

The CIP–Community Facilitation Program was developed to assist local communities bridge the gap between government agencies and private security.[1] As previously mentioned, the program was developed from a research project on the effectiveness of joint cooperation for emergency planning and preparedness. The findings were incorporated in *The Critical Incident Protocol—A Public and Private Partnership* (Jones 2000). The publication includes information on the mutual benefits of joint planning, leadership roles, overcoming roadblocks, critical incident planning, and other pertinent topics. The Protocol identifies the following objectives:

- Create public and private sector understanding of their common goal to protect lives and property while sustaining continuity of community life.
- Encourage the public and private sector entities that may have engaged in the assessment and planning process in isolation to form cooperative partnerships.
- Assist those businesses and communities lacking emergency planning experience in the development of a joint emergency planning process.
- Develop an understanding of mutual or respective goals and understand how public and private resources can complement and support each other.
- Serve as a resource for those engaged in the joint planning process (Jones 2000: 4).

These objectives were incorporated as the goals of the CIP Program that was launched in 2002. Initially, the intent of the program was to work with the first responder community and security directors from a few large businesses in specific locations. However, over time, the list of prospective stakeholders would greatly expand with future communities. Two to three workshops designed by the program staff were offered to each participating community, with each workshop building upon the previous one. The grant funding allowed the program to be offered free to participating communities (cities, counties, and regions) thus facilitating diffusion across the nation.

During the first workshop, participants discuss various crisis management topics and engage in individual/group performance tasks. One fundamental goal of the CIP Program is the creation of stronger relationships between stakeholders. Not surprisingly, the majority of first responders and business leaders may be acquainted, but had little experience, if any, in actually working together. The second workshop, which occurs about four months later, consists of a two-hour group discussion on how to sustain the partnership. A four-hour tabletop exercise, held six months after the second workshop, is used to culminate the facilitation of the partnership as well as development of concrete security plans and processes. A crisis scenario is subsequently designed, which incorporates the local first responder community. During the exercise, the participants not only identify strengths and weaknesses, but simulate a joint Emergency Operations Center (EOC). Thus, businesses and first responder agencies working side-by-side become more effective in response and recovery.

When the CIP Program initially began, four communities requested to work with MSU: Redmond, Washington; Evansville, Indiana; Monroe County, Michigan; and, Richmond, Virginia. Businesses from these communities included Microsoft, UPS, Mead Johnson Nutritionals, Meijer Stores, George Koch Sons Industries, Target Stores, and others. In addition, law enforcement, fire, emergency medical service (EMS), and emergency management officials worked with the business security leadership in the workshops. After four communities completed the CIP Program, an analysis of the participant evaluations in the pilot program indicated that

- 98 percent agreed that the process built teamwork between the public and private sectors;
- 93 percent agreed that the process enhanced individual and collective skills for critical incident preparedness and response;
- 92 percent agreed that the facilitation helped identify roles and clarified responsibilities of public–private partners;
- 90 percent agreed that the process helped assess the capabilities and resources for critical incident preparedness and response;
- 89 percent agreed that the process validated their emergency response plan (Weber 2002: 3).

By late 2009, the four pilot communities expanded to 47 cities, counties, and regions in 24 states and with over 4,200 public and private sector leaders participating in the workshops. However, important changes were made to the program. Initially, security and law enforcement were almost exclusively involved in the partnership. However, this grew to include fire, EMS, and other agencies, since it became apparent that members of other professions and disciplines should be present. For example, when facilitating a tabletop exercise between the local first responder community and the leadership team from an upscale large shopping center, the marketing director was critical to the recovery of the mall in rebuilding relationships with customers, vendors, and partners. When considering the complexities of an organizational or community construct, the list of participants multiplies. In the CIP Program, members of the following units are encouraged to participate:

- Public sector: police, fire, EMS, emergency management, homeland security, health, port security, public works, administrators, public facility management, media (public information officer), water/wastewater, information technology, public transportation, economic development, libraries, parks/recreation, elected officials, and others (local, tribal, state, and federal).
- Private sector: businesses with departments in security, safety, health, environmental, disaster recovery, risk management, business continuity, information technology, marketing, human resources, fleet management, supply chain, facility management, finance, and operations.
- Non-profits and other groups: citizen corps (CERT, MRC, etc.), chambers of commerce, downtown business groups, economic development groups, Red Cross and similar organizations, faith-based disaster recovery groups, airports, railroads, utilities, universities, colleges, hospitals, healthcare, schools, security groups, business trade associations, business continuity groups, contingency planner groups, property managers, and post offices.

Diversity greatly expands individual and group interaction. The more comprehensive the group, the greater the probability that problem-solving will be achieved. One drawback is a potential for confusion or misdirection among larger groups. Therefore, individual and group performance tasks were designed to target all levels. The focus of the workshops is group interaction and partnership sustainability.

Another change that emerged over time, reinforced by events such as Hurricanes Katrina and Rita, was an expanded focus from terrorism to all-hazards. Initially, given the recent September 11 terrorist attacks, the focus of the program was upon security, terrorism, and weapons of mass destruction. However, stakeholders made clear that building capacity to address all types of critical incidents increased preparedness and expanded the relevance of broad participation across sectors. The program concentration quickly moved to an all-hazards approach for crisis management.

Assessing the Program and Identifying Key Dimensions of Preparedness

Over a decade of research, program management, and collaboration with numerous organizations led to the identification of factors that influence partnership building. There are integral concepts to all participants in a partnership group, regardless of their public or private sector alliances. Several elements serve as the building blocks to all sustainable relationships, regardless of type. This includes leadership, clear sense of mission, ongoing communication, and a sense of trust.

All partnerships require leadership and a clear sense of mission to exist and be sustainable: leadership and a clear sense of mission. Thoughtful leadership enables a partnership to move forward. The leadership team must be cognizant of the stakeholders' interests and follow an agenda that is group-oriented to gain commitment. Partnership groups will embrace a formal meeting structure and may create subcommittees. Some tasks of these subcommittees include exercising, resource management, training, partnership sustainability, and reviewing plans and procedures. Equally important to collaboration is the establishment of a clear sense of the partnership mission, goals, or objectives. The goals are developed with input from the membership and provide direction. Like all successful undertakings, the leadership should promote the mission and goals and ensure that they are regularly reviewed for viability.

Leadership and mission, in turn, are essential to the formation of social capital, particularly relationships across sectors. Social capital is defined as the cooperation and collaboration between members of the public and private sectors in order to achieve a common goal. It also refers to the extent and nature of existing relationships in a community. Social scientists use extensive observations and research to examine a network of relationships and measure social capital. One of the challenges for MSU was to develop a method to evaluate the effectiveness of a public–private partnership. Although prior studies on social capital had been conducted in communities, research on the effects of social capital related to critical incident preparedness and response was lacking. To address this gap, a social capital survey was designed and administered to participating communities in order to determine the effectiveness of public–private partnerships for joint crisis management.

Two social capital surveys were administered prior to the orientation/planning session (pre-test) and the tabletop exercise (post-test). The surveys were typically administered ten months apart. This time period intended to capture the activity by the participants when they returned to their respective locations to collaborate on planning, training, and exercising activities. Upon completion of the surveys, an analysis was conducted and reports were distributed to participating communities, as well as the Federal Emergency Management Association, which currently funds the CIP Program. The report highlights information on the demographics of participants, level of participation in critical incident preparedness initiatives, perceptions of critical incident planning and preparedness, and attitudes toward community representatives. A comparison between the pre-test and the post-test results were outlined for each community, and were compared to other communities, respectively.

Perhaps one of the most important findings gleaned from the social capital survey concerns the participants' attitudes toward other community representatives (e.g., police, fire, EMS, corporate security, risk management, media, etc.). Higher levels of contact with members of the other sector led to greater the levels of trust, understanding, cooperation, and empathy toward a particular unit. Thus, in general, individuals who reported frequent contact with other groups also had more positive attitudes toward those groups. In addition, most participants are supportive of planning and preparedness initiatives and believe that joint efforts are necessary to mitigate and recover from the impact of critical incidents. The findings also indicate that when public and private sector leaders collaborate on joint preparedness, mitigation, response, and recovery, they are perceived as being highly effective. This demonstrates that obstacles can be overcome and the sharing of information and resources for joint response and recovery can enhance a community's preparedness capabilities.

The results of the social capital survey indicate the following:

- A 26 percent increase in participation in joint public–private sector meetings focusing on the reduction of risk.
- A 25 percent increase in participation in joint public–private sector meetings during the development of the organization's response plan.
- A 23 percent increase in participation in joint public–private sector exercises.
- A 20 percent increase in participation in a joint incident command post.
- An 18 percent increase in the adoption of the Incident Command System by public–private sectors.
- A 17 percent increase in the transfer of technology from one sector to the other to improve critical incident capabilities.

Understanding the dynamics of relationships between the public and private sectors assists in building effective partnerships to address critical incidents and emergencies in the community. Communities that recognize and utilize social capital are better equipped for emergency response than those who do not have such resources. Equally important, feedback assists stakeholders and communities in identifying strengths and weaknesses in the experiences and relationships of the participants related to joint crisis management. This insight is useful in guiding the development of joint public–private sector strategies within the community.

Benefits and Challenges of Implementing Partnerships

One of the fundamental reasons, and usually the initial one, that people and organizations create a partnership is to mitigate the impact of critical incidents that can negatively affect the community. Partnerships that are formed to solve problems between individuals or organizations are common throughout the country. At the community or regional level, the creation of public–private partnerships provides a forum to foster trust, share information, improve coordination, identify and assess vulnerabilities, address preparedness needs, and develop and implement solutions (TISP 2006). Partnerships inclusive of diverse membership bring an assortment of expertise and resources to mitigate or eliminate problems.

Operation Cooperation describes the following additional benefits for developing partnerships:

- networking and developing the personal touch;
- collaboration on specific projects;

- increased crime prevention and public safety;
- enhanced cross-fertilization of ideas;
- information sharing;
- leveraging of resources (Connors et al. 2000: 2).

An important benefit to a workable partnership is improved communication through the sharing of information, discussion for problem-solving, and learning about their partners' strengths and limitations. A significant challenge in organizations or between entities is communicating from top to bottom, bottom to top, across vertical lines, and with other stakeholders. Communication is important, since perspectives vary regarding the conceptualization of emergency preparedness and crisis management. The public sector is tasked with responding to emergencies, while businesses tend to focus on the costs associated with a particular incident. This profit philosophy is often not a primary concern of law enforcement agencies, since their immediate focus rests with public safety. Therefore, there must be a joint effort to communicate to address respective roles and responsibilities. Partnerships that utilize communication can lead to greater clarity of the issues, retention of valuable information, and collaboration for problem-solving. After a significant event, communication and information sharing between government agencies and the private sector is vital to minimize the loss of potential resources that may be misdirected.

When public and private stakeholders first meet to discuss the challenges they face, this information can enhance a partner's ability to make informed decisions based on the sharing of facts and intelligence. Approximately 85 percent of the nation's critical infrastructure is owned by the private sector (Office of Homeland Security 2002). Therefore, much of the information needed by the public sector to effectively respond to critical incidents is retained by the private sector. When governmental agencies gain a better understanding of the private sector capabilities and needs, public safety agencies can increase their level of preparedness and response. Moreover, businesses and non-profit organizations can improve strategic planning, business continuity, and emergency preparedness plans after learning the public sector's capabilities and responsibilities.

In today's society, we see the public sector facing budget limitations, doing more with less, an increase in citizens' expectations of governmental services, and pressure to keep up with technology, training, and information overload. In a partnership, the public sector benefits in numerous ways, as explained herein.

The identification of new resources is one of the many benefits to the public sector. When members of the public sector are initially exposed to a partnership concept, they immediately recognize the potential benefits in the partnership that a business community may offer. Traditionally, local public safety agencies relied on their own resources to respond to critical incidents. If these resources are exhausted, the public safety agencies will request assistance from adjacent localities, state, or federal agencies depending on the size and complexity of the crisis. However, it may take time for these resources to arrive to their destinations. Therefore, businesses that donate resources can greatly assist their respective public safety agencies and communities in responding to a crisis.

The public sector also possesses a greater understanding of the private sector's risks, threats, resources, and capabilities when engaging in joint planning efforts. A public safety agency's primary responsibility is to ensure the safety of their stakeholders. Often, public safety agencies focus on citizens' needs through planning and response methods with less emphasis on the business community. Learning the critical business functions of an organization can help the public sector devote resources where best needed and aid in the recovery process. Historically, police and fire services spent little time on joint planning in the business community's recovery efforts.

While both sectors should strive for rapid response and recovery, perhaps the largest impact of establishing partnerships is evident with the business community. In partnering with first-line responder agencies the business community can ensure their ability to resume normal operations, which is enhanced through joint planning. The quicker the recovery, the more efficient and effective the capability will be of the private sector to resume operations (Jones 2000). Partnering with government agencies can enhance a business's bottom-line and increase financial profits. One of the most important aspects of partnership building is understanding the "what's in it for me" concept so that prospective stakeholders appreciate the value to their organization.

As previously discussed, an effective partnership can enhance both the public and private sectors' abilities and capabilities in emergency preparedness, mitigation, response, and recovery processes. In turn, this improves a community's quality of life. A partnership built on communication and collaboration that identifies needs, establishes goals, and defines performance tasks to meet the objectives, enables a secure, healthy, and economically robust community. When communities recognize the value of partnerships, the potential for further development becomes realized. Collaboration can lead to other venues, such as economic, health, educational, social, and community planning initiatives.

Despite the numerous benefits of mutual cooperation, there are several reasons why partnerships may not be formed, such as apathy, accountability, denial, and misunderstanding (Jones 2000). Apathy or indifference tends to be more common in the public sector, since governmental employees may be preoccupied with public safety issues and unaware of the business community's needs and concerns. Denial reflects a belief that bad things happen to others but not to us, or in our community. In addition, denial of the possibility of an incident alleviates responsibility when one occurs. Stakeholders have often stated that they do not possess the resources or funding to devote to a partnership. Occasionally, decision-making in the political arena will hinder the development of a partnership as well. Additional reasons for not developing a partnership include a lack of interest, turf protection, and time. Protection of one's domain is commonly referred to as turf protection, in that people will safeguard their job or responsibilities to an unrealistic exclusion of outside influences. Leaders who vigorously protect their turf by restricting collaboration and cooperation will seldom embark on cooperative endeavors. However, many of the obstacles faced by prospective stakeholders can be overcome as the partnership progresses. The following case study illustrates how one group overcame some of the challenges and implemented a successful partnership.

LCRCIP: A Model of Success

Located at the northeastern tip of Illinois, Lake County houses the North Shore Region and is one of the wealthiest counties in the United States. In contrast, Cook County, Illinois is one of the largest counties in the nation, with more than five million residents. It contains several major corporations, prominent medical facilities, and attracts millions of tourists each year who visit the City of Chicago. In 2006, Discover Financial Services approached MSU to initiate a partnership group between the two counties after participating in the CIP Program. Eventually, about 15 communities from the greater Chicago metropolitan area joined together to form the Lake–Cook Regional Critical Incident Partnership (LCRCIP), one of the most progressive, formalized partnership groups in the nation.

During its development stage, there were several organizations that provided the leadership and commitment to move the partnership forward. For example, the Deerfield Police Department, Takeda Pharmaceuticals of North America, and other key organizations initiated contact with other local area businesses when creating the LCRCIP. Previously, the Deerfield Police Department and Takeda Pharmaceuticals security had a long-standing relationship, but the LCRCIP strengthened and expanded it into other areas. As a result, the Deerfield Police Department and Takeda Pharmaceuticals signed a memorandum of understanding that their respective facilities would serve as a back-up EOC.

Joe Sparks, Senior Manager of Security at Takeda Pharmaceuticals and co-chair of the LCRCIP, noted that partnering and networking have led to new resources, improved strategic planning, and expanded the role of security in the company. For example, Takeda Pharmaceuticals has been able to tap into new resources with the Lake County Emergency Management Agency, Lake County Health Department, and the Lake County Transportation Department. Through the Emergency Management Agency, they have greater access to weather related information that the security department shares with campus facility employees. Further, input from the health department on H1N1 has been used for planning purposes.

The partnership with the Lake County Transportation Department had a significant impact upon Takeda employees. When members of the LCRCIP received a tour of the Lake County PASSAGE security system, which monitors major highways and intersections, they were stunned at its capabilities. Sparks and his team returned to Takeda and advised employees that everyone could access the Internet camera system. Employees can now observe traffic and nearby facilities through the camera system to determine which routes to take when leaving the campus. This capability is critical during times of an emergency or crisis.

After attending the CIP Program, Rob Ugaste, Assistant Fire Chief of Lincolnshire–Riverwoods Fire Protection District and co-chair of the LCRCIP, initially speculated on whether the partnership philosophy and program model would be a success in the community. However, Ugaste subsequently noted several rewards for both the department and the district. For example, the traditional focus of the Fire Protection District, which was narrow in scope, is now inclusive of a global perspective. It has expanded the fire district's view in how they can assist area businesses during the recovery process. Finally, the partnership concept has influenced personnel during the decision-making processes and is inherent to the organizational culture.

Currently, there are over 200 members in the LCRCIP from domestic and international corporations, private sector businesses and non-profit organizations. From the public sector, the stakeholders represent local, state, and national members of law enforcement, fire, health, and emergency management departments. Further, there are two steering committees which provide guidance to the group. The private sector steering committee includes Discover Financial Services, HSBC USA, Motorola, Northwest Community Hospital, and Takeda Pharmaceutical Company Limited. The public sector steering committee includes Deerfield Police Department, Deerfield–Bannockburn Fire Protection District, Lake County Emergency Management Agency, Lake County Health Department, Libertyville Fire Department, and the Lincolnshire–Riverwoods Fire Protection District.

Partnership groups require both maintenance and oversight. One of the challenges experienced by the LCRCIP is its rapid growth in membership. To address this issue, the steering committees meet regularly and communicate frequently to manage LCRCIP activities while adhering to their mission and objectives. Another challenge was how to sustain the partnership group in the event that support was lost from participating agencies. Time, resources, and funding can adversely affect the sustainability of partnerships. However, the LCRCIP is committed to providing future services to the communities in which they serve, despite any challenges they may encounter.

There was no major disaster that occurred in the greater Chicago metropolitan area, which sparked the development of the LCRCIP. Rather, the stakeholders recognized the need to establish a formalized group between the two counties for emergency preparedness and joint crisis management. Recently, the IACP presented the IACP Chief Michael Shanahan Award for Excellence in Public–Private Cooperation to the Deerfield Police Department for its outstanding achievements in public safety through joint cooperative efforts. Both Takeda Pharmaceuticals and the Deerfield Police Department were formally recognized at the 2009 IACP International Conference. Since its inception, the LCRCIP has demonstrated the success of public–private partnerships and serves as a model for other communities to emulate.

Conclusion

Today, more leaders and stakeholders are seeking to establish cooperative groups to strengthen their organizations and the communities. In this chapter, we shared information on the rewards and challenges in building sustainable relationships and outlined a case study as a model of success. Public–private partnerships for joint crisis management can benefit both organizations and the community or region as a whole. Partnerships that have leadership, a clear mission, and diverse stakeholders provide the foundation for resilient businesses, public agencies, non-profit organizations, and ultimately, communities. Comprehensive security programs utilize risk and threat assessment to ensure quality and control. Partnering with key stakeholders increases the effectiveness of these programs, policies, and procedures. Through joint preparedness, mitigation, response, and recovery, businesses and government agencies can maximize success through collaboration. Thus, establishing public–private partnerships serves as a community solution for crisis management.

Note

1. The CIP program was developed by Radford Jones under funding to MSU from the Department of Justice and Department of Homeland Security. Jones was a retired special agent in charge from the US Secret Service who later served as global Manager of Safety and Fire Protection at Ford Motor Company and thus brought a perspective shaped by his experience in the public and private sector.

References

Connors, E., W. Cunningham, P. Ohlhausen, P. Oliver, and C. Van Meter. 2000. *Operation Cooperation. Guidelines for Partnerships between Law Enforcement & Private Security Organizations.* Bureau of Justice Assistance, Office of Justice Programs, US Department of Justice.

Cunningham, W., J. J. Strauchs, and C. W. Van Meter. 1990. *Private Security Trends 1970 to 2000: The Hallcrest Report II.* Boston: Butterworth-Heinemann.

Gifford, James P. 1986. "Partnerships and Public Policy Advocacy." *The Academy of Political Science* 36(2):74–86.

IACP (International Association of Chiefs of Police). 2004. *National Policy Summit: Building Public–Private Security/Public Policing Partnerships to Prevent and Respond to Terrorism and Public Disorder.* The US Department of Justice, Office of Community Oriented Policing Services (COPS). Retrieved on September 10, 2009 from: http://www.cops.usdoj.gov/files/ric/Publications/national_policy_summit.pdf.

Jones, R. J. 2000. *Critical Incident Protocol—A Public and Private Partnership.* Office of Justice Programs, US Department of Justice. Retrieved on September 12, 2009 from: http://www.cip.msu.edu/cip.pdf.

Miller, V. 2008, February 25. "Rainy Day Preparations: Disaster Planning Essential to Small and Big Businesses Alike." Las Vegas Business Press. Retrieved on July 11, 2009 from: http://www.lvbusinesspress.com/articles/2008/02/25/news/iq_19819486.txt.

Nemeth, C. P. 2005. *Private Security and the Law.* Boston: Elsevier Butterworth-Heinemann.

Office of Homeland Security (now known as the US Department of Homeland Security). (July 2002). National Strategy for Homeland Security. Retrieved on September 20, 2009 from: http://www.dhs.gov/xlibrary/assets/nat_strat_hls.pdf.

TISP (The Infrastructure Security Partnership). 2006. *Regional Disaster Resilience: A Guide for Developing an Action Plan.* Reston, VA: American Society for Civil Engineers.

US Department of Homeland Security. 2006. *Nationwide Plan Review Phase 2 Report.* In cooperation with the US Department of Transportation. Retrieved on September 28, 2009 from: http://www.dhs.gov/xlibrary/assets/Prep_NationwidePlanReview.pdf.

Weber, B. R. 2002, December 2. "Executive Summary on Development of CIP Program." Unpublished document, Critical Incident Protocol (CIP)—Community Facilitation Program.

Questions

1. How can public–private partnerships increase the effectiveness of each group taking on emergencies on their own?
2. What limitations do these partnerships impose on the clear planning of emergency response and the implementation of responses during an actual disaster?

PART III

CASE STUDIES OF RISK ASSESSMENT AND RISK MANAGEMENT IN AN ALL HAZARDS ENVIRONMENT

We include a number of articles that provide case studies of the ways in which risk assessment and governance have been applied in different contexts. We are particularly interested in including examples that consider risk in a global context. These examples illustrate different ways in which risk assessment and management can be applied to areas that can be impacted by crime and terrorism, but are not often considered in this context, such as global trade (Xu); pharmaceuticals (Fenoff and Wilson); food and food fraud (Spink; Mace); privatized cybersecurity (Yannakogeorgos); and environmental risks (Carter).

The chapters offer an introduction into the more general research areas that are impacted by considerations of risk and represent new thinking about how we might effectively apply this perspective in many different contexts. Given the global nature of these risks, they also suggest the need for broad enterprise risk approaches and cooperation between nations and between public and private sectors.

13. GLOBAL TRADE AND CRIME

Jie Xu

Crime has been a silent partner in modernization. Globalization creates new and favorable contexts for crime. Crimes benefit from opportunities not dissimilar to those enjoyed by multinational enterprise beyond the jurisdiction of the individual state and the limitations of single markets. The globalization of capital from money to the electronic transfer of credit, of transactions of wealth from the exchange of property to info-technology, and the seemingly limitless expanse of immediate and instantaneous global markets, have enabled the transformation of crime beyond people, place, and even identifiable victims.

Globalization brought about a tremendous increase in international business and trade. When more goods are transported by sea, there are more targets for criminal activities. Thus, declining for centuries, maritime piracy has resurged and even increased since the 1990s. It also contributed to the sharp increase in truck cargo theft and container crimes. Global trade brings together people of different traditions, laws, and institutions. Relationships are based on trust. In the event of this trust being misplaced, or of something else going wrong, cultural difficulties between trading partners will be compounded by jurisdictional problems and by the fact that, once a ship has left port, it is not easy to know what she is doing. Fraud has always been known to exist in shipping. In the past, it was generally the underwriter who was the target of fraud, the intent being the collection of insurance on ship and cargo. In some cases the ship was scuttled; in others insurance was affected after the ship was lost. In more recent frauds, however, the victim can be anyone connected with the transaction but usually the importer and particularly the state trading organizations of developing countries. The banks and underwriters are, in general, indirectly involved. The International Maritime Bureau (IMB) estimates that the losses due to maritime fraud are $13 billion per year.

As Beck (1992) pointed out, globalization has contributed to risk with greater temporal and spatial mobility. The universality of global threats is serving to democratize the distribution of risk. On a perceptual plain, this is indicative of a movement away from differential class-consciousness toward a universal risk-consciousness, in essence, the foremost threat of the world risk society—terrorism. The large and more frequent bank transactions brought by global trade are attributed to globalization finance. Almost all banks of the twenty-first century maintain their accounts electronically. This means that all transactions are automatically recorded and so can be traced to their sources. Thus, terrorists and organized crime groups spend much time covering up their money trails, and the globalization of finance in the twenty-first century has made this easier for them. They therefore set up shell companies and shell banks that become correspondent banks with major international and national banks. Terrorists and organized crime groups thus move money through many different banks and venues before it reaches its desired destination. Hence, money laundering becomes a hot topic in dealing with crime in global trade.

Specific Types of Crime in International Trade

The tremendous increase in international trade brought by globalization has facilitated the expansion of transnational crimes: maritime piracy, maritime fraud, cargo crimes, terrorism, and money laundering.

Maritime Piracy

The IMB has adopted a more liberal definition of piracy, as follows: piracy is an act of boarding any vessel with the intent to commit theft or any other crime and with the intent or capability to use force in the furtherance of the act.

The International Maritime Organization (IMO), through their Maritime Safety Committee Circular in 1993, has classified sea piracy into three broad categories:

1. Low-level assault or armed robbery. This is generally carried out from the vicinity of land from small, high-speed craft by groups of petty thieves armed with machetes, clubs, and occasionally low-velocity weapons such as pistols and shotguns.
2. Medium-level armed assault and robbery. This is carried out by people who are better organized and more heavily involved than those in the lower-level category.
3. Major criminal hijacks. These involve extreme violence. In some cases, the cargo is discharged or transferred to another ship. In other cases, the ship is renamed, restaffed, or sold by the pirates. Crew members are sometimes left adrift or killed.

With regard to the location of modern maritime piracy, piracy may well be an international and global phenomenon, impinging on the security of maritime traffic in African, Indian, Asian, and American waters. But the weekly and annual piracy reports from the IMB indicate that nearly 50 percent of such acts are being committed in an area that is known under the name "Maritime Asia"—a maritime region stretching from the Red Sea, the Persian Gulf, the Arabian Sea, and the Bay of Bengal through the Strait of Malacca all the way to the South China Sea and even the East China Sea. Since this maritime region, formerly known as the Maritime Silk Road, is home of some of the world's busiest sea lines of communication, this sorry fact could again be explained from a global perspective by pointing at the huge volume of traffic there. The Strait of Malacca, for example, is even dubbed "the iron highway" because of its very dense traffic: roughly 30 percent of world trade passes through the Strait, two-thirds of the world's supply of liquefied natural gas, and about half of the world's crude oil. And more traffic simply means more potential targets.

Maritime Fraud

The range of maritime fraud is startlingly wide and, with other types of crime, is constantly being expanded in line with the ingenuity and imagination of the people who practice it. Indeed, today's maritime fraud can be defined as a combination of various criminal acts such as forgery, barratry, piracy, theft, arson, etc., where one or more parties end up losing money, goods, or even the vessel. Maritime frauds can be classified into the following types: documentary; charter party; insurance; deviation.

Many maritime frauds are really documentary frauds that involve false or forged documents used to deceive. In the maritime context these documents are usually bills of lading and commercial invoices—or sometimes nothing more elaborate than a telex message. Buyers and sellers are often separated by political, legal, and geographical barriers. Sellers have a natural reluctance to part with goods until the purchase price is paid. Conversely, buyers prefer to pay on delivery of goods. The gap between payment and delivery of goods is bridged by the documentary credit system (UCP 500). The mechanism of this system is well known. The buyer instructs his or her bank (the issuing bank) to open a letter of credit in favor of the seller. The issuing bank instructs the paying bank in the seller's country to hand over the purchase price when the seller tenders documents confirming shipment of the goods. In a collection arrangement, for example, a payment against document transaction, the payment takes place in the buyer's country. The seller instructs a bank in the buyer's country to hand over the documents of title to the goods to the buyer upon receipt of payment. Both the documentary credit and the collection arrangements rely on the honesty of the trading partners and can be easily abused. A fraudulent seller can cash the letter of credit by presenting bogus documents for a non-existent cargo. Alternatively, the cargo can be of lesser quality or quantity. Sometimes, the same cargo is sold to two or more buyers, and the purchase price collected from each. Finally, a fraudulent buyer can present forged bills of lading in a payment against documents transaction and collect the cargo without paying for it (Eric and Campbell 1981).

Charter party fraud is another type of maritime fraud. Many sea voyages are made under the terms of a charter party. The success of this type of voyage depends on all parties, the ship owner, the charterer, and the cargo owner fulfilling their obligations. When one or more of the parties fails to do so, the resulting contractual mess usually requires a great deal of investigation and negotiation if anything is to be salvaged. Failures occur for two reasons, which often become blurred in practice: fraud and business failure. A typical fraud will concern a time charter party, under which a vessel is hired on a premium basis and hire is usually paid in fortnightly installments in advance. The time charterer pays the first hire and the vessel proceeds to the loading port where the cargoes are loaded. The shipowner authorizes the release of freight pre-paid bills of lading to the charterers who pass them on to the cargo owner in exchange for freight. In other words, the shipowner is releasing a document acknowledging the cargo owner's title to the goods on board his ship and that can be negotiated as representing these goods. He does this in recognition of the fact that his freight has been paid, although in fact all that he has received is the first two weeks' hire. What happens next is that the ship sails from the loading port to her destination and the charterers abscond with the entire freight, which they have collected from the cargo owner. The shipowner is contractually bound to deliver the cargo under the bills of lading, which he has issued. But since he is no longer receiving his freight he might not have the financial resources to do so (Eric and Campbell 1981).

Scuttling is the most common type of maritime insurance fraud. Scuttling is defined as the willful casting away of a vessel with the connivance of her owners. Investigations revealed that in the majority of cases the following features could be found: the shipping company was in financial trouble and probably having difficulty in meeting the mortgage payments; the vessel was experiencing difficulties in contracting charters; the freight rates were low; the vessels were grossly over-insured (Kapoor and Gray 1985).

Deviation fraud started as an innocent attempt by a desperate shipowner to avoid bankruptcy but later developed into a premeditated crime. Deviation means departing from the contract route or the usual route. Instead of proceeding to the agreed destination, the vessel deviates from her route to another destination where the cargo is sold illegally. The vessel is then deliberately sunk, is reported to be sunk, or simply vanishes by changing its identity and reappearing under another name (Kapoor 1983).

Cargo Crimes

Cargo crimes include truck cargo theft and container crimes. It is reported that cargo theft is a rapidly increasing crime throughout the world. Truck transportation is vulnerable to theft in many ways. It is highly visible and often highly identifiable; it is relatively slow and easily targeted; trucks may be parked unattended for relatively long periods of time making them easily accessible to professional thieves; trucks are inherently designed to be mobile and easily driven which makes the job of a potential hijacker or thief that much easier (Salzano and Hartman 1997).

Various types of criminal have infiltrated the system of containers. We have the theft of the whole container, either by force or by a manipulation of the system; we have theft of the contents, either by the opportunist or by organized crime. The theft of the whole container is sometimes accomplished by hijacking. It is now becoming increasingly common to be accomplished by altering paperwork or the manipulation of a computer entry. The benefit to this type of theft is that there are no witnesses and sometimes a great deal of time can elapse between the time of the theft and the time that the investigators manage to identify the method. The time lapse often means that the trail of evidence has grown cold.

Theft of part of the contents is far more common both in individual instances and in overall value. This is because it is possible to steal from a container and so disguise the manner of entry that the fact of loss does not become apparent until the container reaches its destination, which might well be both months later and many thousands miles away. The time lapse often militates against effective investigation.

The system of sealing suffered from three fundamental errors. First, many of the early seals could be compromised in that they could be opened and then closed again without any identifiable trace. Second, the locking mechanism of the container could be opened and closed again without trace and without ever touching the seal. Finally, the system of examining and recording seals was quickly eroded by the technicalities of the container handling system. The initial handling system had a requirement that seals should be examined during the transit of the container. This was considered an essential part of the system because the sanctity of the seal was the only replacement for the repeated tally process that existed in the days of conventional cargo. Unfortunately, increasingly sophisticated handling systems, combined with speed of throughout, meant that it was physically impossible to examine seals. This, together with the fact that examination of seals on a busy terminal brought with it a safety risk, has produced a "system by default" in which seals are not always examined when they should be and sometimes never at all. In some areas of operation it is a deliberate management policy not to examine seals because any identified discrepancy could possibly impose on the facility a liability of loss, which it is not obliged to accept (Salzano and Hartman 1997).

Terrorism

Terrorism can be conceptually and empirically distinguished from other modes of violence and conflict by the following characteristics:

- It is premeditated and designed to create a climate of extreme fear.
- It is directed at a wider target than the immediate victims.
- It inherently involves attacks on random or symbolic targets, including civilians.
- It is considered by the society in which it occurs as extra-normal, that is, in the literal sense that it violates the norms regulating disputes, protest, and dissent.
- It is used primarily, though not exclusively, to influence the political behavior of governments, communities, or specific social groups (Wilkinson 2000).

In 2004 the United Nations Security Council unanimously passed Resolution 1566 that defines terrorism and declares that in no circumstances can terrorist acts be condoned or excused for political or ideological reasons:

> Criminal acts, including those against civilians, committed with the intent to cause death or serious bodily injury, or taking of hostages, with the purpose to provoke a state of terror in the general public or in a group of persons or particular persons, intimidate a population or compel a government or an international organization to do or to abstain from doing any act, which constitute offences within the scope of and as defined in the international conventions and protocols relating to terrorism, are under no circumstances justifiable by considerations of a political, philosophical, ideological, racial, ethnic, religious, or other similar nature.

Clarke and Newman (2006) summarized four pillars of terrorist opportunity: targets, weapons, tools, and facilitating conditions. With regard to targets, there are an unlimited number of targets for terrorist attack (subway systems, buses, trains, airliners, power plants, reservoirs, embassies, public buildings, prominent individuals, etc.), but they do not all offer the same opportunity. Terrorists must choose among the various distinguishing characteristics of targets. The attractive characteristics of target include: exposed, vital, iconic, legitimate, destructible, occupied, near, and easy (Clarke and Newman 2006).

Money Laundering

Money laundering is the process of disguising the unlawful source of criminally derived proceeds to make them appear legal. Money laundering begins with the fruits of a crime and ends with funds that can be used safely or at least with minimal risk. Money laundering is usually described as having three sequential elements: placement, layering, and integration. Placement is the introduction of the funds into the financial system, whether through cash deposits or more complex methods. Layering is a set of activities intended to distance the funds from their point of criminal origin. Integration involves converting illegal proceeds into apparently legitimate business earnings through normal financial or commercial operations. Reuter and Truman (2004) found that three offense categories accounted for over 70 percent of entries: drugs, fraud, and other kinds of smuggling.

Criminological Theories and Approaches

Opportunities and Routine Activities Approach

Routine activities theory explains crime in the presence of three factors: motivated offenders, suitable targets, and an absence of capable guardians. Specifically, crime occurs when there are "one or more persons present who are motivated to commit a crime, a suitable target or potential victim is available, and there is an absence of formal or informal guardians who could deter the potential offender" (Cohen and Felson 1979). This theory is a strong theoretical foundation to understand crime and opportunities for crime. The routine activities perspective argues that normal movement and local environment plays a significant role in the interaction between the potential perpetrator and the victim in a crime.

This theory has been applied, with varying success, to such problems as homicide patterns, crime rates, workplace victimization, predatory crime, personal and property crime, and "demand" and "supply" of criminal opportunities, offenders' decisions to commit crimes, and "hot spots" of predatory crime (Messner and Tardiff 1985; Maxfield 1987; Lynch and Cantor 1992; Lasley 1989; Miethe and Meier 1994; Cook 1986; Sherman, Gartin, and Buerger 1989). Most explanations of crime that invoke a routine activities approach are speculative; empirical tests of the theory are not as prevalent as descriptive analyses. Even so, routine activities theory has been described as the most popular theory of victimization (Birkbeck and LaFree 1993).

Crime opportunity is a cornerstone in all criminal behavior, but opportunity alone does not lead to a crime (Felson and Clarke 1998). Perpetrators evaluate the situation or the opportunity by using external and internal variables and then decide how they best proceed. There are ten crime opportunity principles as follows (Felson and Clarke 1998):

1. Opportunity plays a role in causing all crimes.
2. Crime opportunities are highly specific.
3. Crime opportunities are concentrated in time and place.
4. Crime opportunities depend on everyday movements.
5. One crime produces opportunities for another.
6. Some products offer more tempting crime opportunities.
7. Social and technological changes produce new crime opportunities.
8. Opportunities for crime can be reduced.
9. Reducing opportunities does not usually displace crime.
10. Focused opportunity reduction can produce wider declines in crime.

Applying these approaches to maritime piracy, for example, the approach explains the cause of and factors in maritime piracy. Cottrill's investigation of piracy revealed that pirates now attack from speedboats, armed to the hilt with automatic weapons, mortar shells, and even antitank rockets. They are further aided by the Global Positioning System (GPS), and sometimes board ships with computer printouts of the cargo the ship is carrying (Cottrill 1997). When a vessel is carrying millions of dollars' worth of commercial goods, with little or no security measures taken, the motivated offenders component of routine activities scarcely needs discussion.

One trend in the merchant shipping industry in recent years has worked in the pirates' favor. With the advent of modern shipping technologies (e.g., GPS), shippers are employing fewer crew members. With fewer people aboard, would-be attackers are less likely to be deterred. This is compounded by that fact that ships' crew members are rarely armed, trained in piracy countermeasures, or capable of fending off attack.

The Maritime Administration in the US Justice Department has observed that: "Maritime criminals are inclined to operate in waters where government presence is weak, often lacking in both technical resources and the political will to deal effectively with such attacks." (www.globalsecurity.org/military/world/para/pirates.htm) Similarly, "attacks have become especially prevalent in Asian waters, where the absence of naval power, coastal police patrols, cooperative law enforcement and treaty agreements have made the shipping lanes a pirate's dream" (www.globalsecurity.org). Many nations where piracy is a problem also have insufficient resources to mount an effective counter-piracy campaign. The effective reduction of piracy requires something of a naval presence, but many third-world nations are incapable of mustering the materials and manpower to stave off piratical attacks. Mueller and Adler found that several nations, the Republic of Singapore in particular, could simply not fund much of an effort to secure the vast sea lanes around their shores (Mueller and Adler 1985).

The conflicting laws on piracy and complex regulations concerning crime at sea also conspire against effective guardianship. This is particularly true when piratical attacks occur in the territorial waters of one nation, but a pursuit has begun in another nation's territorial waters, or in international waters. Thus, the fact that many nations are clustered together in tight proximity in many regions of the world can be a major impediment to international cooperation in combating crime at sea.

Rational Choice Theory and Situational Crime Prevention

The rational choice theory states that crimes are broadly the result of rational choices based on analyses of anticipated costs and benefits (Clarke and Cornish 1985). Individuals choose to engage in crime in an effort to maximize their benefits and minimize their costs. This choice process occurs in two stages: first, individuals decide whether they are willing to become involved in crime to satisfy their needs. Whether they decide to engage in crime is heavily influenced by their previous learning and experience, including their moral code, view of themselves, personal experience of crime, and the degree to which they can plan and exercise foresight. In turn, the previous learning and experience are heavily influenced by a range of background factors, including individual traits, their upbringing, and their social and demographic characteristics such as sex and class.

Second, once individuals decide to engage in crime, they must decide to commit a particular offense. This decision is heavily influenced by the immediate situation of the individual. The individual may have a desperate need for money or may be out with friends who suggest engaging in crime. The individual then selects a target for the offense based on a consideration of costs and benefits. For example, the individual selects a home to burglarize based on considerations of whether the home is occupied, whether the home is easily accessible, whether there is reason to believe that it contains valuable items. The factors that individuals consider may differ dramatically from one type of crime to another, which is why rational choice theorists argue that crime-specific models must be employed. Different models of decision making are necessary for different types of crime.

The theoretical focus of rational choice theory has important policy implications on how best to reduce crime: situational crime prevention. Situational crime prevention introduces environmental change to reduce the opportunity for crimes to occur. Crime is prevented not by changing offenders but rather by changing aspects of the situations in which offenses occur. The focus is on making crime more difficult to commit or less profitable so that it becomes a less attractive choice. It is difficult to rehabilitate offenders or eliminate the root causes of crime. However, the components of situations are easier to manipulate. The key in doing this situational intervention effectively is to make the choice of crime more difficult or less appealing.

Clarke (1997) provided a useful way of understanding how crime opportunities can be blocked or made less attractive. First, there are strategies that seek to increase the effort needed to commit a crime. These might include using more effective physical barriers to crime. Second, there are strategies to increase the risks of attempting to commit a crime. These might include ways of increasing the chances of detection. Third, there are strategies to reduce the rewards of crime. These might include limiting the cash kept in a store. Fourth, there are strategies to reduce excuses such as set rules, signs. Take cargo theft for example. Since cargo theft is a property crime, situational crime prevention is a useful method to address this problem. Basically this is achieved by applying the following four prevention principles (Clarke 1997):

1. Increase perceived effort (Clarke 1997)—people's motivation to commit a theft is reduced if the perpetrator believes that the crime is too hard to commit. Preventive actions based on this idea can be categorized as physical separation of potential perpetrator and the object of the theft. This can be accomplished through the use of access control and different types of physical barriers (fences, locks etc.).
2. Increase perceived risks (Clarke 1997)—if perpetrators think that they will get away with a theft, it is more likely that they will commit it. By increasing the risk for perpetrators they are less likely to commit a theft. This can be accomplished by using surveillance systems and security personnel and by increasing employees' awareness.
3. Reducing anticipated rewards (Clarke 1997)—people are more likely to commit a theft if they can benefit from it. By making the target for the theft worthless or reducing its resale value it becomes less attractive for the potential perpetrators. This can be accomplished by marking the goods with either unique numbers or a product destroying device. Good examples of this principle are the safety case used in transports of valuables and money and the inc-tags used in fashion stores.
4. Inducing guilt or shame (Clarke 1997)—a theft is more likely to occur if it can be excused by appeal to reasons such as "the company can afford it" or "I've worked hard for the company but they have not thanked me for it" (Tyska et al. 1983). This is a form of ethnical relativity. Companies and organizations can affect this ethnical relativity by using company rules, signs, and regulations that demonstrate the right moral values. When theft is seen as an additional wage benefit for the employees this preventive action has failed (Muir 1996). By appealing to people's moral and making it easy for them to do the right thing, then excuses will be more difficult to use (Clarke 1997).

Applying these approaches to terrorist crimes has been introduced in the book *Outsmarting the Terrorists* (Clarke and Newman 2006). It will be discussed later.

Crime Displacement Theory

Crime displacement predicts that crime prevention in one area may have unintentional consequences for other areas or situations. This means that crime prevention may not lead to an absolute reduction in crime. The theory of crime displacement is based on the rational choice theory, with the following three assumptions concerning the potential perpetrator and the target (Lab 2000).

First, crime displacement assumes that crime is elastic (Reppetto 1976)—this assumption indicates that the demand for crimes is affected by preventive efforts (Hesseling 1994). Second, the perpetrator has mobility (Reppetto 1976)—the perpetrator has flexibility relative to time, place, method, and type of crime to commit. In reality, perpetrators are normally limited in their mobility, adaptability, and flexibility relatively to a particular crime, place, time, and method (Hesseling 1994). Third, there exists unlimited numbers of alternative targets (Clarke et al. 1985)—the perpetrators have unlimited numbers and types of potential targets to choose from.

The theory of crime displacement states that rational thinking perpetrators with crime mobility will alter their criminal behavior in response to crime prevention efforts (Lab 2000). Crime displacement will only occur when the alternative crime has similar cost–benefit structure rationalized within the perpetrators' decision process (Clarke et al. 1985). The key element in all crime is the role that opportunity plays, i.e., if there is no opportunity there is no crime. Based on the ten principles of opportunities and the configuration of the transport network, it is obvious that all opportunities cannot be eliminated but rather they can be reduced by implying substantive preventive countermeasures. The object is to reduce crime opportunities that will lead to a change in all potential situations and therefore crime displacement is a valid theory (Reppetto 1976).

Crime displacement can occur in several ways. Six types of displacement are explained:

1. Crime: transfer to other types of crime, e.g., offenders stop doing robberies and instead commit burglaries.
2. Target: transfer to other types of goods, e.g., offenders stop taking goods and instead target money transports.
3. Method: better locking devices force the offender to be more innovative, e.g., better doors force the offender to break in through the windows.
4. Place: transfer to a less protected target in the same or other areas, e.g., if one area improves security then the offender attacks another area.
5. Time: transfer to different times of the day, e.g., better night security forces the offenders to strike during daytime.
6. Perpetrator: transfer to another perpetrator, e.g., preventing one offender can create an opportunity for another offender.

Applying crime displacement to cargo theft in the transport network reveals that the source of risk for cargo theft in the transport network comes from theft opportunities. These opportunities are highly specific and concentrated in time and place. This is the result of required scheduling and fixed assets of the normal activity in the transport network. The risk for cargo theft can be reduced by initiating resilient countermeasures focusing on reducing theft opportunities. These countermeasures are to induce a change in the characteristics of cargo theft or, more correctly, change the outcome of a potential perpetrator's decision process. If a potential perpetrator can change the target, method, place, or time, they will do so within their own capability. This is crime displacement in the transport network. The key aspect to understanding this effect is the knowledge that the potential perpetrator will only respond to perceived preventive measures.

Risk Balance Theory

In Van Brunschot and Kennedy's *Risk Balance and Security* (2008), a risk balance perspective is offered. The balance derives from the following elements: the values that we retain that shape our views of the dangers and threats that we face, the choices that are made when calculating risks based on information that is available, and decisions regarding allocating resources to mitigate or alleviate harm. Risk is a calculation, based on values, choices, information, and perspectives of the likelihood of harm or benefit. This calculation of harm or benefit takes into consideration the types of hazards faced, along with the resources available to address the likelihood of harm that hazards pose. At the same time, the calculations of harms and benefits depend on the temporal stage of the hazardous event. Security therefore requires balancing between the hazards themselves, the means available to address various hazards, and the stage of the hazardous event. Security may result when hazards are balanced with resources, yet the balance changes depending on the temporal phase of the hazard (Van Brunschot and Kennedy 2008).

However, the characteristics of risk today—its global nature, imperceptibility, and ties to technological advancement—make identification and calculation of future harm difficult. All of these factors make risk balance complicated and inexact. It is not just the uncertainty of harm but also the lack of information relating to certain sources of harm that may confound individuals and institutions in the pursuit of security. Attempts to balance risk become more difficult and less certain in their results (Van Brunschot and Kennedy 2008).

The context of risk society suggests that the means by which we identify harms and threats may be largely influenced by doubt, information gaps, and lack of trust. While the emphasis on experts and expertise grew over the course of modernity, expertise today is increasingly viewed with suspicion. Furthermore, competing claims of experts contribute to security as a social construction. Weaving together various fragments of expert and other knowledge creates many security landscapes that may hinder seeing the forest for the trees (Van Brunschot and Kennedy 2008). A most important impediment to security as it relates to crime is jurisdiction—which jurisdiction harbors information, what kind of information is held, and who is allowed access to it. The threats that are most potentially harmful may be those that require both collaboration and the breaking down of traditional jurisdictional boundaries (Van Brunschot and Kennedy 2008).

Six principles are summarized to guide how risk balance is negotiated—choice, decision making, cooperation, planning, institutional learning, and communication:

1. Choice: we cannot apply infinite resources to address all of the threats that we face in modern life. Priorities are selected for action. Social and political factors as well as risk positions play a role in decision making, and the willingness to prioritize preparedness, response, and prevention measures to identified hazards.
2. Decision making: decision making regarding risk balance is based on three elements: information, trust, and other resources. Information is central to risk

balance: resources are typically allocated on the basis of what is known. Trust in the source of information helps figure into assessments of risk.

3. Cooperation: effective decision making requires cooperation across groups that have not dealt with one another in an easy fashion or that suspect one another in the ongoing battle over jurisdictional turf.
4. Planning: planning for disasters needs to extend beyond treating one type of hazard as completely different from another.
5. Institutional learning: the institutional learning of government agencies can facilitate a more competent response, contain the damage, and assist in a quicker response.
6. Communication: the mobilization of resources to prepare for, respond to, and recover from hazards is dependent on communication both within and between jurisdictions (Van Brunschot and Kennedy 2008).

The application of this approach is broad. One of the applications is discussed below on how to control terrorist and other crimes relating to global trade.

Promising Approaches to Control and Reduce these Threats from Terrorism and Crime

Improve Legal Regimes

Law enforcement is mainly governed by international law. But the existing legal regime has demonstrated a lack of effectiveness in cracking down on maritime piracy and money laundering. There are two major areas that need to be amended and improved: the first is definitional problem and the second is inadequate legal regime.

Definitions of maritime piracy and money laundering vary among nations, and the scope, content, and intent of statutes are sometimes at odds with one another. The divergent definitions of maritime piracy and money laundering can be extremely problematic for two reasons: first of all, it renders prosecution at the national level cumbersome; second, given the transnational character of maritime piracy and money laundering and the need for a unitary and coherent approach, it impacts negatively on the efficiency of the fight against them at the international level. One possible way to address this problem would be to adopt a very broad approach to predicate offenses, to expand their scope as much as possible (Unger 2007).

With regard to inadequate legal regime, international law obligates states to cooperate with the suppression of piracy and money laundering and grants them certain rights to seize pirate vessels and perpetrators and money launderers. However, different countries have different definitions and laws on maritime piracy and money laundering. There are gaps in state practices. Internationally, different organizations have different definitions and regulations, and there are different treaties. For example, no country in East Asia aside from the Philippines actually criminalizes maritime piracy. Also up for debate is how countries implement international treaties at the domestic level (Zou 2005).

In summary, international convergence with regards to maritime piracy and money laundering becomes a must if one wants to fight them effectively. Even if one jurisdiction extends predicate offenses so as to cover all crimes, it could still encounter jurisdictional problems if money laundering and maritime piracy is transnational in character and the other countries involved have not adopted an equally broad approach. Consequently, a move toward convergence would benefit not only investigation and prosecution efforts and lead to efficiency in international cooperation on criminal matters but also have significant beneficial implications for the international exchange of information.

Enhance Regional and International Cooperation

As Van Brunschot and Kennedy (2008) state, a central component of globalization is the breakdown of national borders. Globalization has forced interconnections and partnerships between national and international entities. Effective law enforcement will be very difficult if restricted to a single country. Maritime piracy, money laundering, terrorism, cargo crimes, and maritime fraud are all international crimes often involving several countries.

Take maritime piracy for example. Effective law enforcement is extremely difficult in Southeast Asia areas. A typical example is the South China Sea, which is vast and dotted with numerous uninhabited islands to which pirates can easily retreat. Another key obstacle to effective law enforcement rests with the overlapping territorial claims to some islands. Under such complicated circumstances, cooperation is the only way to maintain regional maritime security. The effective control of maritime piracy at the regional level can benefit not only adjacent countries but also the interests of global society. Multilateral cooperation bears noticeable significance when a coastal state has good reason to believe that a foreign ship has violated its laws and regulations, thus giving that state the right of hot pursuit. However, this right ceases as soon as the ship enters the territorial sea of a third state. Ideally, in this case, the situation would be righted through a regional cooperative arrangement between the affected states. Another critical matter that highlights the importance of cooperative enforcement is that due to domestic laws, many pirates go unpunished (Zou 2005).

References

Beck, U. 1992. *Risk Society: Towards a New Modernity.* London: Sage.

Birkbeck, C. and B. LaFree. 1993. "The Situational Analysis of Crime and Deviance." *Annual Review of Sociology.* 19:113–137

Brunschot, E. Gibbs Van and L. Kennedy. 2008. *Risk Balance and Security.* London: Sage.

Clarke, R. V. and G. R. Newman. 2006. *Outsmarting the Terrorists.* Westport, CT: Praeger Security International.

Clarke, R. V. and D. Cornish. 1985. "Modelling Offenders Decisions: A Framework for Policy and Research". Pp. 147–185 in *Crime and Justice*, Vol. 4, edited by M. Tonry and N. Morris. Chicago: University of Chicago Press.

Clarke, R. V. 1997. *Situational Crime Prevention: Successful Case Studies.* New York: Harrow and Heston.

Cohen, L. E. and M. Felson. 1979. "Social Change and Crime Rate Trends: A Routine Activity Approach." *American Sociological Review* 44:588–608.

Cook, P. 1986. "The Demand and Supply of Criminal Opportunities." *Crime and Justice: An Annual Review of Research* 7:1–28.

Cornish, D. and R. V. Clarke. 1986. Introduction. Pp. 1–16 in *The Reasoning Criminal*, edited by D. Cornish and R.V. Clarke. New York: Springer-Verlag.

Cottrill, K. 1997. "Modern Marauders: Pirates Use High-Tech Weapons." *Popular Mechanics* 170(24):2.

Ellen, E. and D. Campbell. 1981. *International Maritime Fraud.* London: Sweet & Maxwell.

Felson, M. and R.V. Clarke 1998. *Opportunity Makes the Thief: Practical Theory for Crime Prevention.* London: Home office police and reducing crime unit.

Hesseling, René. B. P. 1994. "Displacement: A Review of the Empirical Literature." Pp. 197–230 in *Crime Prevention Studies*, Vol. 3 edited by Ronald V. Clarke. Monsey, NY: Criminal Justice Press.

Kapoor, P. 1983. "Definition and Classification of Maritime Fraud." *Lloyd's Maritime and Commercial Law Quarterly* 29–44.

Kapoor, P. and R. Gray 1985. *Maritime Fraud: An Overview.* Plymouth: Plymouth Polytechnic.

Lab, S.P. 1988. *Crime Prevention: Approaches, Practices and Evaluations.* Cincinnati, OH: Anderson.

Lasley, J. R. 1989. "Drinking, Routines/Lifestyles, and Predatory Victimization: A Causal Analysis." *Justice Quarterly* 6:529–542.

Lynch, J. and D. Cantor. 1992. "Ecological and Behavioral Influences on Property Victimization at Home: Implications for Opportunity Theory." *Journal of Research in Crime and Delinquency* 29(3):335–362.

Messner, S.F. and K. Tardiff. 1985. "The Social Ecology of Urban Homicide: An Application of Routine Activities Approach." *Criminology* 23(2):241–67.

Maxfield, M. 1987. "Household Composition, Routine Activity, and Victimization: A Comparative Analysis." *Journal of Quantitative Criminology* 3:301–20.

Miethe, T. and R. Meier. 1994. *Crime and its Social Context: Toward an Integrated Theory of Offenders, Victims, and Situations.* Albany, NY: State University of New York Press.

Mueller, G.O.W. and F. Adler. 1985. *Outlaws of the Ocean.* New York: Hearst Marine Books.

Reppetto, T.A. 1976. "Crime Prevention and the Displacement Phenomenon." *Crime & Delinquency* 22:166–177.

Reuter, P. and E. Truman. 2004. *Chasing Dirty Money: The Fight against Money Laundering.* Washington, DC: Institute for International Economics.

Salzano, J. and S. Hartman. 1997. "Cargo Crime." *Transnational Organized Crime* 3(1):39–49.

Sherman, L. W., P. R. Gartin, and M.E. Buerger. 1989. "Hot Spots of Predatory Crime: Routine Activities and the Criminology of Place. *Criminology* 27(1):27–56.

Tyksa, L. 1983. *Controlling Cargo Theft: A Handbook of Transportation Security.* Oxford: Butterworth-Heinemann.

Unger, B. 2007. *The Scale and Impacts of Money Laundering.* Cheltenham, UK: Edward Elgar.

Wilkinson, P. 2000. *Terrorism Vs. Democracy: The Liberal State Response.* London: Cass.

Zou, K. 2005. "Seeking Effectiveness for the Crackdown of Piracy at Sea." *Journal of International Affairs* 59(1):117–134.

14.
COUNTERFEIT PHARMACEUTICALS
Examining the Multi-dimensional Risk to African Nations

Roy S. Fenoff and Jeremy M. Wilson

Described as the crime of the twenty-first century, counterfeit products threaten public health and safety, reduce jobs and tax bases, and inhibit corporate innovativeness and profitability. The production, distribution, and consumption of counterfeit pharmaceuticals represent a particularly dangerous public health risk; estimates of the numbers of counterfeit pharmaceuticals range from 10 to 15 percent of the world drug supply. Counterfeit drugs indirectly and directly adversely affect health. Indirectly, false drugs hasten the illness and death of consumers who do not receive the appropriate active agent or dosage to treat their conditions, and sub-potent pharmaceuticals that do not kill a disease-causing pathogen can eventually lead to the development of drug-resistant strains, making even the authentic drugs useless. More directly, toxic ingredients in counterfeit pharmaceuticals can cause serious health problems.

Widespread infectious diseases coupled with pervasive poverty make the developing countries on the African continent particularly vulnerable to harm from counterfeit drugs, which now constitute more than 30 percent of all medicines sold in Africa, including up to 70 percent of drugs sold in Nigeria. Despite the tangible risk the problem represents, little is known about its true prevalence and impact, supply and demand, existing prevention and intervention strategies and their effectiveness, and other issues critical to understanding its nature and to developing evidence-based strategies to address it. While not a definitive account, this chapter is put forward to introduce the topic of pharmaceutical counterfeiting in Africa by highlighting information gathered from open sources, as a means of generating discussion about the next steps to combat the crime, and their consequences.

Introduction

Counterfeiting, perhaps one of the oldest and most profitable occupations, can be traced back to the times of the ancient Egyptians and early Greeks (Craddock 2009; Penn 1979). Advancements in technology have facilitated the production and distribution of larger quantities of counterfeited goods in shorter periods of time, and have made it easier for counterfeiters to profit while successfully escaping detection. The Federal Bureau of Investigation calls counterfeiting the crime of the twenty-first century (UNICRI 2003) and the trafficking of counterfeit goods has become one of the world's largest and most rapidly growing criminal enterprises, with the International AntiCounterfeiting Coalition (IACC) estimating a value of more than US$600 billion annually (IACC 2009). Today, counterfeit goods are found throughout the world; wherever there is a demand for a product, counterfeiters are quick to fill the niche. The kinds of counterfeited products identified by the Organization for Economic Cooperation and Development (OECD) include clothing, auto parts, handbags, DVDs, electronics, food, water, and pharmaceuticals (OECD 2007).

The risks posed to society by counterfeit goods vary in scope and scale depending on the product and its range of distribution. However, of all the different counterfeit goods produced, none presents a greater risk than those affecting public health and safety, such as pharmaceuticals. The World Health Organization (WHO) estimates that of the 1 million malaria deaths that occur each year, 200,000 are the result of counterfeit anti-malarial drugs (WHO 2003). Harris, Stevens, and Morris (2009) state that counterfeit drugs for tuberculosis and malaria kill 700,000 people every year. Although the counterfeit drug trade is wide in scope, it has a more devastating public health effect on developing countries, such as those in Southeast Asia and Africa, where there are intense demands for inexpensive lifesaving drugs. This chapter presents an overview of the current knowledge of the counterfeit drug trade in Africa, with the aim to generate a discussion regarding the risks involved and how to respond to this problem.

Before discussing the damage caused by counterfeit drugs and their potential threat, it is necessary to define the term. Because the extent of the problem differs from country to country, the concept of "counterfeit drug" varies. For the purpose of this chapter we will use the definition developed by the WHO (2009):

> A counterfeit medicine (drug) is one which is deliberately and fraudulently mislabeled with respect to identity and/or source. Counterfeiting can apply to both branded and generic products and counterfeit products may include products with the correct ingredients or with the wrong ingredients, without active ingredients, with insufficient active ingredients or with fake packaging.

The Significant Magnitude of Counterfeiting

Traditionally, the most common types of counterfeit drugs were those described as "lifestyle drugs" such as those used to treat erectile dysfunction or baldness. However, recently there has been an increase in the counterfeiting of "lifesaving" drugs meant to prevent or treat asthma, malaria, cancer, HIV/AIDS, tuberculosis, blood pressure and heart conditions, diabetes, and severe diarrhea. In Africa, the three forms of drugs most frequently counterfeited are those aimed at malaria, HIV/AIDS, and tuberculosis (Amin and Kokwaro 2007; Kapp 2002).

The counterfeit drug epidemic currently threatening many African countries is gaining international attention. World leaders are beginning to realize that the counterfeit drug trade is not a local problem faced by a few developing countries, but a global one with far-reaching consequences. Surprisingly, even with an estimated US$39 billion annual world market (Pitts 2005), little is known about its geographic extent. Organizations such as the United Nations, the WHO, the World Trade Organization, and the OECD, and some government agencies, such as Nigeria's National Agency for Food and Drug Administration and Control (NAFDAC), have collected some startling statistics over the past few years. Armed with these data and following some high-profile arrests, these organizations have made public the impact of the counterfeit drug epidemic in developing countries and the steps that leaders are taking to bring the counterfeiting problem under control.

Recent findings by governmental and non-governmental organizations have identified the production and distribution of counterfeit drugs as a considerable risk to public health throughout the world, particularly Africa. According to the WHO, a significant fraction of the world's drug supply is counterfeit. Estimates of counterfeit drugs range from 10 to 15 percent for the world drug supply, to more than 30 percent of all medicine sold in Africa, and up to 70 percent of the drugs sold in Nigeria (WHO 2008; Primo-Carpenter 2009). In 2005 the estimated US$39 billion global market for counterfeit drugs was growing at 13 percent annually, with a projected increase of 92 percent to reach US$75 billion by 2010 (Pitts 2005). It is important to acknowledge, however, the difficulty of measuring the extent of counterfeiting with any degree of certainty. The clandestine nature of much of the manufacture, distribution, and sale of counterfeits make creating an appropriate sampling frame challenging, thereby inhibiting the accuracy of prevalence estimates. It is difficult to assess how well the sample drawn to test for counterfeits represents the true population of pharmaceuticals in the market.

Because so little is known about the counterfeit drug trade in Africa it is difficult to provide any solid information regarding regional differences. Much of the published information is "recycled" in that organizations often cite each others' statistics. Most of the information has been collected by international organizations and local government personnel and there are few empirical studies. Even with limitations, it is certain that the trade of counterfeit drugs is thriving throughout Africa, and most countries have been unable to effectively combat the threat.

A Model of the Counterfeiting Process

The process of creating counterfeit drugs is similar to counterfeiting any other product and the goal is the same: fabricate false products and/or false labels. Some drugs are produced with no, little, or too much active ingredient while others are composed (intentionally or otherwise) of toxic substances (Bate 2008a). Other methods include forging labels by changing expiration dates, packaging information, holograms, or trademarks (Forzley 2003; Sukhlecha 2007). The imitation or altered product is then produced in various quantities and entered into a need-driven market in which large profits are often made.

Because the trade in counterfeit drugs usually uses complex distribution networks, researchers have been unable to create a distinct criminal profile that identifies counterfeit drug producers and sellers. However, to combat the trade it is necessary to understand how counterfeit drugs are introduced into the distribution system. Figure 14.1 illustrates a three-component counterfeiting process: supply, process, and demand. The model, based on open source documentation, describes how the practice works

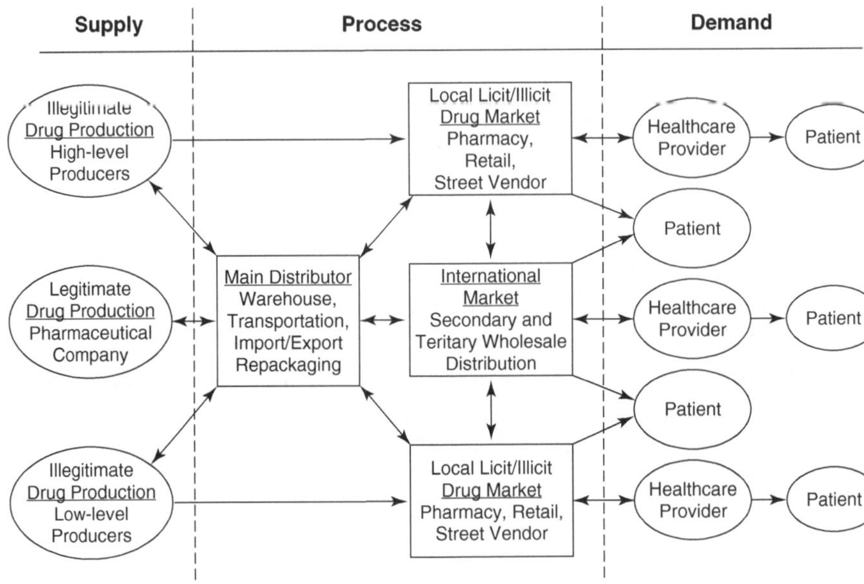

Figure 14.1 Counterfeit drug distribution flowchart

and identifies areas of opportunities for counterfeiters to enter drugs into the African market.

As shown in this model, it is not a single pathway from the manufacturer to the distributor to the customer (for additional depictions of the pharmaceutical supply chain, see Kontnik 2004; OECD 2008; Yankus 2006). The arrows indicate points in the supply chain where drugs exchange hands—each such instance is an opportunity for illegal activity. The process relies on a wholesale distribution network that often includes numerous distributors, intermediaries, and secondary and tertiary wholesalers. Although this process is multifaceted and often may seem confusing, it actually makes the delivery of drugs more efficient and economical for the supplier (Eban 2006). The process allows the supplier to produce drugs in large quantities and then sell them in lots to distributors, who then pass the products along the distribution network where small wholesalers are able to quickly find buyers. The distribution network is a mix of legitimate, semi-legitimate, and criminal players (Phillips 2007). As a result, the wholesale market opens the door for counterfeits and substandard drugs to be introduced into the supply chain (Patton 2007). One way this is accomplished is through diversion, or the manipulation of the legal flow of drugs. This can result in the authentic product being sold outside the intended market while creating an opportunity to introduce fakes into the supply chain. The US Food and Drug Administration claims that every case of counterfeiting is followed by diversion (Daleiden 2009).

As outlined in the model, the process begins with the manufacturer, who can be a legitimate pharmaceutical company or an illegitimate high- or low-level producer. The producers may directly enter the product into the local licit or illicit market or they may pass it on to a distributor who is often connected to a criminal network (Kontnik 2004). The distributor may then directly enter the product into the local licit or illicit market or sell the counterfeit drugs to wholesalers in the international market. These secondary wholesalers may then pass the product on to the local licit and illicit markets, or pass it on to a tertiary wholesaler who continues to forward the drugs along the supply chain; in some cases they may even sell the drugs directly to patients or healthcare providers (Kontnik 2004).

The process also works in reverse. Drugs that are damaged or expired are returned to the producer for destruction. However, many of these drugs never make it back to the manufacturer but are instead diverted, repackaged, and/or relabeled, and then inserted back into the supply chain to be resold to consumers (Yankus 2006).

It has also been claimed that the high-level producers are the ones who sell the counterfeit drugs to criminal networks; some of these producers have been linked to terrorist groups (Bate 2008a; IACC 2005). INTERPOL has identified a relationship between counterfeit drugs and terrorist financing; the relationship is either direct, when a terrorist group produces and sells the goods, or indirect, when the group receives funds from sympathizers (INTERPOL 2009).

The Internet expands opportunities for these diverse groups to connect. It is well known that Africa is becoming a hot spot for Internet scams; some of the better known frauds include phishing scams and the Nigerian 419 scam (BBC News 2003; Namibia Economist 2009). With an increasing number of Africans using the Internet, the number of cybercrimes continues to grow (Wanjiku, 2009). It is reasonable to expect these trends to exacerbate the counterfeiting problem in African countries.

Largely Unknown Supplier Characteristics

Those who take part in the production and distribution of counterfeit pharmaceuticals are extremely diverse (e.g., in terms of role, extent of involvement, and level of organization) and hard to identify, as the flow of these drugs involves numerous people and organizations from different parts of the world. Criminals include individuals, pharmacists, doctors, criminal groups, pharmaceutical companies, corrupt government officials, and terrorist organizations (OECD 2008). Established organized crime groups such as the Russian mafia (OECD 2008) and Chinese triads (Kafchinski 2009) are thought to be one of the main players in the lucrative business of counterfeit drugs (Lybecker 2008). In fact, there have been reports both of drug cartel leaders switching from trafficking in illegal drugs of abuse such as cocaine to counterfeit pharmaceuticals (Yankus 2006) and of cartel leaders adding counterfeiting to their business plans (Patton 2007). While both activities are profitable, the punishment for counterfeiting is relatively minor, with the most common penalty being a fine, while a conviction for dealing cocaine can lead to life imprisonment. Although it is not readily apparent who and how people behind the transportation and distribution of these drugs are organized, the drugs' origins are clearer. Roger Bate (2008a), a director of the non-governmental organization Africa Fighting Malaria states:

> Nearly every observer and researcher on (the) hunt for these dangerous fakes will point to two primary culprits: China and India. My own research on antimalarial drugs suggests that 60 to 80 percent of those fake drugs come from these two countries alone.

Peter Pitts (2005), president of the Center for Medicine in Public Interest, agrees with Bate and testified before the US Senate that China plays a central role in counterfeit drug production, with over 50 percent of the global supply of counterfeit pharmaceuticals having Chinese origin. Recently, Chinese Ambassador to Nigeria, Rong Yansong, issued a formal apology for the counterfeit drugs that have entered Nigeria from China and explained that the Chinese government would ensure the punishment of fake drug exporters (Nwezeh, 2009) (see also Box 14.1).

Box 14.1
Nigeria: Drug Agency Raises Alarm over Illegal Act of Fake Drugs in Kano State, June 22, 2009, Lagos: The Guardian Online (http://guardiannewsngr.com/index.php as of October 2009).

An alarm was at the weekend raised by the National Agency for Food and Drug Administration and Control (NAFDAC) that Kano has assumed the position of a sanctuary where manufacturers of fake food and drugs carry out their heinous activities. The agency's Director of Enforcement, Mr. Hashim Ubale Yusufu, made this known yesterday at a press briefing at the Kano chapter of the Nigeria Union of Journalists (NUJ) secretariat. Reading the Riot Act to distributors, wholesalers, and retailers of pharmaceutical and other NAFDAC-regulated products to stay away from such illegal act; he disclosed that a survey conducted by NAFDAC's officials indicates that counterfeiting activities is shifting from Onitsha and Lagos to Kano.

Yusufu … disclosed the discovery of a warehouse beside Kwari market fully stocked with counterfeit regulated drugs comprising 142 cartons of Tramadol capsules, 574 cartons of Ibumol as DA-ibugold, 244 cartons of chloraphenicol eye drop and packs of Tarivid tablets. He also disclosed the arrest of an illegal manufacturer, who was caught with large quantity of unpacked baking powder called "Super Baking Powder" as well as large quantities of fake curry powder known as "Duros" at 2A Sansusi Street, Sabongari, Kano. He further disclosed the discovery of a warehouse at Unity Road stocked with over 1300 cartons of Husna Kajal made in Pakistan, 118 cartons each containing 12 dozens of "kala-kola hair colour," 24 cartons of different unregistered bleaching creams, which had no manufacturers' name, address, batch number, manufacturing date, expiry date and list of ingredients on the packaging materials or labels.

A Distribution Process with Criminal Opportunity

African countries are attractive targets for trade in counterfeit pharmaceuticals; numerous factors allow criminal organizations to make enormous amounts of money with little risk of being caught. One is the extremely high level of poverty and disease across the continent. According to the WHO, laws and levies added 69 percent to the cost of legitimate drugs in 2003, making it impossible for most Africans to afford medicines from legal sources and further driving illicit drug markets (Bale 2001). Harris et al. (2009) support this contention, arguing that government-imposed distortions of the medicines market through taxes, tariffs, price controls, and licensing regimes contribute to the counterfeit drug epidemic. At least 103 pharmacies in South Africa have closed due to price caps on pharmaceuticals (Tren 2005). A second factor is the ease with which smugglers are able to get their products into different African countries. Harris et al. (2009) explain that the often flawed legal systems of developing countries contribute to this process. The countries also generally lack the necessary resources to invest in technologies to test and identify counterfeit drugs. One of the most common techniques to evade detection is to disguise counterfeit drugs as legal imports, by changing the appearance of the items, providing fraudulent documentation with forged signatures, or both. The number of corrupt and/or politicized legal systems in which informal payments can be easily made in order to pass counterfeits through bureaucratic barriers facilitate this fraud; some contend that many customs agents and police officers are easily bribed or fooled by forged documents (Harris et al. 2009; Anonymous 2008; Buyonge and Kireeva 2008; Edmonson 2005; Beisecker 2005). More sophisticated criminals use small amounts of the drug's active ingredient, referred to as "salting," so when and if a drug is tested it passes as the real product (Mukhopadhyay 2007). Smugglers will also conceal counterfeit drugs within legal shipments or use common delivery systems.

The Increasing Demand for Inexpensive Lifesaving Drugs

Like many other illicit markets such as the illegal trade of wildlife, weapons, and human trafficking, the trade of counterfeit drugs is a lucrative business with a growing demand that continues to fuel the trade. However, unlike the illegal trade of wildlife, weapons, and humans, which tend to have a demand driven by wealthier countries (World Wildlife Fund 2002; Pitts 2005), the demand for counterfeit drugs is highest in impoverished, less developed countries (Khan and Ghilzai 2007; Harris et al. 2009; Taylor 2008). These nations face problems that the developed world is able to escape. For instance, infectious diseases are prevalent across the African continent and a majority of the population is unable to afford or obtain access to lifesaving drugs through legitimate channels (Bate 2008b; Bale 2001) (see also Box 14.2). As a result, many African countries have been bombarded with counterfeit drugs and the governments have had little success in preventing these products from reaching local markets. According to Dr. Taryn Vian (2006), a healthcare expert from Boston University, the market for drugs in the developing world continues to grow with the rate of HIV/AIDS. This increased demand for cheap drugs has resulted in an influx of counterfeits entering the private and public markets in countries whose government regulatory systems are ineffective (Harris et al. 2009). The groups most affected by this are the poor and vulnerable populations (Harris et al. 2009; Vian 2006).

Box 14.2
The Nigerian Meningitis Epidemic, WHO (2003)

In 1995, a meningitis epidemic broke out in Niger. In response to the crisis, Niger authorities organized a country wide vaccination campaign. In an effort to combat the outbreak, Nigeria donated 88,000 vaccines which were given to more than 60,000 people. After the vaccines were administered, more than 25,000 people died. The Niger authorities suspected that the donated vaccines may have caused the deaths and conducted an investigation. The findings confirmed their suspicions that the drugs supplied had been substituted with counterfeit drugs that contained no active ingredients.

The Devastating Impact on Health and the Economy Due to the Risks Associated with Counterfeit Drugs

According to INTERPOL, malaria, tuberculosis, and HIV/AIDS are responsible for the deaths of more than 6 million people a year (Selgelid 2007). Overall, *P. falciparum* malaria kills more than 900,000 Africans annually, mostly children under 5 years old (CDC 2009). It is also estimated that 200,000 children die each year after taking counterfeit anti-malarial medications (INTERPOL 2008a). When the total deaths from malaria and tuberculosis are combined, the estimated number of deaths attributed to counterfeit drugs rises to 700,000 annually (Harris et al. 2009). Although this is an alarming number in itself some researchers claim it is actually a conservative estimate (Harris et al. 2009). Empirical studies in bacteria and parasite drug resistance and tests of active ingredient levels in counterfeit drugs have been conducted. A recent study published in *The Lancet* concluded that up to 40 percent of products labeled as containing the anti-malarial agent artesunate contained no active ingredients (WHO 2006a). A random survey in Kenya found that almost 30 percent of the all drugs were counterfeit, with some drugs being no more than chalk or water (WHO 2006b). Another study analyzed 195 packs of anti-malarials that were sold in six African countries and found that 35 percent of them did not contain efficacious levels of active ingredients (Bate et al. 2008). The danger of receiving drugs with inadequate levels of active ingredient is twofold. First, the individuals receive no therapeutic value from the drug, which increases their risk of death, and second, the low levels of active ingredient assist the microorganisms in adapting to the drugs making the drugs less effective. Regarding the drug resistance of the malaria parasite (*Plasmodium*), malaria expert Dr. Nick White of Oxford University said the following:

> Counterfeit medicine is a major reason why malaria had become, over the past 30 years Africa's biggest child killer, from an illness that used to be easily treated with medicines. There is now just one family of drugs left that malaria has not built up resistance to, Artemisinins—which are being faked. ... Resistance to the Artemisinins would be an absolute catastrophe for our current attempts to try to control malaria. (Shah 2007)

In addition to the devastating effects of drug resistance (Laxminarayan et al. 2006), counterfeit drugs also harm businesses and local economies (see Box 14.3). Counterfeit drugs affect companies that are producing genuine products because the counterfeit products will undersell the real ones. For example, an anti-malarial in Kenya that sells for $5 a dose is competing with counterfeits that are sold for $1 a dose (BBC News 2007). Because it is an illegal trade, the counterfeiters do not pay any import duties when they bring the drugs into the country and they do not pay any sales tax on the drugs that they sell so the local economy also suffers.

Box 14.3
DRC: Health Ministry Warns Citizens against Counterfeit Anti-Malaria Drug, June 12, 2009, Dakar PANA Online (http://www.panpress.com as of October 2009).

Kinshasa, DR Congo (PANA)—The Health Ministry of the Democratic Republic of Congo (DRC) has warned citizens against the purchase of "Quinizen 20%" drug, said to be counterfeit quinine, a government statement, made available to PANA here Friday, disclosed. According to Congolese Health Minister Auguste Mopipi, "the product, sold in the streets of Kinshasa and in some pharmacies in other parts of the country, is dangerous for children and does not heal malaria but rather worsens and complicates the state of patients. ... The 'counterfeit' paediatric quinine was modelled on that of the quinine syrup of 100 ml, manufactured by the Zenuphar firm that denounced the counterfeiting. The analysis of samples of the counterfeit product, made in Kinshasa, at the quality control department of the pharmaceutical firm's laboratory, stated that "Quinizen 20%" did not meet pharmaceutical standards. The country's health ministry has set up a commission of inquiry to dismantle the network of manufacturers of the counterfeit product.

Intervention Efforts

Nigeria appears to have had some success battling this public health crisis. Since 2001, the NAFDAC has reported numerous successes and has been credited with reducing the amount of counterfeit drugs on the market. While no formal estimate should be taken uncritically and further research should attempt to confirm the reduction, progress appears to have been made in curtailing Nigeria's counterfeit drug markets by the NAFDAC's persistence.

The NAFDAC was established in 1993, but for several years the organization was deemed by some to be unable to control the counterfeit drug epidemic (Bate 2008b). In 2001, the WHO estimated that 70 percent of the drugs being sold in Nigeria were counterfeit and little was being done to combat the problem (Primo-Carpenter 2009). In 2002, Nigerian health officials estimated that counterfeits made up 40 percent of the pharmaceutical market (Primo-Carpenter 2009) and a study cited by the American Council on Science and Health estimated that 85 percent of Nigeria's malaria drugs were fake or substandard (Yankus 2006).

Prior to 2001, Nigeria was the main market for counterfeit drugs in sub-Saharan Africa; as a result, neighboring countries responded by banning all pharmaceutical products made in Nigeria (Bate 2008b). This led to the appointment of Dr. Dora Akunyili as director-general of the NAFDAC.

Akunyili restructured the agency and strengthened import and export security by limiting the importation of all drugs and other regulated products to two seaports and two airports, and banning such imports through all land borders (Bate 2008b). The NAFDAC was also given the authority to demand that the Nigerian Ports Authority, independent shipping lines, and airlines release all shipping and cargo manifests to agency inspectors (Bate 2008b).

Because a risk is a perceived hazard (Howes 2005), some people do not see the risks associated with buying and ingesting counterfeit medicines. Some people may buy and ingest medicine knowing that they may be counterfeit because their perception of the risks that come with taking this kind of medicine is distorted, whereas others may be completely blind to the fact that the medicines that they are taking are counterfeit and pose a significant threat to their health. As a result, a large-scale public education campaign was undertaken to increase public awareness of the counterfeit drug problem. Every two weeks the NAFDAC discussed the differences between counterfeit and genuine drugs in Nigerian newspapers and on radio stations. As a result, between 2001 and 2006, Nigerian officials reportedly identified and destroyed more than US$109 million of fake drugs (Edike and Obinwanne 2006). In 2006, a WHO IMPACT report (citing the NAFDAC) announced that the amount of counterfeit drugs had been reduced to 16 percent of the total drug market in Nigeria (WHO 2006b).

Recently, the NAFDAC's new director-general, Dr. Paul Orhii, reported that over the last 8 years more than 120 destruction exercises of counterfeit and substandard drugs have been carried out by the NAFDAC, with 45 counterfeit drug offenders being convicted for their crimes (Obinna 2009). Orhii also reported that regulatory officers would be trained in the use of new technologies that allow them to test and detect the quality and efficacy of drugs coming into the country. In addition, the NAFDAC will soon partner with the private sector to better organize the country's pharmaceutical distribution network (Obinna 2009).

International efforts have been made to curtail criminal attempts to get counterfeit drugs into Africa as well. INTERPOL has been working with numerous police agencies to coordinate operations to track the movement of counterfeit drugs into Africa from other regions of the world. Additionally, in February 2006, the WHO created the first global anti-counterfeiting partnership, the International Medicinal Products Anti-Counterfeiting Taskforce (IMPACT). Together, these efforts have resulted in numerous successes:

- In August 2007, the Chinese pharmaceutical company Holley-Cotec recalled 20,000 doses of the anti-malarial duo-cotecxin from Kenya after testing results revealed low levels of active ingredient (BBC News 2007).

- In September 2008, Belgian authorities confiscated more than 2 million counterfeit anti-malarials and painkillers that were manufactured in India and were being shipped to Togo for distribution (BBC News 2008).
- In September and October 2008, INTERPOL led inspections of 191 locations in Tanzania, including pharmacies, warehouses, and illicit markets, resulting in the seizure of more than 100 different types of counterfeit drugs; law enforcement officials closed four pharmacies and 18 drug shops (INTERPOL 2008b).
- In September and October 2008, INTERPOL led inspections of 45 pharmacies and drug shop locations, 38 of which were investigated further for operating drug businesses without the required licenses and controls to ensure drug safety genuineness (INTERPOL 2008b).
- In November 2008, INTERPOL confiscated US$6.65 million worth of counterfeit HIV/AIDS, malaria, and tuberculosis drugs in Southeast Asia and made 27 arrests (Kaiser Network 2008).

In addition to the international and governmental response, private drug companies have stepped in to reduce this illegal trade. In June 2007, European drug companies announced plans to introduce new security measures, including two-dimensional bar codes and tamper-resistance packaging. In 2008, Pfizer held a conference in South Africa at which customs officials, regulators, the WHO, the Pharmaceutical Security Institute, and representatives from ten sub-Saharan governments came together to discuss the counterfeit drug epidemic in Africa and ways to combat it.

Important Unanswered Questions Remain

Due to the limited amount of information about illicit drug markets and organized crime and their relationship to the pharmaceutical counterfeiting process, there are many questions that need to be answered before evidence-based policy and intervention strategies can be put into place. Among the areas in need of analysis are the following:

1. Identify the level and forms of corruption surrounding the counterfeit drug problem, assess its relation to the counterfeiting process, and, if necessary, consider ways to reduce it.
2. Identify those who are producing, distributing, and selling counterfeit drugs and analyze their characteristics, levels of organization, and processes.
3. Identify the strengths and weaknesses of law enforcement and customs personnel in intercepting and preventing counterfeit drugs from entering their country. Areas of interest might include ports of entry screening procedures, training in forgery and counterfeit document identification, and technology being used to assist in the detection of counterfeit drugs and documents.
4. Articulate how counterfeit drugs are distributed once they pass customs and law enforcement personnel.
5. Identify and evaluate current drug regulatory controls, quality assurance procedures, and intervention strategies to discover weaknesses and propose ways to strengthen them.

6. Identify the ways in which pharmacists, healthcare providers, and consumers authenticate drugs, and the kind of training that would assist in identifying counterfeits.
7. Identify how counterfeit drugs are getting into Africa by investigating the supply chain and the trade routes used by the suppliers and smugglers.
8. Evaluate the current punishments and penalties for drug counterfeiters and determine which penalties may best prevent the crime.
9. Determine what government authorities are doing to stay current on trends and methods used by counterfeiters.
10. Estimate the country-specific quantities and the cost (by type) that counterfeit drugs represent each year.
11. Examine the forms and extent of crimes (and their effects and prevention) associated with the counterfeiting process.
12. Investigate how and the extent to which the Internet facilitates the trade of counterfeit drugs in Africa.
13. Assess potential strategies to reduce the demand for counterfeits.
14. Examine ways to reduce the supply of counterfeit drugs by strengthening law enforcement and rule of law institutions.
15. Examine the effectiveness of educational programs in reducing the purchase of counterfeit drugs.
16. Conduct an international assessment of promising practices for combating pharmaceutical counterfeiting (demand, supply, and process) for use in African countries.

Conclusion

The production of counterfeit drugs has been referred to as the "perfect crime" because a false drug is difficult to detect and the profit for the counterfeiter is high (Kontnik 2004). Because of the challenges in detection, there is much we do not know; societal, economic, and human costs due to deaths and injuries caused by counterfeit drugs may be higher or lower than what has been estimated. The estimates provided above are based on data collected by international organizations and the countries currently dealing with the counterfeiting problem. However, we do know that the risks associated with this illicit activity are becoming more serious. Along with "lifestyle" drugs, such as Viagra, counterfeiters are producing a variety of "lifesaving" drugs such as those used for malaria and tuberculosis. To combat and alleviate the potential threats that counterfeit medicines pose to public health, more research is necessary.

Ultimately, anti-counterfeiting strategies should be evidence-based with a primary focus on understanding and securing the pharmaceutical supply chain from counterfeiters. Any strategy or intervention considered should be supported by objective data and analysis.

References

Amin, A. A. and G. O. Kokwaro. 2007. "Antimalarial Drug Quality in Africa." *Journal of Clinical Pharmacy and Therapeutics* 32(5):1–15.

Anonymous. 2008. "Combating Counterfeit Drugs." *The Lancet*, 371:1551.

Bale, H. E. 2001. "Consumption and Trade in Off-patented Medicines." Commission on Macroeconomics and Health. Retrieved September 28, 2009, from www.icrier.org/pdf/bale65.PDF.

Bate, R. 2008a. "The Deadly World of Fake Drugs." *Foreign Policy* Sept/Oct:56–65.

——. 2008b. *Making a Killing: The Deadly Implications of the Counterfeit Drug Trade.* Washington, DC: AEI Press.

Bate, R., P. Coticelli, R. Tren, and A. Attaran. 2008. "Antimalarial Drug Quality in the Most Severely Malarious Parts of Africa: A Six Country Study." *PLos ONE* 3(5):1–3.

BBC News. 2003, November 26. "Nigeria to Tackle Internet Fraud." Retrieved September 28, 2009, from http://news.bbc.co.uk/2/hi/africa/3241710.stm.

——. 2007, August 17. "Malaria Drugs Recalled in Kenya." Retrieved September 28, 2009, from http://news.bbc.co.uk/2/hi/africa/6951586.stm.

——. 2008, October 3. "Belgians Seize African-bound Drugs." Retrieved September 28, 2009, from http://news.bbc.co.uk/2/hi/europe/7650136.stm.

Beisecker, R. 2005. "Securing and Facilitating Trade: Conflicting Goals under the World Customs Organization." Retrieved September 28, 2009, from http://cns.miis.edu/stories/051222.htm.

Buyonge, C. and I. Kireeva. 2008. "Trade Facilitation in Africa: Challenges and Possible Solutions." *World Customs Journal* 2(1):41–54.

CDC (Centers for Disease Control and Prevention). 2009. "Malaria." Retrieved September 28, 2009, from http://www.cdc.gov/malaria/.

Craddock, P. 2009. *Scientific Investigation of Copies, Fakes and Forgeries.* Burlington, MA: Elsevier.

Daleiden, B. 2009. "Systematic Security Protection through Pharmaceutical Traceability." Retrieved September 28, 2009, from http://secprodonline.com/articles/2009/02/20/systemic-security-protection.aspx?sc_lang=en.

Eban, K. 2006. *Dangerous Doses: A True Story of Cops, Counterfeiters, and the Contamination of America's Drug Supply.* Boston: Houghton Mifflin Harcourt.

Edike, T. and C. Obinwanne. 2006. "NAFDAC Destroys N14bn Fake Drugs." Retrieved September 28, 2009, from http://allafrica.com/stories/200611070673.html.

Edmonson, R. G. 2005. "Potential Barrier: Corruption in Some Countries Could Thwart Goals Set in WCO Framework for Supply-chain Security." *Journal of Commerce*, July 11:16–17.

Forzley, M. 2003. "Counterfeit Goods and the Public's Health and Safety." Retrieved September 28, 2009, from www.iipi.org.

Harris, J., P. Stevens, and J. Morris. 2009. "Keeping it Real: Combating the Spread of Fake Drugs in Poor Countries." Retrieved September 28, 2009, from http://www.ncpa.org/sub/dpd/index.php?Article_ID=18062.

Howes, M. 2005. "The Nature of Risk and Uncertainty." Pp. 3–24 in *Politics and the Environment: Risk and the Role of Government and Industry.* London: Earthscan Publications.

IACC (International AntiCounterfeiting Coalition). 2005. "The Negative Consequences of International Intellectual Property Theft." Retrieved September 28, 2009, from http://counterfeiting.unicri.it/docs/International%20AntiCounterfeiting%20Coalition.White%20Paper.pdf.

IACC. 2009. "Get Real: The Truth about Counterfeiting." Retrieved October 3, 2009, from http://www.iacc.org/about-counterfeiting/the-truth-about-counterfeiting.php.

INTERPOL (International Police Organization). 2008a. "Intellectual Property Crime Training and Operation Workshops." Retrieved September 28, 2009, from http://www.INTERPOL.int/public/ICPO/speeches/2008/IPCrimeNairobi20081121.asp.

INTERPOL. 2008b. "Operation Mamba (IMPACT) – Targeting Counterfeit Medicines in Tanzania and Uganda." Retrieved September 28, 2009, from http://www.INTERPOL.int/Public/News/2008/mamba20081029.asp.

INTERPOL. 2009. "Intellectual Property (IP) Crime." Retrieved September 28, 2009, from http://www.interpol.int/public/FinancialCrime/IntellectualProperty/Default.asp.

Kafchinski, J. 2009. "Global Counterfeit Trade." Retrieved September 28, 2009, from http://policy-traccc.gmu.edu/resources/publications/studentforum/Chinese%20Counterfieting%20Kafchinski.pdf.

Kaiser Network. 2008. "INTERPOL Seizes $6.65M in Counterfeit HIV/AIDS, Malaria, TB Drugs in Southeast Asia." Retrieved September 28, 2009, from http://www.kaisernetwork.org/daily_reports/rep_index.cfm?hint=1&DR_ID=55575.

Kapp, C. 2002. "Counterfeit Drug Problem Underestimated, Says Conference." *The Lancet* 360:1080.

Khan, A. Y. and N. M. K. Ghilzai. 2007. "Counterfeit and Substandard Quality of Drugs." *Indian Journal of Pharmacology* 39(4):206–207.

Kontnik, L. 2004. *Pharmaceutical Counterfeiting: Preventing the Perfect Crime.* Greenwood Village, CO: Lew Kontnik Associates.

Laxminarayan, R., Z. Bhutta, A. Duse, P. Jenkins, T. O'Brien, I. N. Okeke, A. Pablo-Mendez, and K. P. Klugman. 2006. "Drug Resistance." Pp. 1031–1051 in *Disease Control Priorities in Developing Countries.* 2nd ed., edited by D. T. Jamison, J. G. Breman, A. R. Measham, G. Alleyne, M. Claeson, D. B. Evans, P. Jha, A. Mills, and P. Musgrove. Washington, DC: World Bank Publications.

Lybecker, K. M. 2008. "Keeping it Real: Anticounterfeiting Strategies in the Pharmaceutical Industry." *Managerial and Decision Economics* 29:389–405.

Mukhopadhyay, R. 2007. "The Hunt for Counterfeit Medicine." *Analytical Chemistry* 79(7):2623–2637.

Namibia Economist. 2009, March 20. "Another 'Phishing' Scam Hits Bank Clients." Retrieved September 28, 2009, from http://allafrica.com/stories/200903200924.html.

Nwezeh, K. 2009. "Fake Drugs: China Apologizes to Nigeria; as NAFDAC's Drug War Goes Global." Retrieved September 28, 2009, from http://www.thisdayonline.com/nview.php?id=155044.

Obinna, C. 2009. "NAFDAC Introduces Techniques to Combat Fake Drugs." Retrieved September 28, 2009, from http://www.vanguardngr.com/2009/07/09/nafdac-introduces-techniques-to-combat-fake-drugs/.

OECD (Organization for Economic Co-Operation and Development). 2007. "The Economic Impact of Counterfeiting and Piracy: Executive Summary." Retrieved September 28, 2009, from http://www.oecd.org/dataoecd/13/12/38707619.pdf.

OECD. 2008. *The Economic Impact of Counterfeiting and Piracy.* Paris, France: OECD Publishing.

Patton, S. 2007. "Cracks in the Pharmaceutical Supply Chain." Retrieved September 28, 2009, from http://www.cio.com.au/article/170290/cracks_pharmaceutical_supply_chain?fp=&fpid=&pf=1.

Penn, R. G. 1979. "The State Control of Medicines: The First 3000 Years." *British Journal of Clinical Pharmacology* 8:293–305.

Phillips, T. 2007. *Knockoff: The Deadly Trade in Counterfeit Goods.* Philadelphia: Kogan Page Limited.

Pitts, P. 2005. "Counterfeit Drug Sales to Reach $75 Billion by 2010." Retrieved September 28, 2009, from http://www.heartland.org/publications/health%20care/article/17948/Counterfeit_Drug_Sales_to_Reach_75_Billion_by_2010_Report_Says.html.

Primo-Carpenter, J. 2009. "Matrix of Drug Quality Reports in USAID-assisted Countries." Retrieved September 28, 2009, from http://www.usp.org/pdf/EN/dqi/ghcDrugQualityMatrix.pdf.

Selgelid, M. J. 2007. "The Importance of Throwing Money at the Problem of Global Health." *Indian Journal of Medical Ethics* 2:73–75.

Shah, S. 2007. "Drug Giants Accused of Ignoring Fake Medicines that Kill Millions." Retrieved September 28, 2009, from http://www.independent.co.uk/news/world/africa/drug-giants-accused-of-ignoring-fake-medicines-that-kill-millions-444466.html.

Sukhlecha, A. 2007. "Counterfeit and Substandard Quality of Drugs: The Need for an Effective and Stringent Regulatory Control in India and Other Developing Countries." *Indian Journal of Pharmacology* 39(5):255.

Taylor, L. 2008. "UK Launches New Drug Access Alliance for Third World." Retrieved September 28, 2009, from http://www.pharmatimes.com/Article/08-06-11/UK_launches_new_drug_access_alliance_for_Third_World.aspx

Tren, R. 2005. "Price Controls Reduce the Availability of Medicines." Retrieved October 2, 2009, from http://www.freemarketfoundation.com/ShowArticle.asp?ArticleType=Publication&ArticleID=1093.

UNICRI (United Nations Interregional Crime and Justice Research Institute). 2003. "Global Counterfeiting: Background Document." Retrieved September 28, 2009, from http://counterfeiting.unicri.it/docs/Global%20counterfeiting%20background.pdf.

Vian, T. 2006. *Anti-corruption in the Health Sector: Preventing Drug Diversion through Supply Chain Management.* Bergen, Norway: Chr. Michelsen Institute.

Wanjiku, R. 2009. "Online Scams Up as More Africans Use the Internet." Retrieved September 28, 2009, from http://www.thestandard.com/node/136796.

WHO (World Health Organization). 2003. "Substandard and Counterfeit Medicines." Retrieved September 28, 2009, from http://www.who.int/mediacentre/factsheets/2003/fs275/en/.

WHO. 2006a. "Counterfeit Medicines." Retrieved September 28, 2009, from http://www.who.int/medicines/services/counterfeit/impact/ImpactF_S/en/index.html.

WHO, IMPACT. 2006b. "Counterfeit Medicines: An Update on Estimates." Retrieved September 28, 2009, from http://www.who.int/medicines/services/counterfeit/impact/TheNewEstimatesCounterfeit.pdf.

WHO, IMPACT. 2008. "Counterfeit Drugs Kill!" Retrieved September 28, 2009, from http://www.who.int/impact/FinalBrochureWHA2008a.pdf.

WHO. 2009. "General Information on Counterfeit Medicines." Retrieved September 28, 2009, from http://www.who.int/medicines/services/counterfeit/overview/en/.

WWF (World Wildlife Fund). 2002. "Switching Channels." Retrieved September 28, 2009, from www.wwf.org.uk/filelibrary/pdf/switchingchannels.pdf.

Yankus, W. 2006. "Counterfeit Drugs: Coming to a Pharmacy Near You." Retrieved September 28, 2009, from http://www.acsh.org/docLib/200608171_counterfeitdrugw.pdf.

15. CASE STUDY
Counterfeit Product Risk Model
John Spink

As described in Chapters 7 and 14, food counterfeiting represents a crime, a business risk, and a risk to public health. With the goal of better assessing the nature of these risks, and thereby improving risk management, a food ingredient company (referred to as "CSFI") was confidentially engaged to conduct this case study to demonstrate the use of the Counterfeit Risk Model. A series of interviews was conducted to solicit background information, to gain insight from a company expert panel, and to discuss both the working of the model and the results.

The company manufactures branded food ingredient products in a single-use "form" combined in a multiple-unit primary package selling unit. The product is sold by business-to-business (B2B), business-to-consumer (B2C), and also there is evidence of selling by consumers-to-consumers (C2C) on websites such as "e-Bay." The privately held company was started between 1985 and 1995 and has revenue in the $5M–$20M range. The company has global distribution of their product in both the bulk and finished form, and manufactures 95 percent of their product in one, US-based facility.

Methods

The researcher visited the company and conducted follow-up interviews with key personnel to gather background and specific information on the CSFI operations (procurement, manufacturing, distribution, customer service, and others) and the marketing (competitive products, sales process, distribution, retailing, and others). The researcher and the key company contact (the vice-president responsible for brand integrity) developed the preliminary CFSI data that would be entered in the Counterfeit Risk Model (also referred to as the Model).

The researcher and key contact developed the first versions of the completed Model, which were presented to the company expert panel (the full leadership team) and used this larger corporate group as the gathered expert panel. The expert panel discussed the sub-factors, sub-factor derivation, the summary formulas, and the overall results of the Counterfeit Risk Model.

The expert panel chose to narrow the details of the product, geography, and channel:

- Product: a specific CSFI-branded, packaged product that is sold by anyone directly to consumers (B2B, B2C, or C2C), including all forms of the product and all package styles including promotional items or special packs.

- Market geography: the specific geography was for sales in the US.
- Market channel: the specific market channel was actually all channels including wholesale, retail, on-line retail, and auction sites (e-Bay, etc.).

Results

The Model was used with CFSI data and presented in an overall counterfeit risk in Figure 15.1. To protect the identity of the company, the sub-factor derivation descriptors (identifying why a sub-factor was very high, high, medium, low, or very low) was

	Probability					
	Very high	*High*	*Medium*	*Low*	*Very low*	
Total rank				X		(1) 95+% of product is manufactured at one proprietary location; (2) the company has comparably low volume combined with advanced packaging and product features that are somewhat complex to duplicate; (3) the company monitors product in the marketplace including online C2C
Factors						
Summary function: A + B + C + D + E =F						All equal, manual derivation
1. Counterfeit-History	0	0	0	X	0	No known counterfeits, little diversion, but related lower premium product are faked
2. Counterfeit-Ability	0	0	X	0	0	The product and from are frequently counterfeited, but unique feature reduce risk
3. Counterfeit-Attractiveness	0	0	0	X	0	This is a premium product with the price two to five times the generic product pricing, but tight supply chain controls
4. Counterfeit-Hurdles	0	0	0	X	0	High hurdles for counterfeiters, in terms of packaging features, tight supply chain, and monitoring
5. Market Profile	0	0	0	X	0	Brand loyal consumers, price insensitive, pursue reputable sources, and in-house production

Figure 15.1 Case study overall counterfeit risk for CFSI

recorded as "tbd" (to be determined) and the sub-factor derivation comments were modified.

The result of the Counterfeit Risk Model was that CSFI has a "low" probability of counterfeit risk because: (1) 95+% of product is manufactured at one proprietary location on proprietary equipment; (2) the company has comparably low volume combined with advanced packaging and product features that are somewhat complex to duplicate; and (3) the company monitors product in the marketplace including on-line C2C.

Regarding the factors and sub-factors, it should be emphasized that it is up to the expert panel to define how sub-factors are arranged and weighted. Unique production, supply chain, marketing, retailing, or consumer factors may not be apparent to someone outside the company.

Details of the "factors" are included here:

- Counterfeit-History. This Total Rank was "low" since there are no known counterfeit products, there is little evidence of diversion of product outside the legitimate supply chains but there is some risk since related, lower premium priced products, are known to be counterfeited and diverted. These points lead to a "low" rank because CSFI is very diligent in monitoring the market and has no evidence of counterfeit product and very little diverted product.
- Counterfeit-Ability. This total risk was "medium" since the basic product form is frequently counterfeited but not "high" since CSFI uses very unique and complex packaging that would increase challenges for counterfeiters. These points lead to the increased "Medium" rank because even though there is a no evidence of counterfeiting of this product, there is a high likelihood that counterfeiters have the capabilities to attack.
- Counterfeit-Attractiveness. The total rank was "low" even though the product has a premium price of two to five times more than the generic version of the product. Tight supply chain monitoring and supply chain relationships make it hard to place counterfeit product in the legitimate supply chain. These points lead to a "low" rank because even though there is a high profit-margin between the cost of generic goods and the CSFI product, there are many supply chain actions that make it difficult for counterfeiters to get, and keep, product in the legitimate supply chain.
- Counterfeit-Hurdles. The total rank was "low" since there are many functional packaging features (tamper-evidence, child-resistance, print carton tape, and others), as well as other supply chain controls. These points lead to a "low" rank because these actions lead counterfeiters to need more complex production or packaging, or they must overcome supply chain controls. (For continuity in the model, "low" is good, so "low" in this rating means that there are a many or complex hurdles, which lead to a "low" risk.)
- Market Profile. The total rank was "low" since very brand loyal consumers are fairly price insensitive and pursue reputable sources of supply, and 95+% of the production and warehousing is in-house. These points lead to a "low" rank

because the consumer does not go looking for good deals from disreputable sources, and with in-house production there is little risk of unauthorized production or co-mingling by contract manufacturers.

Summary

The Counterfeit Risk Model was successfully used to review a company's specific counterfeit risk attributes. The process of completing the Counterfeit Risk Model provided a mechanism to gain corporate consensus and set clear direction for effective countermeasures in product development, regulatory affairs, customer service, and legal strategy. The sub-factor derivation discussion and analysis was valuable to allow reproducing the analysis in the future and to consider how new knowledge of emerging threats change the overall risk.

Reviewers not involved in the process may perceive that some factors or sub-factors overlap, but there was agreement by the expert panel.

Bringing the expert panel together was useful since it brought issues to light/ discussion in a very focused review of the counterfeit threat, so that the overall company more clearly understands their actual risks. CFSI considered the process efficient to conduct the risk assessment and as an internal communication tool.

Although the development of this risk model was intended to primarily benefit a specific company's in its strategy to reduce the risk of food counterfeiting, the process appears to have broader applicability. From a corporate security perspective, the risk management model provides a mechanism for assessing levels of risk for varying products, with varying supply chains, and for domestic and international markets. For regulatory agencies (e.g., US Food and Drug Administration, Environmental Protection Agency, Homeland Security), wide application of these models could provide assessment of relative risk across various sectors. For police and public sector law enforcement and security agencies, the utilization of expert panels offers an approach for systematic risk assessment and management.

16.
FOOD FRAUDS
Counterfeit and Adulteration in Food Supply Chain Networks
Robyn R. Mace

Food and agriculture are critical national infrastructure elements, depending on water, energy, and transportation infrastructures for successful production and distribution. Production and consumption of foods is fraught with numerous risks: inclement weather, diseases or pests, product quarantine, environmental contamination, civil unrest, port closures, market competition, consumer preferences, capitalization, and trade barriers. Certain risks, such as dependence on weather or susceptibility to rust, are inherent vulnerabilities in food commodities; these risks may be intensified or managed and minimized by human agricultural choices and necessities. Other risks, such as spoilage in port, customs seizure, and opportunities to perpetrate food adulteration and fraud, are introduced or amplified through processing, packaging, and transportation involved in supplying international global food markets.

Counterfeit food and food adulteration pose public health risks from ingestion, social (and legal) risks from consumer illness and death, and economic losses to food producers, suppliers, distributors, and retailers, as well as the governments that tax them. Recent international food scandals and customs seizures indicate that counterfeit and adulterated food and consumer products should be a serious international public health concern. Despite sensationalized public attention to shortcomings in industrialized food production systems through food scares and recalls over the past few years (dairy products, meats, spinach, lettuce, chili, tomatoes, peanut butter, pet foods, cereals, etc.), little is known about the prevalence, characteristics, or impacts of food fraud and adulteration or the types and effectiveness of various prevention and intervention strategies. Like other disturbances to food supplies, food frauds are likely to disproportionately affect poor consumers who have fewer food choices and shopping alternatives, less disposable income, and generally less social recourse to remedy. This chapter will describe the range of food adulteration and fraud, elucidate public health and economic impacts from various types of food fraud using domestic and international examples, and discuss recommendations to minimize the impact of food fraud on consumers, producers, and markets.

Counterfeit food refers to raw, whole, or manufactured food whose origin or contents are misrepresented through presentation, labeling, and other forms of misidentification. There is ample historical evidence that food fraud and swindling have always been an unsavory aspect of food markets, and are still more pervasive than a casual modern eater might reasonably expect. Sturgeon roe (caviar), truffles, and rare wines have long been coveted by gastronomes and kings, with counterfeits substituted by opportunists able to successfully misrepresent them as the actual food product. Unscrupulous merchants in the middle ages reportedly sweetened beer with lead salts and "bittered" it with strychnine (Barnes 1996). During early industrialization, bread, milk, candy, and spices were routinely adulterated by merchants to extend supplies and increase profits (Wilson 2008). The scope and scale of the modern economy generates unprecedented opportunities for the production and distribution of fraudulent foods.

International Foods Markets

Food is necessary to sustain human life, and specific foods can be imbued with tremendous cultural and social significance. Food is also big business. The US food industry accounts for almost 20 percent of US gross national product and directly employs about 14 million workers and another four million indirectly (Strongin 2002). The US agricultural products market is mature, growing slowly but steadily, and has annual revenues of approximately US$147 billion, about 17 percent of the global market (Datamonitor 2006). In 2009, the global agricultural product market continued to grow at a decelerating rate and generated revenues of approximately US$1.37 billion; the Asia-Pacific region accounts for almost 60 percent of this value (Datamonitor 2010).

The agricultural products market consists of fresh foods for human consumption and food products processed for human and animal consumption. For many comestibles, the food supply chain network is high velocity and high volume, featuring perishable commodities with small profit margins. The growing prepared food market demonstrates similar characteristics with higher margins, while the processed food market has the advantage of better perishability management through storage, processing, and preservation conditions and techniques. Due to globalized commodity markets, food travels more miles than ever before from farm to fork: consequently, in the US as in many countries, food imports continue to increase while food self-sufficiency at the national level declines. Growing markets for specialty products,[1] enhanced production techniques, inexpensive shipping, and consumer demands shape international markets for food and food products and generate opportunities for food fraud.

Continuum of Food Frauds (Adulteration to Misrepresentation)

Recent infrastructure protection activities have been informed by recognition of the interdependencies of infrastructure sectors, the threat of catastrophic disruption due to terrorism and natural disasters, and increasingly, a risk management approach that identifies and prioritizes assets, threats, and vulnerabilities and organizes activities to protect (differentially) critical, prioritized assets. From a military and national security perspective, deliberate adulteration of the food supply has long been recognized as a threat to specific actors and human assets (including military units). From a risk management perspective, more proximate vulnerabilities for widespread harm are related to the safety and authenticity of food products.

Table 16.1 Food Crimes: Continuum of Adulteration and Fraud

Act	*Description*	*Examples*
Ingredient adulteration/ substitution	Unannounced introduction or substitution or adulteration with inferior/restricted ingredients	Melamine in dairy products and pet foods Unlabeled allergens or prohibited products in foods
Up-labeling	Deliberate misrepresentation of product as being higher quality	Non-organic produce sold as organic Common fish sold as up-market species Transshipment for processing to disguise origin
Product substitution or repackaging	Substitution of product similar in appearance but inferior in quality in branded packaging	Refilling branded soda or alcohol bottles
Expiration date misrepresentation	Removing or changing labels to obscure proper expiration of sales or use date	Perishables, dairy products to extend sales life
Volume misrepresentation	Using filler, packaging, or other items to increase the apparent weight of the product	Pre-packaged foods Specialty and bulk items

The following typology of food frauds identifies five types of deliberate and direct action on the part of the perpetrator to misrepresent a food product (see Table 16.1). Malicious food adulteration and tampering are deliberate actions designed to harm or kill;[2] these *mala in se* acts are not included in the food fraud continuum as they are categorically different than a producer or vendor who offers adulterated or purloined food without the intent to harm, despite the fact that the possibility may exist or even be reasonably foreseeable. In fact, most individual and commercial-scale counterfeiters, presumably motivated by ongoing profits, would seem rather unlikely to want to mortally poison potential repeat customers.

International and Domestic Food Frauds

Food fraud raises significant challenges for regulators and public health officials as well as food producers and marketers as these products are apparently increasingly distributed through legitimate markets and outlets. Food frauds are perpetrated at the individual, group, corporate, and institutional levels. The examples discussed below were selected to highlight the range of food frauds, their impact on health and economic indicators, and key components in food frauds. According to news reports, these basic food frauds seem to occur in almost every country.

Ingredient Adulteration/Substitution: Melamine in Chinese Dairy and Pet Food Products, 2007–2008

China has become the world's largest exporter of food, food additives, and food products, with a market valued at US$54 billion (Ellis and Turner 2008). International and domestic concerns about the quality and integrity of Chinese-produced food products were heightened after a series of food product scares culminating in the 2007–2008 Chinese melamine scandals. Over an approximately two-year period, melamine was introduced as a food additive in numerous products in order to boost measurable nitrogen (a surrogate protein marker). Adulterated products included pet food, milk products, eggs, candy, seafood, and others in the international animal food market and domestic and international human food supply chain networks. The number of companies and products implicated in these adulterations combined with reports of animal feed-related deaths in Asia in 2003–2004 and human dairy product induced health issues in China in 2003 suggests that melamine boosting has been a common industrial production practice in China for some time.

Domestically, more than 30 Chinese dairies were implicated in the 2007–2008 adulteration, over the course of official Chinese investigations in 2008–2009. Local government officials were fired; at least 60 persons were arrested and 21 tried. Sentences included two executions (a farmer-producer of melamine, and a broker-adulterator) and life imprisonment for the former chairperson of Sanlu. Culpable dairies were ordered to pay US$16 million in cash settlements to the families of affected children and to fund medical care until the children turn 18. This intentional, industrial-scale contamination with melamine was the second major Chinese infant formula quality scandal this decade; in 2003, widespread unsafe practices and sub-standard nutritional value of 55 different infant formulas from 40 companies were found responsible for 12 infant malnutrition deaths (Chan, Griffiths, and Chan 2009).

Up-labeling: Fish and Specialty Foods

International fish demand and consumption is at an all-time high. International fish stocks are overexploited and threatened by international competition, improved boating and catch-storage technology, and ecological conditions. The depletion of certain stocks of fish and limited availability of high-demand species facilitates numerous opportunities for fraud.

Many Americans are familiar with the concept of fish fraud, thanks to a news story in 2008 about two thoughtful and enterprising New York City high school students who scientifically challenged the content claims of a small convenience sample of sushi restaurants and groceries (Schwartz 2008). The Trinity School students visited four restaurants and ten groceries; using DNA sequencing analysis,[3] they found that two of the restaurants (50 percent) and six of the groceries (60 percent) were selling mislabeled fish (overall 8/14 samples or 57 percent). Ensuing media coverage appears to have temporarily raised public awareness about the possibility of food fraud.

Fish fraud may occur in one of several ways—substituting a less desirable fish for a more palatable, expensive species; including ice or chilling materials in the weight of the seafood; or misrepresenting the place of origin, cultivation/capture methods, or other properties of the product (including prohibited species for capture/consumption, contamination with prohibited matter, preparation and preservation methods, and spoilage). Over the past decade, there have been intermittent reports of action against seafood wholesalers, restauranters, and retailers by United States Attorneys or State Attorneys General (Consumer Protection) over the fraudulent representation of fish to consumers in restaurants, wholesale markets, and grocery stores. In January 2007, the US District

Court in Florida sentenced a seafood importer to 51 months imprisonment for mislabeling Vietnamese seafood for import to avoid US Department of Commerce anti-dumping duties, and fraudulently labeling fish as grouper for resale in the US. In May 2007, the US Food and Drug Administration posted a consumer update warning consumers not to buy or consume imported monkfish from a lot of 282 imported in September 2006, as the lot was puffer fish, containing deadly textrodotoxin, that had been mislabeled; the mislabeled fish was not prepared in accordance with US standards.

This problem of up-labeling occurs with all types of food products—wines and spirits, prized fruit varietals, truffles, meats, etc., and may be especially common with specialty foods whose authenticity or provenance (source and contents) is difficult or impossible to verify using organoleptic techniques (sensory methods—smell, taste, appearance). Unsuspecting consumers seeking organic, halal or kosher, or allergenic foods, may be exposed to ingredients, chemicals, or products from undesirable locations while paying to avoid exactly those product characteristics. The health effects of this practice are unknown, although loss of consumer trust is certainly a byproduct of discovery of this type of fraud. Several major US grocery retailers are under federal and state regulatory and popular scrutiny for the fraudulent labeling of non-organic foods as organic, although up-labeling also occurs in smaller markets. Whether deliberately or unwittingly as the result of unprincipled distributors, it is likely that this type of fraud will continue as opportunity proliferates.

Product Substitution and Repackaging: Branded Beverages

Alcohol has been a prized (processed) agricultural product since the dawn of civilization, and general production techniques for beer, wine, and liquor are widely known and practiced. Unfortunately, in numerous countries, unscrupulous and informal producers capitalize on the demand for alcohol by providing a potentially toxic brew to unsuspecting customers. Illicit and informally produced alcohol causes thousands of deaths annually, often due to methanol poisoning from adulterants used improperly in the distillation process. One common distribution technique for fraudulent spirits is to refill branded bottles with illicit alcohol; others include diluting or tipping existing contents, and the manufacture of new packaging and fake contents. In June 2006, the Russian Interior Minister announced that counterfeit alcohol kills 42,000 Russians each year (Eke 2006). Mass casualties from adulterated alcohol have also been reported in Indian and Chinese media. In these incidents, consumers are either unable or unwilling to recognize cost, sensory, or seller clues regarding the provenance and quality of the alcohol, and potential severity of harm. The World Health Organization (WHO) has identified illicit and informally produced alcohol as a major health threat to developing countries and one of its priority areas in strategies for reducing the harmful use of alcohol (WHO 2010).

Expiration Fraud: Perishables

Expiration fraud involves changing or updating labels to obscure proper expiration of use or sales dates. Depending on the type of product, this practice could result in illness for consumers; it also conveys a decided disinterest in consumer safety, although it may be profitable in the short-term. In July 2006, a federal grand jury indicted executives at two now defunct food companies for misleading a third company into applying re-printed expiration labels to 1.6 million bottles to Henri's salad dressing to extend sell-by dates. Halliburton is currently under investigation for selling the US Army expired rations. Expired and spoiled rations have even been delivered as part of humanitarian aid, adding an additional, potentially life-threatening indignity to the already dispossessed.

Volume Misrepresentation: Fish

Finally, one of the simplest food frauds is to charge for more product than is sold; most customers have no ability to verify purchase weights and relatively few sources from which to buy commodities in bulk. These frauds may be accomplished by manipulating scales or thinning (adulterating) the commodity with materials similar in appearance or texture; this apparently common historical practice with grains, coffee, and other bulk items is still a popular and low-risk method to increase profits. In March 2010, the National Conference on Weights and Measures (NCWM) announced that over the course of a two-month investigation in 17 states, its inspectors discovered and removed over 21,000 packages of seafood that included the weight of the ice or chilling glaze in the cost of the seafood (NCWM 2010). This report also highlights the complex web of regulatory agencies assigned responsibilities for food authentication and safety, and reinforces the necessity to include a broad array of agencies and actors in developing and promulgating prevention, detection, and intervention strategies for adulterated foodstuffs.

Governing Food Fraud: Harm and Risk Reduction Strategies

Producers, consumers, and their exchanges comprise the vast web of production and consumption of global food supply chain networks. Historically, markets and transactions were regulated by participants in informal, ongoing interactions. As agriculture and economies expanded, complex and variable systems of voluntary and mandated domestic and international product standards, processing capacities, and labeling evolved and now characterize the international trade in agricultural commodities and food products. Within these systems of interaction, there are numerous agencies, interests, and conditions that influence the global and local risks inherent in and associated with the production and distribution of ingestible products. Constituent members of these extended supply chain networks and markets have vested interests in the satisfactory outcomes of food market exchanges, and consequently, if properly motivated and constructively engaged, can exert influence to reduce risks from food-related adulteration and counterfeiting.

It is possible and reasonable to reduce specific and identifiable risks from food fraud through systematic applications of risk management principles, technology, and good production and management practices, however, it is unlikely that any combination of law, producer oversight, and consumer awareness can entirely eliminate the conditions that facilitate food fraud. Identifying biggest health risks, most likely products, victims, and market channels can assist in targeting limited social resources to reduce public health and economic consequences of food counterfeit and fraud in international food supply chain networks.

Awareness, accountability, and transparency in food markets are essential for product assurance and safety. Awareness and detection are two of the most critical factors in undermining fraud in food products. Variable temporal lags from the discovery of food fraud to an epidemiological understanding of its source and cause and justifiable public health foci of response strategies complicate the detection and prevention of these problems.

Recognizing that the fundamental quality of a product should be a determining factor in consumer purchasing decisions, legitimate producers have few incentives to repeatedly cheat local, long-term consumers. However, if customers have little direct knowledge of the product and its production methods, they may be unable to determine the authenticity or quality of a particular food. Therefore, both producers and consumers have a vested interested in increasing awareness of the potential for food fraud. For consumers especially (as well as public health officials and social scientists), critical impacts of food frauds relate to product ingestion, essential or discretionary product demand, and volition in selecting product attributes (i.e., expectation of ingredients and materials). The willingness of consumers to purchase potentially counterfeit goods due to cost, false brand preference, or other reasons, such as a lack of alternatives, is an important subject for future research and policy development.

For producers, accountability and transparency in production systems are essential to provide product safety and assurance, and to insure ongoing public confidence and profitability. Self-regulation contributes to brand and product market share protection through consumer confidence and satisfaction. Best agricultural and manufacturing practices (BAP and BMP respectively) and industry benchmarking assist in the iterative ongoing process of awareness, education, and training about safe food production and handling procedures for producers (and indirectly consumers). These conjointly designed practices reflect regular, meaningful communication by regulators and industry producers with stakeholders and at-risk groups.

Industry, regulatory, and third-party efforts to enhance operational controls over inadvertent or deliberate production contamination are also critical to improving food quality and assurance. While inspections to catch violations of quality or administrative errors in paperwork generally only address a small percentage of product volume, inspection efficacy can be enhanced by using risk-driven or criteria-based inspection, rather than a random or percentage-based approach. Although post-hoc inspection does not constitute prevention, the selective use of authentication and traceability technologies can contribute to effective pre-identification and integrity assurance programs during production and distribution.

Authentication and Traceability Technologies

Technology can be particularly helpful in the authentication and traceability of food products. Authentication technologies and techniques include spectroscopy, isotopic analysis, chromatography, and thermal analysis among others. Currently, access to most of these authentication technologies is limited to food scientists and commercial and research laboratories and by their costs, although public–private partnerships are pursuing the commercialization of inexpensive point-of-sale and customer test kits. Traceability technologies, such as radio frequency identification (RFID) and global positioning service (GPS), have demonstrated great utility in assisting in early detection, recalls, and limitation of food related illnesses or health concerns; these powerful tools are equally applicable in situations of deliberate, fraudulent and unintentional, inadvertent food adulteration.

Staying ahead of counterfeiter innovations is challenging. There are numerous ways to bypass internal production and third-party quality controls and regulations (i.e., falsification of data, substitution of inferior products) in the absence of continual, direct oversight and a cultivated awareness of the issues and consequences. In 2003, British Food Safety Agency conducted DNA testing to determine the authenticity of Basmati rice; only 54 percent of the 363 samples contained purely Basmati rice. Publicity and merchant shaming seemed to reduce adulteration in future surveys, albeit temporarily, since fraudsters quickly began using (less expensive) Basmati hybrid varietal mixes to fool the authentication assays (Wilson 2008).

Increasingly sophisticated authentication, tracking, and monitoring systems are heavily reliant on technologies. Due the expense and effort required to realize them, technological investment must be leveraged by integration with other operations level processes and functions without introducing undue delays or disruptions. Technology can facilitate vital business functions, but it cannot perform them. While technology presents a powerful tool to address product, supply chain, and security management issues, technology is not stand-alone security solution and requires ongoing investment and employee training.

Conclusions

Reducing food fraud is a multi-pronged process involving consumer awareness, producer-certified standards, and effective regulation. Intervention strategies can and should be developed, targeted, and enforced for specific products, producers, consumers, and distribution networks and outlets based on consideration of social and economic harm.

Solutions to counterfeiting will involve an ongoing and evolving mix of policy and program development supported by technology. Regulatory and supranational agencies must continue to establish and enforce punitive penalties against counterfeiting organizations and nations, particularly with respect to comestible and medical products. Strengthening existing legislation and penalties to better discourage and punish offenders may assist in disrupting trade by interdicting the methods of production (equipment, skilled workers, raw materials). Public information on food seizures and food frauds schemes will increase public awareness of product authenticity and the potential for food frauds.

Producers must develop incremental, layered plans to address the risks and threats posed by food counterfeiting and fraud. Using enhanced cargo and supply chain tracking and inspection to identify illicit goods, their transit methods and routes, and method of entry into the system may also reduce the number of available fake and sub-standard goods available to consumers. Depending on the product and the type of threat from counterfeiting, regulatory agencies and affected companies should consider using in-field and in-bound product authentication in their standard operating procedures as well as source/production inspections and controls. Regulators, producers, and consumer groups must also continue to collaborate to educate consumers about the availability and potential safety concerns of counterfeit foods.

Notes

1. Market channels and consumer preferences include organic, fair-trade, non-genetically modified organisms (GMO), allergenic, religiously-sanctioned (kosher or halal), vegan, and farm dynamics (free-range, feed type, etc.).
2. Besides the widely known 1982 Tylenol tampering that established crisis management as an essential corporate function (with Tylenol's handling of the response emerging as a revered standard for crisis management), there are surprisingly frequent news reports of stories of responsible adults or police officers presented with foods "spiked" with narcotics, laxatives, or other contaminants for a variety of reasons—desire for harm or revenge, anger, misguided humor, and humiliation.
3. Conducted by a graduate student at the University of Guelph in Ontario and facilitated by the DNA scientist-father of one of the teens.

References

Barnes, P. J. 1996. "Adulteration and Fraud Threaten Safety." *Nutrition & Food Science* 96(4):23–26.

Chan, E. Y. Y., S. M. Griffiths, and C. W. Chan. 2008. "Public-health Risks of Melamine in Milk Products." *The Lancet* 372(9648):1444–1445, 25 October.

Datamonitor. 2006. *Agricultural Products in the United States: Industry Profile.* Datamonitor.

——. 2010. *Global Agricultural Products: Industry Profile*, Datamonitor.

Eke, S. 2006. "Fake Russian Alcohol 'Kills Many'." *BBC News*, June 23. Retrieved from http://news.bbc.co.uk/2/hi/5111762.stm.

Ellis, L.J. and J. L. Turner. 2008. *Sowing the Seeds: Opportunities for U.S. China Cooperation on Food Safety.* Woodrow Wilson International Center for Scholars, China Environment Forum. Retrieved fromhttp://www.wilsoncenter.org/index.cfm?topic_id=1421&fuseaction=topics.publications&group_id=476235.

NCWM. 2010. "National Investigation Exposes Fraud in Frozen Seafood Labeling." Press release, 29 March.

Organization for Economic Cooperation and Development. 2008. *The Economic Impact of Counterfeiting and Piracy.* Paris: OECD.

Schwartz, J. 2008. "Fish Tale Has DNA Hook: Students Find Bad Labels." *The New York Times*, August 22. Retrieved from http://www.nytimes.com/2008/08/22/science/22fish.html..

Strongin, R. J. 2002. "How Vulnerable Is the Nation's Food Supply? Linking Food Safety and Food Security." *National Health Public Forum Issue Brief* 773(May 17):22.

WHO (Report by Secretariat). 2010. *Strategies to Reduce the Harmful Use of Alcohol: Draft Global Strategies* (A63/13). Sixty-third Global Health Assembly, Provision Item 11–10, March 25. Geneva: World Health Organization.

Wilson, B. 2008. *Swindled: The Dark History of Food Fraud, from Poisoned Candy to Counterfeit Coffee.* Princeton, NJ: Princeton University Press.

See also

Mace, R. and J. Spink, 2007. *The Business Case for Food Counterfeit: Final Project Report.* East Lansing, MI: Food Safety Policy Center, National Center for Food Safety and Toxicology, Michigan State University.

17.
PRIVATIZED CYBERSECURITY AND THE CHALLENGES OF SECURING THE DIGITAL ENVIRONMENT

Panayotis A. Yannakogeorgos

Cyberspace continues to elicit a sense of a fundamental departure from business as usual. Dynamic new technologies and new applications are the epitome of a global project on which financial transactions, infrastructure controls, and communications are increasingly reliant on to function. The processes of securing cyber assets that are critical for the functioning of the developed world, as well as the risks of utilizing cyberspace, are surprisingly familiar. A common methodology of assessing risk in cyberspace is lacking, and efforts to mitigate risk in this domain are hampered by ineffectual cooperation between public security and private actors who are responsible for computer and network security. Although private technological innovation in the development of the Internet is unquestioned, corporations have repeatedly been the victims of cyber attacks. Efforts to reduce threats and vulnerabilities in cyberspace have proven elusive in the face of actors who can attack from anywhere in the globe with true anonymity. In sum, private efforts to secure their digital dominions have led to a succession of dead ends. In this chapter I examine the obstacles for reducing risk in cyberspace. It will be shown that an inability to generate a common framework for managing cyber risk has only enhanced insecurity for all actors as failure to agree how to protect a non-excludable good has led to a negative-sum outcome. In other words, private sector attempts to secure their own networks have led to everyone being worse off. Creating the Internet, and policing it, are two very different problems.

What is Cyberspace?

Understanding the cyber environment is important in understanding the nature of the threat. The strategic definition of cyberspace is "a domain characterized by the use of electronics and the electromagnetic spectrum to store, modify and exchange data via networked systems and associated physical infrastructure."[1] Contrary to the popular metaphor, this definition indicates that cyberspace is a physical domain composed of electronic hardware and software internetworked via, and powered by, the electromagnetic spectrum to communicate and store information. Thus, cyberspace is more than the Internet or World Wide Web (WWW), although both comprise its most recognizable elements. In the modern era, cyberspace is an extension of Earth's environment just as sea, air, and outer space are. Like these other spaces, it is considered a global commonage.[2]

The technical complexities of the Internet, which are now developed by the private sector, rely on the Transmission Control Protocol and Internet Protocol (TCP/IP) and (increasingly) wireless communications methods, both of which lack secure identity management protocols, and present challenges to the international communities' efforts to prevent, identify, and prosecute perpetrators of cybercrime, cyber terror, and cyber war.[3] As a result, the Information Society lacks the critical elements of trust and security required for it to reach its full potential.

Computer networks are dependent on the good use of internationally standardized communications protocols, such as the TCP/IP, to send and receive data packets and information. TCP/IP allows for the flow of data-packets and information across computer networks, including the Internet. TCP/IP is standardized by the International Organization of Standards for the Open Systems Interconnection (OSI) model as the basis of Internet networking. To better understand the significance of TCP/IP, a brief description of how information is sent across networks is necessary. Data-packets are the basic units of network traffic. They are the standard way of dividing information into smaller units when sending information over a network. A significant component of the computer networks is the IP header, which contains information pertaining to the source and destination addresses. Machines require these strings of numbers to connect with other computers on the Internet or other networks.[4] All networked hardware must have a valid IP address to function on a network. Data-packets are recreated by the receiving machine based on information within a header of each packet that tells the receiving computer how to recreate the information from the packet data. Without international standards, such as TCP/IP, there would be no assurance that packets could be read by a receiving machine.[5]

The Domain Name System (DNS) allows people to use Uniform Resource Locators (URLs) to communicate with other machines on the Internet. So instead of having to type in the numerical address of a website, which might look like 123.1.2.1, a person can type in a plain language URL (such as www.google.com) into a web browser to reach an Internet address. This makes the WWW user friendly. IP addresses reside on DNS databases on root servers that allow for the translation of URLs into IP addresses.[6] The top-level domain names, such as .com or .org, are maintained and updated by the Internet Corporation for Assigned Names and Numbers (ICANN), which was once under the Department of Commerce (DOC). Now operating under a memorandum of understanding with the DOC, ICANN is a private entity responsible for governing and maintaining the DNS database's 13 root servers allowing for global Internet communications.[7]

Vital computer networks that are part of the DNS are open to electronic attacks. As will become apparent, the trend with computer security officers is to not take the necessary steps to safeguard the DNS.[8] Common exploits include Denial-of-Service (DoS) and Distributed-Denial-of-Service (DDoS). In both types of attacks, an enormous amount of useless data is sent to a server in an attempt to overload the system and render it inoperable. DDoS attacks are more sophisticated in that thousands of computers controlled by malicious software installed on computers around the world allow an attacker to mount DoS attacks on a grand scale. The consequences of such attacks were observed in February 2007 when attackers from the Asia-Pacific region launched a DDoS attack on the Internet's 13 DNSs.[9] Six root servers were affected.[10] The Department of Defense "g-root" and the ICANN "i-root" DNS servers were severely crippled during the attack. Neither of these critical infrastructures were outfitted with new security technology that had been implemented on the other attacked servers. Quick recovery was possible due to the use of "new technology, combined with the speed, skills and experience learnt by root server operators over the years."[11] Even so, the tangible consequence of a successful attack against the DNS would mean that the Internet would not be operational until its operators could recover from the attack. However, there is a low probability that this could occur given the redundancy of the DNS.

Internet protocols, such as the Internet Protocol version 4 (IPv4), DNS, and the Border Gateway Protocol (BGP) are identified by the National Security Strategy for Cyberspace as being prone to security problems.[12] This is significant since critical infrastructures are networked via this protocol so that key functions may be accessed remotely. Tom Leighton, a professor of Mathematics at MIT and co-founder and chief scientist of Akamai Technologies, argues that beyond worms, viruses, and Trojan horses, the fundamental protocols on which the Internet runs are too weak to provide reliable security mechanisms.[13] According to Leighton, the DNS, ports, and IP address systems are plagued by flaws that "imperil more than individuals and commercial institution. Secure installations in the government and military can be compromised" as well.[14] The reason for the current flaws in the network architecture of the Internet is that they still rely on the protocols that were built 35 years ago when the Internet was not a global entity, but a closed research network. When it did become global, there was no shift to create stronger security mechanisms.

The challenge of securing cyberspace rests in the technical complexities described above. Cyberspace is a dynamic environment where no defense will be perfect, and attackers will have a variety of means available to deny a network's services to users. In addition, if targeting a specific network proves too difficult, taking out its supporting subsystem might prove just as effective.

Tools and Tactics of Exploiting Threats and Vulnerabilities in Cyber Systems

The changing nature and variety of the network vulnerabilities, tools, and tactics available to cyber criminals and terrorists makes it impossible to document the full range of threats in cyberspace. Dorothy Denning's classification of tools and tactics that computer criminals, terrorists, and information warriors might use in their information warfare campaigns is still relevant 11 years after its publication. The tools and tactics developed since the article was published fit neatly into Denning's following classifications:

- eavesdropping and packet sniffing;
- tampering or data diddling;

- snooping and downloading;
- spoofing;
- jamming or flooding;
- injecting malicious code;
- exploiting flaws in design, implementation, or operation;
- cracking passwords, codes, and keys.[15]

These archetypical types of cyber attacks have become enhanced and more widespread in the past 11 years in that the cyber attacks are more complex. A sophisticated attack today may use all of the above attack methods to penetrate an information system. This complexity is due to overlap between categories, and will be noted below. When a group of highly sophisticated hackers, possibly using state resources, are involved in a highly complex attack, all of the basic elements converge.

Eavesdropping and Packet Sniffing

These techniques refer to the capturing of data-packets en route to their destinations without altering information. They are typically used to capture user IDs and passwords. Technological changes that have occurred since Denning's first writing have made such attacks more consequential. Whereas in the past data were transmitted over copper wire, the mass adoption of wireless technologies such as WiFi and Bluetooth enable hackers to sniff out data from the electromagnetic ether rather than having to find a way to have direct access to a data source.

Tampering or Data Diddling

Denning defines tampering or data diddling as "making unauthorized modifications to software stored on a system, including file deletions."[16] This corruption of data can have profound effects in the financial world where a digit in a financial report is changed to reflect less than stellar profits, or a remote sensor on critical infrastructure is reprogrammed to display false pressure data.

For example, the threat to the US power grid is very tangible. During a Department of Homeland Security (DHS) exercise, hackers were tasked with hacking into the information system of a power generator. Succeeding in gaining remote access to the generator's SCADA control system, the hackers were able to physically destroy the generator. The Aurora vulnerability, as this exploit is called, lends credence to the suggestion that the manipulation of computer code can be just as effective in destroying critical infrastructure as would a missile. In a letter to Congressman John Dingell (D-MI), Chairman of the Committee on Energy and Commerce, Congressmen Bennie Thompson (D-MS) and James Langevin (D-RI) emphasized that an attack exploiting the Aurora vulnerability "could enable a targeted attack on infrastructure connected to the electric grid, potentially destroying these machines and resulting in catastrophic losses of power for long periods of time."[17] The cost for a cyber attack resulting in a three-month power outage is cited at over $700 billion.[18] Using the same techniques required to enter a computer network without authorization, hackers can inject false information into a system with tragic results. Utility companies have been the targets of such attacks. Recently, the Central Intelligence Agency warned that: "cyber attackers have hacked into the computer systems of utility companies outside the United States and made demands, in at least one case causing a power outage that affected multiple cities."[19] Since utility companies now use SCADA to remotely monitor and access controls for their services, and because SCADA is a ubiquitous technology, US utilities are just as vulnerable to these sorts of attacks. There is circumstantial evidence that Chinese hackers may have been responsible for past power blackouts in the United States, including the widespread power loss that occurred on August 15, 2003 affecting the East Coast. During this event, over one hundred power plants were shut down in part due to the disruption of communications lines used to manage the power grid. This disruption was attributed to a computer virus in circulation at the time.[20] It has been noted by "one security analyst in the private sector with close ties to the intelligence community … that some senior intelligence officials believe that China played a role in the 2003 blackout that is still not fully understood."[21] Further investigation of why there is not a full understanding of China's role is not possible due to limitations on the information available.

Snooping and Downloading

Snooping and downloading does not involve capturing information that is in transmission. Rather, attackers obtain access to files and folders on a given computer and are able to browse without restriction. Files of interest may then be downloaded to the attacker's data storage device, such as a flash drive.[22] To conduct such an attack, physical access to the information system is required. Indeed, outright theft of an information system constitutes such an attack. The consequences of snooping and downloading are not light. In 1996 "United Nations officials reported that four computers containing data on human rights violations in Croatia were stolen from their New York offices, dealing a heavy blow to efforts to prosecute war crimes."[23] Thus, in at least one case, evidence critical to the reconciliation process of populations in post-conflict zones suffered setbacks after such an attack.

Key loggers are tools that can be used by unauthorized persons to gain access to information on a computer. Software and hardware key loggers exist, and both types record all keystrokes made on a computer. They are used to capture passwords and other information required for login. A criminal can glean other sensitive information so that he or she might access a computer network. Software key loggers are installed directly on a computer, and therefore may be identified and removed by a suspicious skilled person, or by anti-malware software, which most users rely upon to detect malicious code. However, hardware key loggers are undetectable by such software since they are physically attached to the cable that connects the keyboard to the computer. These data can be read if the device is retrieved and the captured keystroke data are output to a computer. The only way to discover hardware key loggers is by physically examining the keyboard-to-computer connection.

Spoofing

Spoofing attacks strike at the root of trust in identity and security in cyberspace. An individual may manipulate various layers of the OSI in order to create the false appearance of a user, device, or website. Given the global nature of the Internet, it is possible that hackers can exploit software vulnerabilities, spoofing their location and making the attack to appear to have originated anywhere in the world. This technique permits skilled computer criminals to thwart cybercrime investigations. Denning aptly points out that in order to "trace an intruder, the investigator must get the cooperation of very system administrator and network service provider on the path."[24] This time- and patience-consuming task is complicated by weak domestic legislation and regulation. Attackers bouncing their attacks through a maze of servers in different countries increase their chance of getting away with a crime due to the lack of international cooperation for political and technical reasons.

Jamming or Flooding

This method is used to disable or tie up system resources. DoS and DDoS attacks are the primary examples of these sorts of attacks. Hackers targeting information systems with DoS or DDoS have been able to prevent legitimate consumers from accessing e-commerce websites, slowed down the DNS, and even shut down the country of Estonia during a massive cyber attack in May 2007. What is notable about the Estonian incident is that the Kremlin-backed youth group that claimed responsibility for the attack asserts that they did not use any software programs to conduct their operations.[25] Instead, they claim, they used the Internet to organize thousands of dispersed youth to make legitimate requests on specific Estonian information systems. The high volume of legitimate requests precipitated the outage of the information systems.

Injecting Malicious Code

A computer virus is one form of software that contains malicious code. These functions can range from simple annoyances to the user, such as resetting the computer's clock, to completely erasing all the data on the drive. Viruses can replicate, and have the ability to spread to other computers. Worms are similar to viruses, however, they spread by scanning networks and replicate themselves on networks that have specific security holes. Trojan horses are malicious codes disguised as legitimate programs. When run, they perform some malicious task instead of the task that a user was expecting the program to perform. Further threats are root kits, which are stealthy software programs that conceal their presence from security and other software by modifying parts of the target computer's operating system and/or computers kernel (core), thereby preventing the system owner from discovering, managing or getting rid of the root kit.[26] In all instances, compromised computers are known as bots or zombie computers.

While a single malicious program (or malware) installed on a single computer can compromise data on that computer, cyber criminals are now using malware to potentially take over thousands of computers infected with malware. In doing so, the criminals create expansive networks of computers, called botnets, from which cyber attacks can potentially be launched.[27] A botnet containing a large number of bots can be used to send spam, commit click-fraud, and steal information via computers the attackers do not own, unbeknownst to the legal user, granting them distributed computing power dispersed geographically allowing diligent attackers time to cover their tracks.[28] Thus, what has changed since Denning's taxonomy in 1998 is that the current trend is that there is now less of a propensity to make the user aware of the presence of malicious code on a computer, and more of a will to have the code run silent and deep so that the attacker can remotely control the targets computer to launch massive attacks or exfiltrate data from a sensitive network.

Although a variety of software exists on the market that aims to protect information on a computer from malicious code, encrypting and compressing (packing) malware are two popular methods that terrorist programmers use to conceal malware from security software and/or computer network security analysts. Packers and cryptors disguise malware code by compressing or encrypting the program or file in which malicious code has been inserted. This transforms the code into random byte sequences, and has the effect of making the malicious code appear harmless to malware detection programs. Approximately 90 percent of malicious code is delivered in packed or encrypted form, and the level of sophistication increases each year, posing new challenges to computer security officers.[29]

Exploiting Flaws in Design, Implementation, or Operation

Poor programming of computer software and operating systems creates flaws that hackers exploit "to gain access to systems, files, accounts, or root privileges, or to sabotage the system or its files."[30] On the one hand, the companies that supply software are under pressure to release their latest software package without rigorously checking their product for programming errors. The danger of not identifying these errors, or when discovered not making the errors public, can make information systems insecure, unbeknownst to trusting users. Since most users lack the knowledge to discover bugs on their own, they rely on patches and updates from software manufacturers. They then have to have to know how to fix the problem themselves, and vigilantly keep tabs on whether or not their information systems are running software with the latest security patches installed. This focus on the user protecting his or her own information system by vigilantly patching their systems is perhaps the greatest pitfall of current cybersecurity systems, which is discussed in greater detail below.

Cracking Passwords, Codes, and Keys

These methods entail either a brute force or guess and check method of guessing a user's password, or the use of sophisticated software designed to decrypt password files and keys. Denning notes that: "It is often possible to crack a key much faster than would be expected by exploiting a hole in the implementation of the encryption algorithm or key management function."[31] Thus, even with long encryption chains that in theory cannot be broken, the risk of an attacker successfully cracking a password is increased when individuals or enterprises do not ensure that their software is current.

Attacks Targeting TCP/IP Network Architectures

Vital computer networks that are part of the DNS are open to electronic attacks. In some cases, computer security officers do not take the necessary steps to safeguard the DNS.[32] Common exploits include DoS and DDoS. The consequences of such attacks were observed in February 2007 when attackers from the Asia-Pacific region launched a DDOS attack on the Internet's 13 DNS servers.[33] Six root servers were affected including the DOD's "g-root" and the ICANN "I-root" DNS servers were severely crippled during the attack. Neither of these critical infrastructures were outfitted with new security technology that had already been implemented on the unaffected servers.[34] This demonstrates the importance installing software updates meant to fix security holes in software. Further root servers, which host a company's DNS, are vulnerable to complicated attacks known as "pharming."[35] Pharming occurs when an attacker replaces the IP address of a company's server with the address of a machine he or she has set up, thus hijacking the companies' website traffic. When pharming occurs, users access their bank's website IP address and unwittingly send their personal information to hackers, rather than to the intended online banking site. Thus, DNS has much potential for misuse by a variety of actors with differing objectives. Some might want to bring down the Internet, while others might want to preserve the Internet in order to carry out the equivalent of high-robbery in cyberspace.

While attackers can exploit DNS, they face enormous redundancy by design. The DNS is composed of tens of thousands of systems through which information can be routed.[36] However, although it might be difficult to bring down the Internet, insecurities in network infrastructure lead to a reduction in people's trust of the technology.

Challenges for Developing a Common Methodology in Managing Cyber Risk

National and international policymakers have recognized that in order to prevent the misuse of cyberspace, especially its ubiquitous elements of the Internet and WWW all interested stakeholders must cooperate to harmonize their national legislation within the framework of existing efforts undertaken by the international community.[37] However, the framework as it stands does not emphasize the importance of a strong public law enforcement presence in cyberspace. Instead, private architectures of control that has government playing a lawmaking role, while law enforcement and intelligence services role in securing private networks are discouraged. Relying on the private sector to provision cybersecurity undermines trust and security in cyberspace, and hampers the distribution of justice. National and international law enforcement capacity building must take place in order to all for public law enforcement entities to take the lead and enforce the dictates and laws create by national and international bodies.

The Information Society depends on the free flow of information over computer networks. The utility of computer networks rests with on the academic concept of network-effects. That is, "the greater the size of a network, the greater the benefits."[38] A networked community benefits each time a new user joins that network. People tend to join networks if they can trust that they will benefit as members of the network.

Neal K. Katyal in *The Dark Side of Private Ordering: The Network/Community Harm of Crime* identifies the main reason for current cybersecurity strategies as being the wrong approach in addressing injustice in cybserspace. This, he argues, is in part due to the focus of criminal justice on the individual impact of a crime, rather than on the harm of a crime to the community at large.[39] In his view, the focus on the harm to the community rather than the individual impact of a crime is especially important in cyberspace some each instance of a cybercrime, no matter how trivial it is, leads to the network's users to not trust the network.[40] With each intrusion, mistrust increases, and the number of users using the intruded network decreases. As a result the utility of that network to the remaining users decreases.[41]

Katyal argues against the common perception that "the strong arm of law enforcement has no business when the only harm to a victim is remote and intangible."[42] This view is fallacious, since it focuses only on the individual harmed rather than the harm done to the computer network. Katyal argues that even breaches of computer networks motivated by curiosity seriously damage a community's trust in the network, thereby decreasing the principle of network-effect. It follows that justice in the Information Society, can be achieved only if the injustice to the community is considered in addition to the injustice to the individual. In order for the Information Society to reach its full potential, all cyber criminals should be tracked and punished to the fullest extent possible under law so as to prevent the loss of trust and increase network usage.

Even if laws exist, this does not mean that the crime will be prevented since law enforcement authorities must have the capabilities and procedures to prevent, investigate prosecute cybercrime.[43] Contrary to the views of international bodies within the United Nations system dealing with cybersecurity, such as the General Assembly and the International Telecommunications Union, Katyal argues that is the responsibility of law enforcement organizations, and not private individuals or corporations, to enforce cyber law and prosecute all infringements. Private ordering efforts, such as proprietary anti-virus or firewall software, will not prevent computer crime simply because this software or hardware is purchased and installed by a user, and the interests of private industry is to

> promote sales of anti-virus software, intrusion systems and the like. Yet, the ability to afford and the knowledge to use such technologies will not be distributed equally. Those with fewer resources will not be able to adopt them in the same way that richer individuals and institutions can. Further, because these technologies are often complicated, there will be some who have the resources to purchase them but lack the skills necessary to use them effectively.[44]

Thus since not all computer users will be able to afford or know how to use protection software, their computers will be prey to attackers. This gap will result in those without protective measures to use the network less because they lack the trust in the network to allow for more migration of human activity into cyberspace.[45] Thus, these constraints indicate that private efforts cannot protect every user. Hence, governments must steer clear from the current approach and bear greater responsibility for network protection in order to assure the Information Society is a secure environment in which information flows freely.

A further consequence of private industry being given the responsibility of securing cyberspace is that in their effort to assure they meet benchmarks set by the government, connectivity to their networks may be diminished.[46] Put otherwise, if Internet service providers are burdened with the responsibility to assure that their services are not being used for criminal acts, then they might react to any suspicious activity (now matter how slight it is) by purging users for such actions. This too harms the open networks overall.

Threat awareness by both federal and private actors is thus a linchpin in securing cyberspace via networked efforts. However, a culture of secrecy surrounds the full extent of private sector vulnerabilities. According to Verton, this dynamic persists largely because:

> The private companies that own and operate the bulk of the nation's most critical infrastructure system continue to balk at sharing with the government the lion's share of information about cyber-vulnerabilities and security incidents. Most fear that the government, through Freedom of Information Act requests, will inadvertently disclose proprietary company data to competitors.[47]

To demonstrate his point, Verton describes the financial industry as one example where the private sector is reluctant to share data on cyber intrusions with outsiders. Beyond the protection of proprietary data, private industry is also concerned that if the full extent of their vulnerabilities is made public, people will not trust their services and stop using them.

The federal government also appears to be less willing to share sensitive security information with both the private sector and within the interagency since "interagency rivalries and distrust have too often slowed progress" in organizing a networked response to security threats.[48] This is emphasized by Beckstrom, who when comparing his role in DHS to the capabilities of National Security Agency noted that "intelligence culture is very different than network operations or security culture."[49] Thus, without changes in organizational cultures aiming to reduce distrust amongst all stakeholders, and promote the sharing of data related to cyber attacks in the public and private sectors, malicious actors will continue to benefit from the stovepiping of information.

Official sources of data do exist. This includes Computer Emergency Response Teams and the Common Vulnerabilities and Exposures database that are sponsored by DHS and based on expert opinion. Such databases however focus on ways that one might detect a malicious code, and how one might go about fixing a system to patch vulnerability. In addition to these, organizations such as the European Institute for Computer Antivirus Research (EICAR) are examples of organizations that are networks of independent information security that monitor and provide on cybersecurity. With such lists available online, unsophisticated cyber criminals have a list of common tools that they may utilize in order to infiltrate an information system. Since preventing the use of these tools and tactics depends on an individual user securing his or her own system, there are a great number of computers that are unsecured due to individuals not patching their systems and remaining alert. Further, although such active monitoring programs exist, and common attacks are made public, it is the uncommon attacks that are the greatest threat to government and corporate information systems.

So-called zero-day attacks are those that are not listed in databases as the vulnerability is not known to anyone other than the hacker, privateer hacker network, or information warrior who has discovered the flaw, and not publicized it or exploited it.

Conclusion

Unanimous agreement exists on the issue of the private-ordering of cybersecurity in both national and global strategies. This guiding tenet of global and national cybersecurity efforts is flawed since public–private partnerships in which there is no federalized effort to guarantee the security of privately owned networks will erode trust in the networks over the long term. Foreign companies offering their services to consumers may not follow their legal obligations in cases where the security of an information system is breached. One might suggest that this model of cybersecurity applies better in the US case where many corporations, including those overseeing the day-to-day functioning of the Internet, interact with the government in formal meetings of an advisory nature. This chapter emphasizes the point that if private industry is relied on to secure information systems, there will be an increased risk in the exploitation of cyberspace, and thus, the full utility of those networks will not be realized. Users will not trust systems that they cannot be expected to know how to secure. Preventing a computer from being hijacked by malicious code is not as simple as putting a lock on one's house. This is not to say that private industry plays no role in securing cyberspace, only that law enforcement organizations should take on the brunt of the effort while carefully balancing privacy concerns. With rampant criminal activity and the militarization of cyberspace, the consequences of not having more proactive efforts at the individual, organizational, and national levels to secure cyberspace on the basis of international cooperation paint a bleak picture of the future of cyberspace.

Notes

1. US Department of Defense (2008). It should also be noted that the Department of Defense dictionary definition for cyberspace is a "global domain within the information environment consisting of the interdependent network of information technology infrastructures, including the Internet, telecommunications networks, computer systems, and embedded processors and controllers" (http://www.dtic.mil/doctrine/dod_dictionary/data/c/10160.html).
2. Rattray, Evans, and Healey (2010); Kamal (2005).
3. Hancock (2006).
4. Molyneux (2003: 85–86).
5. Ibid., 27.
6. Ibid., 86.
7. ICANN (2000).
8. Sherman (2007).
9. ICANN (2007).
10. Ibid.
11. Ibid., 2.
12. White House (2003: 30).
13. Leighton (2006: 44).
14. Ibid.
15. Denning (1998: 29–55).

16. Ibid., 33–34.
17. Thompson and Langevin (2008).
18. Ibid.
19. Nakashima and Mufson (2008).
20. Harris (2008).
21. Ibid.
22. Ibid., 33.
23. Wall Street Journal (1996). See also Denning (1998: 33).
24. Denning (1998: 35).
25. Grant (2009).
26. Cass (2006).
27. Markoff (2007).
28. Ianelli and Hackworth (2005).
29. Lyda and Hamrock (2007).
30. Denning (1998: 38).
31. Ibid., 40.
32. Sherman (2007).
33. ICANN (2007).
34. Ibid., 2.
35. Sherman (2007: 41).
36. Lewis (2006: 332).
37. International Telecommunication Union (2006).
38. Ibid.
39. Katyal (2006).
40. Ibid., 197.
41. Ibid.
42. Ibid., 196.
43. Ibid., 194.
44. Ibid., 199.
45. Ibid., 194.
46. Ibid., 214.
47. Verton (2006: 25).
48. Arquilla and Ronfeldt (2001: 364).
49. Beckstrom (2009).

References

Arquilla, John and David Ronfeldt. 2001. "Afterword September 2001: The Sharpening Fight for the Future." Pp. 363–371 in *Networks and Netwars*, edited by John Arquilla and David Ronfeldt. Santa Monica, CA: RAND.

Beckstrom, Rod. 2009. *Letter of Resignation to DHS Secretary Janet Napolitano*. March 5.

Cass, Stephen. 2006. "Antipiracy Software Opens Door to Electronic Intruders: Sony BMG Shoots Itself—and its Customers—in the Foot." *IEEE Spectrum* January 12–13.

Denning, Dorothy E. 1998. "Cyberspace Attacks and Countermeasures." Pp. 29–55 in *Internet Besieged: Countering Cyberspace Scofflaws*, edited by Dorothy E. Denning and Peter J. Denning. New York: ACM Press.

Grant, Ian. 2009. "Kids Responsible for Estonia Attack." *Computer Weekly*. March 13. Retrieved from http://www.computerweekly.com/Articles/2009/03/13/235262/kids-responsible-for-estonia-attack.htm.

Hancock, Bill. 2006. "How to Stop Talking About—and Start Fixing—Cyber Security Problems." *Cutter IT Journal* May 6–11.

Harris, Shane. 2008. "Chinese Hackers Pose a Clear and Present Danger to U.S. Government and Private-Sector Computer Networks and May be Responsible for Two Major U.S. Power Blackouts." *National Journal Magazine* May 31, 2008. Retrieved from http://www.national-journal.com/njmagazine/cs_20080531_6948.php.

Ianelli, Nicholas and Aaron Hackworth. 2005. *Botnets as a Vehicle for Online Crime*. CERT Coordination Center, Carnegie Mellon University. December 1.

ICANN. 2000. "Memorandum of Understanding Concerning the Technical Work of the Internet Assigned Numbers Authority." March 1. Retrieved from http://www.icann.org/en/general/ietf-icann-mou-01mar00.htm.

——. 2007. *Factsheet: Root Server Attack on February 7*. Retrieved from http://www.icann.org/announcements/factsheet-dns-attack-08mar07.pdf.

International Telecommunication Union. 2006. *Research on Legislation in Data Privacy, Security and the Prevention of Cybercrime*. Geneva: ITU.

Kamal, Ahmad. 2005. *The Law of Cyber-space: An Invitation to the Table of Negotiations* Geneva: United Nations Institute for Training and Research. Retrieved from http://www.un.int/kamal/thelawofcyberspace/The%20Law%20of%20Cyber-Space.pdf.

Katyal, Neal K. 2006. "The Dark Side of Private Ordering: The Network/Community Harm of Crime." Pp. 193–217 in *The Law and Economics of Cyberspace*, edited by Mark Grady and Francesco Parisi. Cambridge: Cambridge University Press.

Leighton, Tom. 2006. "The Net's Real Security Problem." *Scientific American* September:44.

Lewis, James A. 2006. "Cybersecurity and Critical Infrastructure Protection." Pp. 324–338 in *Homeland Security: Protecting America's Targets. III: Critical Infrastructure*, edited by James J. F. Forest. Westport, CT: Praeger Security International.

Lyda, Robert and James Hamrock. 2007. "Using Entropy Analysis to Find Encrypted and Packed Malware." *IEEE Security and Privacy* 5(2), March/April:41–45.

Markoff, John. 2007. "Attack of the Zombie Computers is a Growing Threat Experts Say." *The New York Times* January 7, section 1, column 5.

Molyneux, Robert E. 2003. *The Internet under the Hood: An Introduction to Network Technologies for Information Professionals*. Westport, CT: Libraries Unlimited.

Nakashima, Ellen and Steven Mufson. 2008. "Hackers Have Attacked Foreign Utilities, CIA Analyst Says." *Washington Post* Saturday, January 19:A04.

Rattray, Greg, Chris Evans, and Jason Healey. 2010. "American Security in the Cyber Commons." Pp. 137–176 in *Contested Commons: The Future of American Power in a Multipolar World*, edited by Abraham M. Denmark and Dr. James Mulvenon. Washington, DC: Center for a New American Security.

Sherman, Erik. 2007. "DNS: Definitely Not Safe, New Attacks on the Internet's Domain Name Systems Keep CISOs Guessing." *CSO* February: 38–41.

Thompson, Bennie and James Langevin. 2008. "Letter from Congressmen Bennie Thomson and James Langevin to Congressman John D. Dingel." May 29. Retrieved from http://homeland.house.gov/SiteDocuments/20080530130810-85574.pdf.

US Department of Defense. 2008. "Fighting in Cyberspace Means Cyber Domain Dominance." Retrieved from http://www.af.mil/news/story.asp?id=123042670.

Verton, Dan. 2006. *Black Ice: The Invisible Threat of Cyber-terrorism*. Emeryville, CA: McGraw-Hill.

Wall Street Journal. 1996. "The Risk: The Usual Hazards of not Having a Good Data Backup Plan?" January 17 [Brian_Mulvaney@intersolv.com via risks-digest 17, Issue 65].

White House. 2003. *National Strategy to Secure Cyberspace*. February. Retrieved from http://www.dhs.gov/xlibrary/assets/National_Cyberspace_Strategy.pdf?.

18.
THE RISK POSED BY ZEBRA MUSSELS

David L. Carter

Risks can take many forms. In the case of terrorism risks, there are some unusual types of risks that are not particularly intuitive and certainly not found in the traditional literature of terrorism. A case in point is zebra mussels: a threat that is currently spreading across the United States that has a significant impact when it strikes and, in an unusual case, is not only an ecological and economic threat, they were also an instrumentality for one type of terrorism.

Zebra mussels (*Dreissena polymorpha*) are an invasive species native to the Black Sea and Caspian Sea regions of Eurasia. Zebra mussels get their name from the striped pattern of their shells, though not all shells bear this pattern. They're usually about fingernail size but can grow to a maximum length of nearly 2 inches. The mussels live 4 to 5 years and inhabit fresh water at depths of 6 to 24 feet. A female zebra mussel begins to reproduce at 2 years of age, and produces between 30,000 and 1 million eggs per year. About 2 percent of zebra mussels reach adulthood. A species that multiplies quickly, even in less than ideal conditions, the fast growing collection of zebra mussels will aggressively consume underwater vegetation, overwhelming and eventually killing aquatic species that are indigenous to North America.[1] These characteristics are indicative of the type of massive risk zebra mussels pose to North American ecosystems where there is no natural predator or ecological control system to balance their growth.

In 1988 zebra mussels were introduced to US fresh water in Lake St. Clair—between Lake Erie and Lake Huron on the Michigan, US–Ontario, Canada border—through ballast water discharges from transoceanic vessels traveling through the St. Lawrence Seaway. The zebra mussel aggressively competes with native species of mussels and fish for vegetation as a food source. Beyond its consumption of food, the zebra mussels' capacity to reproduce in large numbers in a short period of time has become a significant problem to communities by clogging pipes, valves, and drains that affect drinking water, hydroelectric plants, and a wide variety of manufacturing interests that rely on fresh water pumped from lakes. In some bodies of water where there is a zebra mussel infestation, water intake pipes have to be cleaned on a weekly basis. According to the Nonindigenous Aquatic Species Program of the US Geological Survey: "Zebra mussels can have profound effects on the ecosystems they invade … and represent one of the most important biological invasions into North America."[2] Zebra mussels are small and easily transported by almost any kind of container such as a plastic bag, jar, or bucket. They can stay alive out of the water for several days in cool, humid conditions by simply closing their shell tight. Under the right environmental conditions, it would take as few as three zebra mussels to begin an infestation in a body of water.[3]

Zebra mussels upset ecosystems, threaten native wildlife, damage structures, and cause other serious problems—they pose a series of threats:

- Threat to other species. Zebra mussels can severely affect native mussels and clams by interfering with their feeding, growth, movement, respiration, and reproduction.
- Threat to navigation, boating, and industry. They may colonize water intake pipes, severely restricting the water flow to power plants or other municipal or private facilities that rely on fresh water. Impacts include damage to the facilities as well as the cost of removing or controlling the mussels. Zebra mussels may also foul beaches and create boating and navigation hazards.
- Threat to outdoor recreation. Shells of zebra mussels foul beaches and near-shore swimming areas. Bare feet are at risk from the sharp shells, and clean up costs are high. Due to changes in fish populations, zebra mussels also adversely impact recreational fishing.

The risk associated with zebra mussels is aggravated because they are difficult to eradicate and control. Once zebra mussels become established in a water body, they are impossible to eradicate with the technology currently available. Many chemicals kill zebra mussels, but these exotics are so tolerant and tough that everything in the water would have to be poisoned to destroy the mussel. Most commercial water users rely on chemicals such as chlorine, filters, or mechanical scraping to remove mussels from their intake pipes and facilities.[4] Millions of dollars are spent each year in attempting to control these small but numerous mollusks.[5] Clearly zebra mussels pose risks to communities—their presence in the United States has already caused significant economic loss and has damaged the environment.

Based on the discussion to this point, it is clear that zebra mussels pose significant risks to communities and the environment, albeit at this point, the risk is not criminal. However, increasingly law enforcement agencies are being relied upon to assist in controlling the infestation based on their public safety, public order, and public welfare responsibilities. For example, increasingly officers working in rural and aquatic areas have been asked to examine the hulls of boats being towed to ensure there are no zebra mussels attached that could infect a different body of water. The process of using law enforcement authority to stop a vehicle for a non-criminal public order purpose poses a delicate balance for public safety risk control versus civil rights and constitutional protections of the driver.

In a recent case, however, the risk caused by zebra mussels had a new dimension. A Congressman from a Midwestern state was a vocal supporter of legislation to ban Internet gaming in the US. An individual who opposed this legislation made a threat to the Congressman's Office that if the Congressman voted for the legislation, the individual would introduce zebra mussels into some of his state's lakes. The mussels were now being used as an instrumentality of a threat to a member of the United States Congress.

The Congressman voted for the ban on Internet gaming. Of recent, zebra mussels have been newly appearing in local lakes in the Congressman's state. The immediate question is whether the presence of the newly discovered zebra mussels in the Congressman's state was a product of the threat.

Intentional transportation or use of zebra mussels as an injurious species that can cause damage in various forms as outlined by law are explicitly included in the United States Code[6] as crimes. Mere transportation of zebra mussels is a federal misdemeanor. If an individual intentionally causes damage or loss of property as a result of the introduction of zebra mussels, or conspires to do so, this can be the federal crime "Animal Enterprise Terrorism," punishable as a felony, depending on the value of property loss. Other possible federal and state crimes include extortion, terroristic threat, and criminal environmental law violations. As a result, this case represents a criminal risk by a form of domestic terrorism.

Beyond the criminal risk, the presence, and possible intentional infestation of zebra mussels, poses other risks to the community and economy. Are threats posed to other bodies of water as a result of this act? What preventive/protection measures should local critical infrastructure or key resources vulnerable from this threat take? Intelligence requirements need to identify persons with zebra mussels in their possession and determine the reason. Businesses and government entities whose operations could be effected by the zebra mussels must be identified and notified.

Risk and Intelligence Requirements

To assess risk, there must be a constant information flow to update the risk environment and that new information must be analyzed in the context of all previously received information. In the discipline of intelligence, this ongoing information need is referred to as "intelligence requirements." Requirements articulate specific types of information that need to be collected in order for the analysis to provide the most accurate picture of the threat environment.

- What information is available about the individual who made the threat?
  - Has the Congressman received threats in the past? If so, collect all related information.
  - Are there vocal activists against the ban on Internet gaming who could be reasonably tied to the Congressman and/or the state?
  - Are there any linkages between these individuals and environmental issues?
- How can zebra mussels be introduced into a new environment?
  - What do zebra mussels look like?
  - What are the different methods/processes that might be used for introduction?
  - What are the indicators of zebra mussels being introduced?
- What evidence is needed to prove that
  - the zebra mussels were intentionally introduced?
  - there was intent to cause damage or a loss of property?

- What damage was caused by the zebra mussels?
  - What is the evidence that supports this?
- If someone is identified with zebra mussels in their possession:
  - Identify the individuals and the reasons for their possession of the zebra mussels.
  - Document precautions taken to avoid introduction of the zebra mussels to the local environment.
  - Document any evidence to support the elements of applicable state and/or federal laws.
- What additional evidence may be needed?
- Is there a need for the fusion center to forecast their spread and impact or can this be handled more effectively by another agency?
  - If so, which agency?
  - Is there a memorandum of understanding in place to work with this agency?
- What are the characteristics of the new host environment that would help target places the zebra mussels may be introduced and flourish?
- Do any of the identified host environments have characteristics that increase the seriousness of the invasion (e.g., public water supply, hydroelectric plant, manufacturing, commercial or recreational body of water)?
- Given the spread of zebra mussels, what is the probability their presence in the Congressman's state is a product of natural transmission rather than an intentional act?
- What methods can be used to control zebra mussels?
- What resources are available to assist in handling possession and possible introduction of zebra mussels (e.g., National Oceanographic and Atmospheric Administration, United States Geological Society, Environmental Protection Agency, state natural resource agencies)?

Assessing risk and responding to risk requires the collection of comprehensive information in the risk environment. Risk is static—it changes—hence the need for a constant flow of information to ensure risk-related judgments are accurate with regard to the nature of risk at any given point in time. These intelligence requirements—that is, the answers to this risk-related questions—illustrate the types of information that are needed to make reasonable judgments of risk reduction in the current scenario.

Notes

1. Http://www.fws.gov/midwest/endangered/clams/zebra.html.
2. Http://nas.er.usgs.gov/queries/factsheet.asp?speciesid=5.
3. Personal correspondence, Nonindigenous Aquatic Species Program biologist, US Geological Survey, Gainesville, FL.
4. Http://www.nationalatlas.gov/articles/biology/a_zm.html.
5. Http://www.nationalatlas.gov/articles/biology/a_zm.html.
6. 18 USC §42–§43; 18 USC §3371–§3378; and the *Lacey Act*.

PART IV

SUMMARY AND CONCLUSIONS

The goals for this volume were ambitious. We sought to connect a systematic approach to risk analysis and risk management to our scholarly area of inquiry: crime and the criminal justice system. Additionally, we not only sought to connect issues of crime to issues of terrorism but we also applied an "all hazards" approach that connected these areas of foci to natural hazards, financial risks, environmental emergencies, and so on.

The danger with such an approach, of course, is that the volume becomes "an inch deep and a mile wide." Indeed, we acknowledge such a "risk" (no pun intended) and believe that each sub-section of this book could profitably be expanded into a distinct area of inquiry. We hope that this volume will generate theory and research that will result in precisely these types of inquiry.

More fundamentally, the inclusion of both risk assessment and risk management, the focus on crime, terrorism, and their relationship to other hazards, as well as the examination across public and private sectors represents a belief that such a multi-disciplinary, comprehensive perspective is critical if the risk framework is to provide practical benefit to police and security officials and policy makers. It also represents a belief that research and theory development across all of these domains is necessary to inform policy and practice and that such scholarship can benefit from such cross-fertilization. In many respects, the decision to include this broad framework represents a desire to learn from the experience of risk scholarship, whereby risk assessment has developed well beyond the technological assessment of risk to include the social, cultural, and organizational dimensions of risk management.

The chapters in this volume suggest the value in the attempt to learn across sectors and domains. Further, such learning can move back and forth across disciplinary boundaries. As several chapters suggest, criminologists have long experience in examining risk of criminal activity and vulnerability of victimization at individual, group, neighborhood, and community levels. Similarly, although not using the risk terminology, criminologists have experience in assessing the effectiveness of risk management strategies. Examples include the chapters on the reduced risk of homicide associated with a "pulling levers" strategy (Corsaro and McGarrell, this volume) and the reduced risk of victimization by employing spatial prediction techniques (Caplan, Moreto, and Kennedy, this volume). Criminologists have long theorized and studied the tensions between the goals of security and liberty, crime control and due process, and these theoretical perspectives resonate with a variety of value and policy conflicts in risk governance.

Similarly, the field of criminology and criminal justice has much to learn from scholarship on risk assessment and governance developed in other domains. Van Brunschot and Kennedy (this volume) illustrate this potential by noting the value of considering risk from a probabilistic perspective that creates a metric for planning and assessment. Similarly, they borrow from a well-established public health risk approach and call for surveillance systems to be tested in crime contexts. Similarities can be drawn to the impacts of fear of terrorist incidents and pandemics. Yet, criminologists have rarely drawn on the tradition of risk communication to consider fear reduction effects along with prevention and mitigation. There is a call for improved collaboration across firms and agencies. Certainly there are lessons to be learned from corporations that have transformed their financial risk management as well as their security operations into multi-sector enterprise risk models.

This comprehensive approach to risk assessment and risk management was also reinforced through series of events that occurred recently. On November 5, 2009, Nidal Malik Hasan, a major in the US Army, entered the Fort Hood, Texas Army base and opened fire on a group of fellow soldiers. Hasan fatally shot 13 service members and injured 32 others, most of them military personnel. Subsequent investigation revealed several risk indicators ranging from mental health concerns to electronic communications with alleged extremists in high-risk locations. Significant issues were raised in the aftermath about the risk analysis approach being employed with respect to military personnel. Similarly, risk management issues including inherently value-laden issues such as profiling based on ethnicity and religious affiliation will continue to be debated as various risk management approaches are considered.

On Christmas Day, 2009, Umar Farouk Abdulmutallab attempted to detonate a bomb on an airliner traveling from Amsterdam to Detroit. Similar to the Fort Hood shooting, a variety of risk assessment and management issues arose and will continue to be debated. From an assessment standpoint, there appear to have been a series of intelligence issues that precluded Abdulmutallab to have either been placed on a no-fly list or subject to increased screening despite a number of risk indicators. From a risk management perspective, policy debate has focused on the lessons to be gleaned from this attempted bombing. Should investment be made in full body screening technology? In all airports of the world? If not airport technology, should the focus be on improving intelligence profiles? Should greater screening be applied to individuals of certain racial, ethnic, and religious origin or for travelers from certain countries? Who decides? What are the costs and associated benefits? How much loss of privacy are citizens willing to tolerate? What is the impact on commerce?

This time period has also been one when US governors, legislatures, and corrections officials have struggled with the costs of correctional systems at the time of severe budget shortfalls. A variety of strategies and policies are being considered to reduce correctional populations. However, such decisions involve a great deal of uncertainty. What type of risk assessment instrument will be used to make these decisions? What degree of risk will policy makers tolerate? Are there risk communication strategies that could be used to gain the public's support? Will the management structures and processes surrounding these risk decisions be considered as legitimate by the public and by key stakeholders (e.g., victim advocates, police, prosecutors)? None of these issues have easy answers but the risk assessment and risk management framework offers fresh perspectives for addressing these inherently conflicting decisions.

Finally, in January 2010 Haiti suffered a devastating earthquake. Exacerbated by conditions of severe poverty, the death and destruction illustrated the need for hazard mitigation planning before such catastrophes occur. Although as of this writing the story remains to be told, the situation also points to risk for the type of "spin-off" impacts creating the need for public health and security risk management. As was seen in the aftermath of Katrina in New Orleans, the police are a crucial resource when such incidents occur and an ineffectual law enforcement response increases the risk of significant harm. The lessons learned point to the need for greater cross-sector collaboration and planning before such incidents occur as well as the critical role of public–private, multiple sector response for recovery and mitigation.

We believe that the risk assessment and risk management framework offers utility within the criminal justice system as well as for building capacity across sectors and domains. This ranges from risks emanating at individual, family, neighborhood, community, regional, and global scales; ranging across the stages of prevention, preparedness, response, and recovery; and including traditional crimes, emerging crimes, terrorism, pandemics, and natural disasters. We hope that this volume stimulates research and theory within the various topics covered as well as across these dimensions and, ultimately, that this research and theory informs policy and practice in a way that reduces harm to humans and that maximizes the benefits and reduces the costs through effective governance.

INDEX

9/11 *see* September 11, 2001 attacks

Abdullahu, Agron 188–9
Abdulmutallab, Umar Farouk 187, 274
Adler, Freda 219
AIVD (Dutch Intelligence Service) 83
Akamai Technologies 256
Akunyili, Dora 236
Al-Aulaqi, Anwar 191, 193n4
American Civil Liberties Union 180
American Society for Industrial Security International 199
American Terrorism Study (ATS) 139–41
anti-abortionists 58, 59, 63, 64, 66*f*, 70, 74*f*, 77
Arrestee Drug Abuse Monitoring survey 22
Association of Chief Police Officers (UK) 135
Audit Commission 145n2

Baker, T. 3, 32
Bale, H. E. 233
Bate, R. 231
BBC 7, 236, 237
Beck, U. 35, 213
Beckman, R. C. 102
Beckstrom, Rod 264
Bedell, John Patrick 187
Berk, Richard 26n2
Bhatt, A. 83, 84, 87
Bialik, Carl 118
Booth, K. 100–1
Boston: Operation Ceasefire 18–19, 20, 137–8, 144
Braga, Anthony et al. 19, 21, 25, 26n1
Bureau of Justice Assistance (BJA; US) 136, 144, 171

cargo crimes 216–17, 222
Carlson, Joseph 59, 77
Carter, D. L. 136, 144, 186
Carter, J. G. 186
CAT (Communities Against Terrorism) 171, 172
CCTV (closed-circuit television) 10–11, 131
Central Intelligence Agency (CIA) 259
charter party fraud 215–16
Chermak, Steven M. 22, 23
China: adulterated alcohol 250; counterfeit drugs 231–2; food fraud 248–9
CIB (Counterfeit Intelligence Bureau) 117, 118
CIP (Critical Incident Protocol) Community Facilitation Program 200–2
CISCO Systems 10
Clarke, R. et al. 217–18, 219, 220, 221, 222
closed-circuit television (CCTV) 10–11, 131
Cohen, L. E. 218
Collins, David 118
Combined Joint Task Force Horn of Africa (CJTF-HOA) 103
Combined Task Force 150 (CTF-150) 103
Combined Task Force 151 (CTF-151) 103
Common Terrorism Information Sharing Standards (CTISS) program 179
Communities Against Terrorism (CAT) 171, 172
community-oriented policing (COP) 136, 143
Community Oriented Policing Services (COPS) 57, 78
Compstat programs 9, 137, 144, 145, 154
Congo: counterfeit drugs 235
Connors, E. et al. 199, 204–5
Conway, Sheila 197
Cope, N. 144
Corsaro, Nicholas 23, 26n6
COSO financial risk management model 2
Cottrill, K. 219
counter-terrorism 2–4, 43 *see also* intelligence-led policing (ILP); law enforcement intelligence; terrorist threat: law enforcement view
Counterfeit Intelligence Bureau (CIB) 117, 118
counterfeit pharmaceuticals 227–38; counterfeiting process model 229–31, 230*f*; definition 228; demand for lifesaving drugs 233–4; distribution process 232–3; impact on economy 235; impact on health 234; intervention efforts 235–7; significant magnitude 228–9; supplier characteristics 231–2; and terrorist

financing 231; questions remaining 237–8; *see also* product counterfeiting
Counterfeit Product Risk Model 121–6, 125*f*; Counterfeit-Ability 123; Counterfeit-Attractiveness 123–4; Counterfeit-History 123; Counterfeit-Hurdles 124; counterfeit risk sub-factors 124–5; CSFI case study 242–5, 243*f*; Market Profile 124; model development 125–6; risk factor selection analysis 121–5
crime displacement theory 221–2
criminological theories and approaches 218–25; crime displacement theory 221–2; enhancing regional and international cooperation 225; improving legal regimes 224–5; opportunity 218–20; rational choice theory and situational crime prevention 220–1; risk balance theory 223–4; routine activities approach 218
Critical Incident Protocol (CIP)—Community Facilitation Program 200–2
CTF-150 (Combined Task Force 150) 103
CTF-151 (Combined Task Force 151) 103
CTISS (Common Terrorism Information Sharing Standards) program 179
Cunningham, W. et al. 198
Customs and Border Protection Bureau (US) 119
cyber warfare 7, 256
cybercrime 7, 255–66; attacks targeting TCP/IP network achitectures 262; Aurora vulnerability 258–9; bots 260–1; cracking passwords, codes, and keys 261; cyberspace 255–7; Denial-of-Service (DoS) 256, 260, 262; Distributed-Denial-of-Service (DDoS) 256–7, 260, 262; Domain Name System (DNS) 256, 257, 262; eavesdropping and packet sniffing 258; exploiting flaws 261; injecting malicious code 260–1; Internet protocols 256, 257; jamming or flooding 260; key loggers 259; pharming 262; risk management methodology 262–5; snooping and downloading 259; spoofing 260; tampering or data diddling 258–9; TCP/IP 256, 257, 262; threats 7, 10, 231; tools and tactics 257–62; Uniform Resource Locators (URLs) 256
cyberspace 255–7

data diddling 258–9
Davos World Economic Forum 7
Deerfield Police Department 207, 208
Delaware Information Analysis Center 173
Denning, Dorothy E. 257–8, 260, 261
Department of Defense (US) 265n1
Department of Health and Human Services (HSS) (US) 38–9
Department of Homeland Security (US) 42*f*, 43, 134, 167, 168, 258, 264
Department of Justice (DOJ; US) 57, 135, 167, 168, 199, 200, 219
deviation fraud 216
Dingell, John 258
Discover Financial Services 207, 208
documentary frauds 215
downloading 259
Duka, Dritan "Tony" 188–9
Duka, Eljvir "Elvis" 188–9
Duka, Shain 188–9
Dutch Intelligence Service (AIVD) 83

eavesdropping 258
eco-terrorists 63, 67*f*, 69, 71, 73*f*, 77; *see also* zebra mussels
Economically Motivated Adulteration (EMA) *see* product counterfeiting
Edward Byrne Memorial Grant Program 78
Elkins, D. 132
emergency response *see* public–private partnerships for crisis management
enterprise risk governance *see* risk management
environmental criminology 101–2
ethnic supremacists 59, 63, 64, 66*f*, 70, 73*f*, 77
European Institute for Computer Antivirus Research (EICAR) 264
European Institute for Crime Prevention and Control 9–10
European Space Agency (ESA) 39, 40
European Union 119
event-based models 2

Failed States Index 48, 104, 105
Federal Bureau of Investigation (FBI) 57, 81, 82, 117, 134, 139, 146n10, 189, 190, 198, 227
Federal Law Enforcement Training Center (FLETC) 199
Felson, M. 218, 219
financial suspicious activity reporting 168
flooding (cybercrime) 260
Food and Drug Administration (FDA; US) 117, 119, 120–1, 230, 250
food frauds 246–54; adulteration–misrepresentation continuum 247–8, 248*t*; authentication and traceablity technologies 252–3; counterfeit food 246–7; expiration fraud: perishables 250–1; harm and risk reduction strategies 251–3; ingredient adulteration/substitution:

food frauds (*cont.*):
China 248–9; international foods markets 247; product substitution and repackaging: beverages 250; up labeling: fish and specialty foods 249–50; volume misrepresentation: fish 251
Food Safety Agency (UK) 253
foreign terrorist groups 63, 64, 69*f*, 70, 75*f*
Fort Dix, New Jersey 82, 188–90
Fort Hood, Texas 190–3, 274
fraud *see* food frauds; maritime fraud
Freilich, J. et al. 141
Friedman, T. L. 191
Fund for Peace and Foreign Policy 104

GAO (Government Accounting Office) 76–7, 118
GBLAAC (Global Business Leaders Alliance Against Counterfeiting) 119, 120*t*
GDACS (Global Disaster Alert and Coordination system) 40–1
Gibbs, N. 191
GIVAS (Global Impact and Vulnerability Alert System) 49
Global Intelligence Working Group 135, 168
Global Justice Information Sharing Initiative 175–6
Global Pulse 41
global risk assessment 30–51; analysis structure 33–4, 33*f*, 45, 45*t*; composite measures of risks 46, 48; decision-making 34–5, 37–8, 40–1, 42–3, 44; methods 35–6, 45–6; natural disasters (floods) 39, 40–1; public health (pandemic influenza) 36–9, 37*t*, 38*t*, 39*t*; risk assessment 34, 36–7, 40, 42, 43–4; risk assessment and decision-making 46–9, 47*f*; risk management 35, 38–9, 38*t*, 39*t*, 41, 43, 44–5; terrorism 41–3, 42*f*; uncertainty 33
Global Risk Network 30, 31, 49
global trade and crime 213–25; cargo crimes 216–17, 222; criminological theories and approaches 218–25; legal regimes 224–5; maritime fraud 215–16; maritime piracy 98–113, 214, 219–20, 224–5; money laundering 218, 224–5; regional and international cooperation 225; terrorism 217–18
Goldstein, H. 136, 144
Government Accounting Office (GAO) 76–7, 118
Grant, J. 151

Haiti 275
Hale, C. et al. 145n2, 146n13–14
Hallcrest Report II 198, 199
Hamm, M. 140
Harris, J. et al. 228, 233
Hasan, Nidal Malik 190–1, 192–3, 274
Her Majesty's Customs and Excise (HMCE; UK) 144
Her Majesty's Inspectorate of Constabulary (HMIC; UK) 135, 145n2
Hesseling 1994 221, 222
Home Office (UK) 43, 135
homicide risk reduction 17–26; awareness 22–3; Boston's Operation Ceasefire 18–19, 20; effectiveness 23–4, 24*f*; Indianapolis, Indiana 19; IVRP intervention 19, 20–3, 21*t*; IVRP "pulling levers" strategy 20–2, 26n3; law enforcement response 19–20; networks of capacity 25; Project Exile 138, 145n6
Hoover, J. Edgar 198
Horgan, J. 87
House of Commons Transport Committee (UK) 102
Howick, Susan et al. 126
HSS (Department of Health and Human Services; US) 38–9
Hurricane Katrina 198, 275

I&A (Information and Analysis Directorate; DHS) 168
IACC (International AntiCounterfeiting Coalition) 228
IACLEA (International Association of Campus Law Enforcement Administrators) 61
IACP *see* International Association of Chiefs of Police
IACP Tribal Police 61
ICANN (Internet Corporation for Assigned Names and Numbers) 256
ILP *see* intelligence-led policing
IMB *see* International Maritime Bureau
IMO (International Maritime Organization) 214
IMPACT (International Medicinal Products Anti-Counterfeiting Taskforce) 236
India: adulterated alcohol 250; counterfeit drugs 231
Indianapolis: Brightwood investigation 22, 23
Indianapolis Violence Reduction Partnership (IVRP) 19, 20–3, 21*t*
individual risk 8
Information and Analysis Directorate (I&A; DHS) 168
information sharing environment-suspicious activity report (ISE-SAR) 168, 178–9

institutional risk 8–9
insurance fraud 216
intellectual property 117, 121–2
intelligence-led policing (ILP) 133–46, 186–7; Boston Ceasefire program 137–8, 144; challenges and key components 142–5; community-oriented policing (COP) 136, 143; international terrorism 142, 144; origins of ILP 135–6; problem-oriented policing (POP) 136–7, 143; Project Safe Neighborhoods (PSN) 137, 143, 144, 145n6; responding to terrorism 138–9; Strategic Approaches to Community Safety Initiative (SACSI) 137, 138, 144; terrorism and far-right extremists 139–41; *see also* law enforcement intelligence; Turkish National Police (TNP)
Intelligence Reform and Terrorism Prevention Act (2004) 168, 178
International AntiCounterfeiting Coalition (IACC) 228
International Association of Campus Law Enforcement Administrators (IACLEA) 61
International Association of Chiefs of Police (IACP) 57, 60, 61, 135, 199, 208
international cooperation 225
International Maritime Bureau (IMB) 102, 105, 213, 214
International Maritime Organization (IMO) 214
International Medicinal Products Anti-Counterfeiting Taskforce (IMPACT) 236
international terrorism 142, 144
Internet *see* cyber warfare; cybercrime
Internet Corporation for Assigned Names and Numbers (ICANN) 256
INTERPOL 118, 231, 234, 236, 237
ISE-SAR (information sharing environment-suspicious activity report) 168, 178–9
Islamist terrorist groups 63, 64, 68*f*, 69, 70, 71, 72*f*, 77
IVRP *see* Indianapolis Violence Reduction Partnership

jamming 260
Janis, I. L. 88
John, T. 153
Joint Strategic Plan for Intellectual Property (US) 117
Jones, R. J. 197, 200, 206

Kaczynski, Theodore "Ted" 187
Kahl, Gordon 141
Kaiser Network 237
Kamarck, E. C. 145n3
Katyal, Neal K. 263
Kennedy, David et al. 22, 26n1, 26n3
Kennedy, L. W. 14, 154, 223–4, 225
Kent Constabulary (UK) 135

Lab 2000 221, 222
Lake–Cook Regional Critical Incident Protocol (LCRCIP) 207–8
Langevin, James 258
law enforcement: Compstat programs 9, 137, 144, 145, 154; following 9/11 81–2, 134, 144, 184; risk assessment 8–9; risk management 151; *see also* cybercrime; law enforcement intelligence; terrorist threat: law enforcement view; Turkish National Police (TNP)
law enforcement intelligence 184–93; crime and terrorism risk: determination and prevention 184–6, 185*f*; Fort Dix, New Jersey 82, 188–90; Fort Hood, Texas 190–3, 274; intelligence-led policing (ILP) 186–7; law enforcement cycle 186, 187*f*; likelihood of prevention of crime and terrorism 192, 192*f*; opportunity and time 185, 186, 186*f*; self-radicalization of terrorists 187–8; suspicious activity reports (SARs) 186–7; *see also* intelligence-led policing (ILP)
LCRCIP (Lake–Cook Regional Critical Incident Protocol) 207–8
left-wing terrorist groups 63, 64, 68*f*, 70, 75*f*
legal regimes 103–4, 105–7, 106*f*, 107*f*, 108*f*, 224–5
Leighton, Tom 257
Lobe, J. 104
Local Law Enforcement Block Grant 78
lone-wolf terrorism 187, 193n2
Lum, C. et al. 81, 139

McCauley, C. 86–7
McGarrell, E. F. et al. 22, 23, 26n6, 154
McVeigh, Timothy 187
Maguire, M. 153
Major Cities Chiefs Association 176, 179
Mandel, D. R. 83
maritime fraud 215–16
maritime piracy 98–113; causes and correlates 102–3; classification 214; definition 102, 214; maritime chokepoints 104–5, 109, 110*f*; maritime security 100–1, 113n3, 225; opportunity 219–20; results 109, 110, 111*f*, 111*t*; risk factors 98–9, 105, 219–20; risk terrain modeling (RTM) 99–100,

maritime piracy (*cont.*)
101–2, 111–13, 113n3; safety risk management 113n3; security dilemmas 100–1; shipping routes 104, 107, 108, 109*f*, 214; state piracy 113n4; state status 103–4, 105–7, 106*f*, 107*f*, 108*f*, 224–5
maritime terrorism 100
Marquise 10 191
Marsh, M. H. 131
Matthews, Robert 141
Mazerolle, L. 102
Merari, A. 139
Michigan State University (MSU) 200
money laundering 218, 224–5
Morgenstern, Brian 189, 190
Moskalenko, S. 86–7
Mueller, G. O. W. 219
Murphy, M. N. 100, 103, 113n4

Naim, Moises 118
Nassau County (New York) Police Department 173
National Advisory Commission on Criminal Justice Standards and Goals (US) 135
National Agency for Food and Drug Administration and Control (NAFDAC; Nigeria) 229, 232, 235, 236
National Center for Ecological Analysis and Synthesis 107–8
National Conference on Weights and Measures (NCWM) 251
National Crime Squad (UK) 144
National Criminal Intelligence Resource Center (NCIRC) 180
National Criminal Intelligence Service (UK) 135
National Criminal Intelligence Sharing Plan (US) 135, 136, 144
National Infrastructure Advisory Council 2
National Intelligence Model (NIM; UK) 135, 146n11, 153
National Security Strategy for Cyberspace 256
National Sheriff's Association (NSA) 60–1, 199
National Youth Gang Survey (1988) 19
Nationwide Suspicious Activity Reporting Initiative (NSI) 179
natural disasters (floods) 39, 40–1
NCIRC (National Criminal Intelligence Resource Center) 180
NCWM (National Conference on Weights and Measures) 251
Nelson, Frank 141
New York Partnership 199
New York (City) Police Department (NYPD) 137, 172, 173
Newman, G. R. 217–18, 221
Nigeria: counterfeit drugs 229, 232, 234, 235–6
NIM *see* National Intelligence Model
Noonan, T. E. 131
Norris, Ed 57
Northumbria Constabulary (UK) 135
NSA (National Sheriff's Association) 60–1, 199
NSI (Nationwide Suspicious Activity Reporting Initiative) 179
NYPD *see* New York (City) Police Department

OECD (Organization for Economic Co-operation and Development) 122–3, 228
offender decision-making *see* homicide risk reduction
Omar, Mahmoud 189
Oots, K. L. 87
Open Systems Interconnection (OSI) model 256
opportunity for crime 185, 186, 186*f*, 218–20
Orhii, Paul 236
OSAC (Overseas Security Advisory Council) 199

packet sniffing 258
Peel, Sir Robert 174
Peele, R. B. 104
Pfizer 237
pharming 262
Phillips, Tim 119
Pitts, P. 232
PM-ISE (Program Manager's Office for the Information Sharing Environment) 168, 176
police *see* intelligence-led policing (ILP); law enforcement; law enforcement intelligence; Turkish National Police (TNP)
POP (problem-oriented policing) 136–7, 143
Posse Comitatus Act 57
Post, J. et al. 89
precautionary prinviple 3
Pressman, D. E. 83
Private Sector Liaison Committee (PSLC) 199
Private Security Advisory Council 199
problem-oriented policing (POP) 136–7, 143
product counterfeiting 116–26; Counterfeit Product Risk Model 121–6, 125*f*; scale of threat 117–19, 119*t*; scope of threat 119–21, 120*t*; *see also* counterfeit pharmaceuticals; food frauds
Program Manager's Office for the Information Sharing Environment (PM-ISE) 168, 176
Project Exile 138, 145n6
Project Safe Neighborhoods (PSN) 25, 137, 143, 144, 145n6

Psarros, G. et al. 113n3
PSLC (Private Sector Liaison Committee) 199
public health (pandemic influenza) 36–9, 37*t*, 38*t*, 39*t*; *see also* counterfeit pharmaceuticals
public–private partnerships for crisis management 196–209; benefits and challenges of partnerships 204–6; CIP—Community Facilitation Program 200–2; crisis management 197; critical incidents 197; definitions 197; historical aspects 198–200; Lake–Cook Regional CIP (LCRCIP) 207–8; leadership 203; mission 203; social capital 203–4

RAND Corporation 2, 59, 77
Ratcliffe, J. 153, 185
rational choice theory 220–1
regional cooperation 225
Reppetto 1976 221, 222
Reuter and Truman 2004 218
right-wing extremists 58, 63, 67*f*, 70, 74*f*, 86, 139–41
Riley, K. J. et al. 146n8, 146n12
Ripley, A. 184
risk analysis 2; scalability 29; *see also* global risk assessment; homicide risk reduction
risk assessment 1–4; cybercrime 7; definition 46; framework 6–15; global aspect 10; individual risk 8; institutional risk 8–9; law enforcement 8–9; nature of threat 10–12; prospective risk assessment 97; responsibility for threat 12; societal risk 9–10; three stages of risk 12–14, 13*t*; *see also* global risk assessment; maritime piracy; product counterfeiting
risk balance theory 223–4
risk communication 4
Risk Correlation Matrix 48
risk management 1, 2–3, 131–2, 151; cybercrime 262–5; *see also* global risk assessment; intelligence-led policing (ILP); suspicious activity reporting; Turkish National Police (TNP)
risk terrain modeling (RTM) 99–100, 101–2, 111–13, 113n3
Rodrigue, J. P. 104–5
Rong Yansong 232
Rosenfeld, Richard et al. 26n2
routine activities theory 218–20
Royal Canadian Mounted Police 83
Russia: counterfeit alcohol 250

Sacco, V. 14
SACSI *see* Strategic Approaches to Community Safety Initiative
Sageman, M. 84–5
Sampson, D. W. 103
SARA (scanning, analysis, response, and assessment) 136
September 11, 2001 attacks 81–2, 134, 144
Shah, S. 234
Shahzad, Faisal 187, 193n4
Shanahan, Michael 80 199
Shnewer, Mohamad Ibrahim 188–9
Silber, M. 83, 84, 87
Silke, A. 81, 139
Simon, J. 3, 32
situational crime prevention 220–1
SLATT (State and Local Anti-Terrorism Training) 171
Slovic, P. 14
Smith, B. L. et al. 139–40
Snell, Richard 141
snooping 259
social capital 203–4
social construction of risk 4
societal risk 9–10
Sodipo, Bankole 121–2
Somalia 102, 103, 104
Southern Poverty Law Center (SPLC) 91
Sparks, Joe 207
spoofing 260
Sprinzak, E. 83, 85–6, 87
State and Local Anti-Terrorism Training (SLATT) 171
Strategic Approaches to Community Safety Initiative (SACSI) 137, 138, 144
suspicious activity reporting 167–81, 186–7; "all crimes" SAR 168; establishing controls on SARs/information 174–6; financial SAR 168; information sharing environment (ISE-SARs) 168, 178–9; nationwide SAR initiative 179–80; overcoming dogmatism 178; private sector and community engagement 172–4; reporting system recommendations 176–7; risk relationship 174; SAR processes 169–72; three-part model: (Observe 170–1; Document 171; Report 171–2)

TADOC (Turkish International Academy Against Drugs and Organized Crime) 155–6
Takeda Pharmaceuticals 207, 208
tampering 258–9
Tatar, Serdar 188–9
TCP/IP (Transmission Control Protocol and Internet Protocol) 256, 257, 262

technology 144–5, 146n13; *see also* cyber warfare; cybercrime
terror groups and radicalization 80–95; delegitimization 86; disengagement 87; group dynamics 88–9; group extremity shift 89; group radicalization 85–6; "groupthink" 88; leadership 88–9, 93–4, 93*t*; links with other hate groups 92–3, 92*t*; networks 85; radicalization: definitions 83; radicalization process 84–8, 187–8; reasons for radicalization 89–90; recruitment strategies 91–2, 92*t*; violent and non-violent groups 90–4
terrorism: characteristics 217; definition 217; and far-right extremists 139–41; ILP response to 138–9; international 142, 144; Islamist groups 63, 64, 68*f*, 69, 70, 71, 72*f*, 77; left-wing terrorist groups 63, 64, 68*f*, 70, 75*f*; lone-wolf terrorism 187, 193n2; maritime terrorism 100; opportunities and targets 185, 186, 186*f*, 217–18; risk assessment 41–3, 42*f*; self-radicalization 187–8; threat 11; *see also* counter-terrorism; terror groups and radicalization; terrorist threat: law enforcement view
terrorist threat: law enforcement view 54–79; findings 62–76; group fear 69–71, 70*t*, 72–5*f*, 72*t*; group presence 62–5, 63*t*, 65*t*, 66–9*f*, 72*t*; group variations 58–9; national frequencies 69–71, 70*t*, 76; predictive model 71–2; previous research 59; projected terrorist activity 65; role of armed forces 57; role of federal agencies 57, 81; role of local agencies 57–8; survey and dataset 60–2; terrorist attack forecasting 77–8; terrorist group typologies 60; *see also* law enforcement intelligence
Tew, J. 132
Thomspon, Bennie 258
threat: dynamic nature of 12; nature of 10–11; responsibility for threat 12
Tilley, N. 136, 145n3
TNP *see* Turkish National Police
Toch, H. 151
trade *see* global trade and crime
Transmission Control Protocol and Internet Protocol *see* TCP/IP
Trinity School, New York City 249
TRIPS (Trade Related Aspects of Intellectual Property Rights) 121–2
Turkish International Academy Against Drugs and Organized Crime (TADOC) 155–6
Turkish National Police (TNP) 151–65; ACTION risk model 154–5, 157–63: (assessment of risk 157–8; connections 158–9; tasks 159 60; information 160 1; organization 161–2; notifying others 162–3); Al-Qaeda suicide bombings, Istanbul 142, 144; intelligence-led policing and risk assessment 153–4; organizational structure 155; POLNET (Intranet Police Network in Turkey) 158–9, 162–3; research design 155–6; research questions 157

Ugaste, Rob 207
UN/ISDR (Inter-Agency Secretariat of the International Strategy For Disaster Reduction) 46, 50–1
Uniform Resource Locators (URLs) 256
United Nations: ISDR (Inter-Agency Secretariat of the International Strategy For Disaster Reduction) 31; Office for the Coordination of Humanitarian Affairs (OCHA) 41; Security Council 217
Urban Environmental Research 197
US Geological Survey 268

Van Brunschot, E. 154, 223–4, 225
Verton, Dan 264
Vian, Taryn 06 233

Washington Law Enforcement Executive Forum (WLEEF) 199
WCO *see* World Customs Organization
Wheeler, N. J. 100–1
White, Nick 234
white supremacists *see* ethnic supremacists
WHO (World Health Organization) 36–7, 38, 38*t*, 120, 228, 229, 233, 234, 235, 236, 237, 250
Winship, Christopher 25
Withington, Neil 118
Wooldridge, M. 32
World Customs Organization (WCO) 118, 119, 120*t*
World Economic Forum 30, 31, 32, 46, 48, 49
World Health Assembly 37
Wortley, R. 102

Yagar, Loren 118
Yankus, W. 236
Yusufu, Hashim Ubale 232

zebra mussels 268–71
Zimmermann, C. 143, 145
Zionist Occupation Government 86